MATH.
STAT.
LIBRARY

Encyclopaedia of Mathematical Sciences

Volume 45

Editor-in-Chief: R.V. Gamkrelidze

Springer
*Berlin
Heidelberg
New York
Barcelona
Budapest
Hong Kong
London
Milan
Paris
Santa Clara
Singapore
Tokyo*

Yu. V. Prokhorov A. N. Shiryaev

Probability Theory III

Stochastic Calculus

Springer

Consulting Editors of the Series:
A. A. Agrachev, A. A. Gonchar, E. F. Mishchenko,
N. M. Ostianu, V. P. Sakharova, A. B. Zhishchenko

Title of the Russian edition:
Itogi nauki i tekhniki, Sovremennye problemy matematiki,
Fundamental'nye napravleniya, Vol. 45,
Teoriya veroyatoastej – 3,
Publisher VINITI, Moscow 1989

Mathematics Subject Classification (1991):
60Hxx, 60G40, 60G44, 60G48, 60F17, 60J60, 60J65

ISSN 0938-0396
ISBN 3-540-54687-1 Springer-Verlag Berlin Heidelberg New York

This work is subject to copyright. All rights are reserved, whether the whole or part of the material is concerned, specifically the rights of translation, reprinting, reuse of illustrations, recitation, broadcasting, reproduction on microfilm or in any other way, and storage in data banks. Duplication of this publication or parts thereof is permitted only under the provisions of the German Copyright Law of September 9, 1965, in its current version, and permission for use must always be obtained from Springer-Verlag. Violations are liable for prosecution under the German Copyright Law.

© Springer-Verlag Berlin Heidelberg 1998
Printed in Germany

Typesetting: Camera-ready copy produced from the translators' input files
using a Springer TeX macro package.
SPIN 10020167   41/3143 - 5 4 3 2 1 0 – Printed on acid-free paper

List of Editors, Authors and Translators

Editor-in-Chief

R. V. Gamkrelidze, Russian Academy of Sciences, Steklov Mathematical Institute, ul. Gubkina 8, 117966 Moscow; Institute for Scientific Information (VINITI), ul. Usievicha 20a, 125219 Moscow, Russia; e-mail: gam@ipsun.ras.ru

Consulting Editors

Yu. V. Prokhorov, Steklov Mathematical Institute, ul. Gubkina 8, 117966 Moscow, Russia, e-mail: prokhoro@genesis.mi.ras.ru

A. N. Shiryaev, Steklov Mathematical Institute, ul. Gubkina 8, 117966 Moscow, Russia, e-mail: shiryaev@genesis.mi.ras.ru

Authors

S. V. Anulova, Institut problem upravleniya RAN, ul. Provsoyuznaya 65, 117342 Moscow, Russia

N. V. Krylov, Department of Mathematics, University of Minnesota, Minneapolis, MN 55455, USA, e-mail: krylov@math.umn.edu

R. Sh. Liptser, Institute for Problems of Transmission of Information of the Russian Academy of Sciences, ul. Ermolovoj 15, 101447 Moscow, Russia; Department of Electrical Engineering-Systems, Tel Aviv University, Ramat Aviv 69978, Israel, e-mail: liptser@eng.tau.ac.il

A. N. Shiryaev, Steklov Mathematical Institute, ul. Gubkina 8, 117966 Moscow, Russia, e-mail: shiryaev@genesis.mi.ras.ru

A. Yu. Veretennikov, Inst. of Information Transmission Sciences, Russian Academy of Sciences, 19 Bolshoy Karetnii, 101447 Moscow, Russia, e-mail: ayu@sci.lpi.ac.ru, veretenn@ippi.ac.msk.su

Translator

P. B. Slater, Community and Organization Research Institute, University of California, Santa Barbara, CA 93106-2150, USA, e-mail: slater@sbitp.ucsb.edu

Stochastic Calculus

S.V. Anulova, A.Yu. Veretennikov, N.V. Krylov, R.Sh. Liptser, A.N. Shiryaev

Translated from the Russian
by P.B. Slater

Contents

Preface 5

Chapter 1. Introduction to Stochastic Calculus (N.V. Krylov) 7

§1. Brownian Motion and the Wiener Process 7
§2. Probabilistic Construction of the Solution of the Heat Equation.
    Relation Between the Wiener Process and the Laplace Operator . 15
§3. The Itô Integral and Differentiation Rules for Composite
    Stochastic Functions 19
§4. Stochastic Differential Equations and Diffusion Processes.
    Girsanov's Theorem 26
§5. Stochastic Differential Equations with Boundary Conditions 32
References 36

Chapter 2. Stochastic Differential and Evolution Equations 38

I. Stochastic Differential Equations (SDEs) (S.V. Anulova,
A.Yu. Veretennikov) 38

§1. Strong Solutions of Stochastic Differential Equations 38
§2. Weak Solutions of Stochastic Differential Equations with
    Non-Smooth Coefficients in E^d 50
§3. Differentiation of Solutions of SDEs with Respect to the Initial
    Data 54

§4. Invariant Measure of a Diffusion Process 57
§5. Support of a Diffusion 59
§6. Stochastic Differential Equations in Domains 63
References 72

II. Stochastic Evolution Equations (A.Yu. Veretennikov) 76

§1. Introduction 76
§2. Martingales and Stochastic Integrals in Hilbert Space 77
§3. The Itô Formula for the Square of the Norm 81
§4. Stochastic Differential Equations of Monotone Type in Banach
     Spaces 82
§5. Stochastic Partial Differential Equations I. The First
     Boundary-Value Problem for Parabolic-Type Non-Linear
     Equations 85
§6. Stochastic Partial Differential Equations II. The Cauchy
     Problem for Second-Order Linear Equations 88
References 90

III. Stochastic Calculus (Malliavin Calculus). Applications to Stochastic Differential Equations (A.Yu. Veretennikov) 91

§1. Introduction 91
§2. Stochastic Derivatives 92
§3. Rules of the Malliavin Calculus 96
§4. Smoothness of the Density (Scheme of the Proof) 98
§5. The Bismut Approach 1 99
§6. The Bismut Approach 2. Stochastic Differential Equations 101
§7. Stochastic Differential Equations (Smoothness of the Density
     with Respect to Inverse Variables) 107
References 108

Chapter 3. Stochastic Calculus on Filtered Probability Spaces (R.Sh. Liptser, A.N. Shiryaev) 111

I. Elements of the General Theory of Stochastic Processes 111

§1. Kolmogorov's Axioms and the Stochastic Basis 111
§2. Stopping Times, Adapted Stochastic Processes, Optional and
     Predictable σ-Algebras. Classification of Stopping Times 112
§3. Martingales and Local Martingales 116
§4. Increasing Processes. Doob-Meyer Decomposition. Compensators . 118
§5. Random Measures. Integral Random Measures 120
§6. Locally Square-Integrable Martingales. The Quadratic
     Characteristic 123
§7. Decomposition of Local Martingales 123

II. Semimartingales. Stochastic Integrals 125

§1. Semimartingales. Quadratic Variation. Quasimartingales 125
§2. Construction of Stochastic Integrals with Respect to
   Semimartingales 127
§3. The Itô Formula 129
§4. Construction of Stochastic Integrals with Respect to Random
   Measures 131
§5. Characteristics of Semimartingales. The Triple of Predictable
   Characteristics $T = (B, C, \nu)$. Martingale and Semimartingale
   Problems. Examples. 133
§6. Integral Representation of Local Martingales 137
§7. Stability of the Class of Semimartingales with Respect to a
   Series of Transformations 138

III. Absolute Continuity and Singularity of Probability Distributions . 140

§1. Local Density. Lebesgue Decomposition 140
§2. Girsanov's Theorem and its Generalization. Transformation of
   Predictable Characteristics 141
§3. The Hellinger Integral and the Hellinger Process 143
§4. General and Predictable Criteria of Absolute Continuity and
   Singularity of Probability Measures 147
§5. Particular Cases 148
Commentary to Chapter 3 152
References 154

Chapter 4. Martingales and Limit Theorems for Stochastic Processes
(R.Sh. Liptser, A.N. Shiryaev) 158

I. Theory: Weak Convergence of Probability Measures
on Metric Spaces 158

§1. Introduction 158
§2. Different Types of Convergence. Skorokhod Topology 160
§3. Brief Review of a Number of Classical Limit Theorems of
   Probability Theory 166
§4. Convergence of Processes with Independent Increments 178
§5. The Convergence of Semimartingales to Processes with
   Independent Increments 188
§6. Relative Compactness and Tightness of Families of Distributions
   of Semimartingales 200
§7. Convergence of Semimartingales to a Semimartingale 202
§8. The Martingale Problem 210

II. Applications: The Invariance Principle
and Diffusion Approximation 213

§1. The Invariance Principle for Stationary and Markov Processes ... 213
§2. The Stochastic Averaging Principle in Models without Diffusion . 226
§3. Diffusion Approximation of Semimartingales. The Averaging
    Principle in Models with Diffusion 229
§4. Diffusion Approximation for Systems with Physical White Noise . 233
§5. Diffusion Approximation for Semimartingales with Normal
    Reflection in a Convex Domain 237
Commentary to Chapter 4 243
References 244

Author Index 249

Subject Index 251

Preface

In the axioms of probability theory proposed by Kolmogorov the basic "probabilistic" object is the concept of a probability model or probability space. This is a triple $(\Omega, \mathcal{F}, \mathbf{P})$, where Ω is the space of elementary events or outcomes, $\mathcal{F}$ is a σ-algebra of subsets of Ω announced by the events and $\mathbf{P}$ is a probability measure or a probability on the measure space $(\Omega, \mathcal{F})$. This generally accepted system of axioms of probability theory proved to be so successful that, apart from its simplicity, it enabled one to embrace the classical branches of probability theory and, at the same time, it paved the way for the development of new chapters in it, in particular, the theory of random (or stochastic) processes.

In the theory of random processes, various classes of processes have been studied in depth. Theories of processes with independent increments, Markov processes, stationary processes, among others, have been constructed. In the formation and development of the theory of random processes, a significant event was the realization that the construction of a "general theory of random processes" requires the introduction of a flow of σ-algebras (a filtration) $\mathbf{F} = (\mathcal{F}_t)_{t \geq 0}$ supplementing the triple $(\Omega, \mathcal{F}, \mathbf{P})$, where $\mathcal{F}_t$ is interpreted as the collection of events from $\mathcal{F}$ observable up to time t.

It is this assumption of the presence on $(\Omega, \mathcal{F}, \mathbf{P})$ of a flow $\mathbf{F} = (\mathcal{F}_t)_{t \geq 0}$ that has given rise to such objects as Markov times or stopping times, adapted processes, optional and predictable σ-algebras, martingales, local martingales, semimartingales, the stochastic integral, the Itô change of variables formula, etc., which are the ingredients of the theory of stochastic calculus.

Thus, stochastic calculus axiomatizes the concept of a stochastic basis

$$\mathcal{B} = (\Omega, \mathcal{F}, \mathbf{F} = (\mathcal{F}_t)_{t \geq 0}, \mathbf{P}),$$

which lies at the basis of our entire discussion. Here $(\Omega, \mathcal{F}, \mathbf{P})$ is a probability space, and $\mathbf{F}$ is some distinguished flow of σ-algebras.

The formation of the concept of a "stochastic basis" as one of the specifications of a probability space passed through many stages of discussion of particular cases, refinements, generalizations, etc. Here the theory of stochastic integration for Brownian motion and a centered Poisson measure, developed by Itô, was crucial.

In the first chapter of this volume, an introduction is given to stochastic calculus which is called upon for presenting various aspects of Brownian motion and its connection with the theory of partial differential equations, the latter being the fundamental principle of Kolmogorov's classical paper "Analytical Methods in Probability Theory".

Stochastic integration theory has been brought to perfection for the most part in random processes that are solutions of stochastic differential equations which, as we have said, are a particular case of semimartingales — that wide class of random processes for which stochastic calculus gives a powerful method of analysis.

The second chapter is devoted specifically to the theory of stochastic differential equations as well as stochastic evolution equations and the stochastic calculus of variations (or the Malliavin calculus), a highly effective probabilistic apparatus for study in the theory of partial differential equations, theoretical physics, and ergodic theory.

The third chapter is devoted to the general theory of stochastic calculus proper on probability spaces with filtrations. It presents the basic elements of the general theory of random processes, stochastic integration over semi-martingales, and a number of their applications.

The ideas and methods of stochastic calculus traditionally have found and still find application in diverse sections of probability theory and mathematical statistics. This is illustrated, in particular, in the fourth chapter, in which the methods of martingale theory and stochastic calculus are applied to study questions of the weak convergence of random processes considered as random elements with values in metric spaces.

The contributions made by the team of authors of this volume, S.V. Anulova, A.Yu. Veretennikov, N.V. Krylov, R.Sh. Liptser and A.N. Shiryaev are as follows: Chap. 1 was written by N.V. Krylov; Chap. 2, Part I, §§1, 3, 4; Chap. 2, Part II; Chap. 2, Part III were written by A.Yu. Veretennikov; Chap. 2, Part I, §§2, 5, 6 were written by S.V. Anulova; Chap. 3 was written by R.Sh. Liptser and A.N. Shiryaev; Chap. 4, Part II was written by R.Sh. Liptser; Chap. 4, Part I was written by A.N. Shiryaev.

A.N. Shiryaev

Chapter 1
Introduction to Stochastic Calculus

N.V. Krylov

§1. Brownian Motion and the Wiener Process

1.1. In 1826 the English botanist Brown observed that the microscopic particles found in a liquid move (as afterwards became clear, under the influence of impacts of the liquid's molecules) in a disordered fashion. As has been clarified, the characteristics of this motion, for example, the mean-square deviation from the initial position after a unit of time, depend on the liquid's temperature, its viscosity and other physical parameters. This phenomenon discovered by Brown and thereupon called Brownian motion, is of interest from the viewpoint of physics. Brownian motion was studied by Bachelier (1900), Smoluchovski and Einstein (1905) at the level of rigour typical of modern physics. A mathematical model of Brownian motion was first rigorously constructed in 1923 by Wiener, in honour of which the corresponding random process is also called a *Wiener process*. Wiener constructed a model of Brownian motion as a measure in the space of continuous functions, although 10 more years were required before measure theory, thanks to Kolmogorov's axioms, became the universally recognized basis of probability theory.

Real Brownian motion and its model the Wiener process is three-dimensional. Each of the coordinates of a three-dimensional Wiener process is called a one-dimensional Wiener process.

The mathematical definition of a one-dimensional Wiener process is given in the following manner. Let $(\Omega, \mathcal{F}, \mathbf{P})$ be a probability space and for $t \geq 0$ let $W_t(\omega)$ be a Gaussian process defined on Ω that is continuous with respect to t for almost all ω and is such that $\mathbf{E}W_t = 0$ and $\mathbf{E}W_t W_s = t \wedge s$ for all $t, s \geq 0$. One then says that W_t is a *one-dimensional Wiener process*. Sometimes we add that W_t is a *standard* process, when we want to emphasize that $\mathbf{E}W_t W_s$ is not only proportional to $t \wedge s$, but also equal to it. When the conditions of the definition are satisfied only for $t, s \in [0, T]$, where the constant $T \in (0, \infty)$, one speaks of a Wiener process on $[0, T]$. In what follows, unless specifically stated to the contrary, when we speak of a Wiener process, we will always have in mind the one-dimensional (standard) Wiener process.

The above definition of a Wiener process accords well with the physical nature of Brownian motion on a small time interval or in a volume that is infinite in all directions. Indeed, the Gaussian property is naturally explained by the effect of the large number of "equivalent" actions of the molecules on the suspended particle. Its mean deviation from the starting position must, of course, equal zero. Furthermore, it is natural to take the initial position of the particle as the origin: $W_0 = 0$. Since the increments of the position of the particle over disjoint time intervals of identical length must naturally be independent and have identical variance, in view of the temporal and spatial homogeneity of the liquid under consideration, $\mathbf{E}|W_t - W_s|^2$ must be proportional to $|t - s|$. Choosing the coefficient of proportionality equal to 1 (for example, by means of a change of scale of time or length), we obtain $|t - s| = \mathbf{E}|W_t - W_s|^2 = \mathbf{E}W_t^2 - 2\mathbf{E}W_tW_s + \mathbf{E}W_s^2$, which for $s = 0$ gives $\mathbf{E}W_t^2 = t$. Hence, $2\mathbf{E}W_tW_s = \mathbf{E}W_t^2 + \mathbf{E}W_s^2 - |t - s| = t + s - |t - s| = 2(t \wedge s)$ for any $t, s \geq 0$.

1.2. In order to convince oneself of the *existence of a Wiener process*, it is convenient to use Fourier series and a very special case of Besov's theorems on the imbedding of classes (see Nikol'skii 1969, p. 279).

Lemma 1.1. *Let $\alpha > 1/p$, $p \in [1, \infty)$. Then there exists a constant c_0 such that for any function $f(t)$ measurable on $[0, \pi]$ we have*

$$|f(t) - f(s)| \leq c_0 |t - s|^{\alpha - 1/p} \left(\int_0^\pi \int_0^\pi \frac{|f(x) - f(y)|^p}{|x - y|^{1 + \alpha p}} dx\, dy \right)^{1/p} \quad (1.1)$$

for almost all $t, s \in [0, \pi]$.

Theorem 1.1. *Let $\eta_0(\omega), \eta_1(\omega), \ldots$ be independent normal random variables on some probability space $(\Omega, \mathcal{F}, \mathbf{P})$ such that $\mathbf{E}\eta_k = 0$, $\mathbf{E}\eta_k^2 = 1$. Then for some sequence of integers $N(k) \to \infty$ the functions*

$$W_t^k := \frac{1}{\sqrt{\pi}} t \eta_0 + \sqrt{\frac{2}{\pi}} \sum_{n=1}^{N(k)} \eta_n \frac{1}{n} \sin nt \quad (1.2)$$

converge uniformly for $t \in [0, \pi]$ and their (a.s. [almost surely] continuous) limit is a Wiener process on $[0, \pi]$.

Proof. It follows from Parseval's equality for the Fourier series expansion on $[-\pi, \pi]$ of the functions $I(|x| < t)$, $I(|x| < s)$ as functions of x that for $t, s \in [0, \pi]$

$$\frac{1}{\pi}(t - s)^2 + \frac{2}{\pi} \sum_1^\infty \frac{1}{n^2}(\sin nt - \sin ns)^2$$
$$= \frac{1}{2} \int_{-\pi}^\pi [I(|x| < t) - I(|x| < s)]^2 dx = |t - s|. \quad (1.3)$$

We now take $\alpha \in \left(\frac{1}{4}, \frac{1}{2}\right)$ and set

$$c(n) = \frac{4}{\pi^2} \int_0^\pi \int_0^\pi \frac{1}{|t - s|^{1 + 4\alpha}} \left(\sum_{k=n}^\infty \frac{1}{k^2}(\sin kt - \sin ks)^2 \right)^2 dt\, ds.$$

It follows from (1.3) that

$$c(1) \le \int_0^\pi \int_0^\pi \frac{1}{|t-s|^{4\alpha-1}}\,dt\,ds < \infty.$$

Consequently, $c(n) \to 0$ as $n \to \infty$, and we can define $N(k)$ so that $c(N(k)) \le 2^{-k}$. We now set $\delta_t^k = W_t^{k+1} - W_t^k$. Then by Lemma 1.1 for $p=4$, $s=0$ and taking into account the simple relation between the second and fourth moments of Gaussian variables, we conclude that

$$\mathbf{E}\sup_{t\in[0,\pi]} |\delta_t^k|^4 \le c_0^4 \pi^{4\alpha-1} \int_0^\pi \int_0^\pi \frac{1}{|t-s|^{1+4\alpha}} \mathbf{E}|\delta_t^k - \delta_s^k|^4\,dt\,ds$$

$$= c_1 \int_0^\pi \int_0^\pi \frac{1}{|t-s|^{1+4\alpha}} \left(\sum_{N(k)+1}^{N(k+1)} \frac{1}{n^2}(\sin nt - \sin ns)^2 \right)^2 dt\,ds$$

$$\le c_2 c(N(k)) \le c_2 2^{-k}, \tag{1.4}$$

where c_2 does not depend on k. It follows from this and Hölder's inequality that the quantity

$$\sum_{k=1}^\infty \sup_{t\in[0,\pi]} |W_t^{k+1} - W_t^k|$$

has finite expectation and is therefore finite (a.s.). Thus we have proven that the sequence (1.2) indeed converges uniformly in t (a.s.) to some continuous process, which we denote W_t. Clearly W_t is a Gaussian process, $\mathbf{E}W_t = 0$. Finally, from (1.3) we obtain $\mathbf{E}|W_t - W_s|^2 = |t-s|$, $\mathbf{E}W_t^2 = t$, $\mathbf{E}W_s^2 = s$, $\mathbf{E}W_t W_s = t \wedge s$, as required. □

One can, with the help of the Wiener process on $[0,\pi]$, construct the Wiener process on $[0,\infty)$ by gluing together independent replicas of the processes W_t^n, $t \le \pi$, $n \ge 0$ (each of which is given for example, by a series of the form (1.2)) by the rule

$$W_t = \begin{cases} W_t^0, & 0 \le t < \pi, \\ W_\pi^0 + W_{t-\pi}^1, & \pi \le t < 2\pi, \\ W_\pi^0 + W_\pi^1 + W_{t-2\pi}^2, & 2\pi \le t < 3\pi, \end{cases}$$

and so on.

1.3. The Wiener process plays a highly important role in the theory of random processes, and therefore, considerable attention has been paid to studying its properties. We list just a few of them. The reader can find a vast amount of information regarding the Wiener process in, for example, the book (Itô and McKean 1965).

Theorem 1.2. *The following assertions are equivalent:*
1. *W_t is a Wiener process;*

2. W_t is a continuous process a.s. such that a) $W_0 = 0$ a.s., b) $W_t - W_s \sim \mathcal{N}(0, |t-s|)$,[1] c) $W_{t_1}, W_{t_2} - W_{t_1}, \ldots, W_{t_n} - W_{t_{n-1}}$ are independent for any $0 \leq t_1 \leq \ldots \leq t_n$;

3. W_t is a continuous process a.s. such that a) $W_0 = 0$ a.s., b) $W_t - W_s \sim \mathcal{N}(0, |t-s|)$, c) for any $0 \leq t_1 \leq \ldots \leq t_n$, the random variable $W_{t_n} - W_{t_{n-1}}$ does not depend on $(W_{t_1}, \ldots, W_{t_{n-1}})$.

Khinchin and Lévy found deep properties demonstrating the strong *irregularity* of trajectories of the Wiener process.

Theorem 1.3. (Khinchin's law of the iterated logarithm). *With probability 1 we have*

$$\limsup_{t\downarrow 0} \frac{W_t}{\sqrt{2t \ln \ln \frac{1}{t}}} = 1, \quad \liminf_{t\downarrow 0} \frac{W_t}{\sqrt{2t \ln \ln \frac{1}{t}}} = -1.$$

Theorem 1.4. (on the modulus of continuity, P. Lévy). *With probability 1 we have*

$$\limsup_{\substack{0 \leq s < t \leq 1 \\ u = t-s \to 0}} \frac{|W_t - W_s|}{\sqrt{2u \ln \frac{1}{u}}} = 1.$$

Theorems 1.3, 1.4 show that

$$|W_r| \leq C(\omega)\sqrt{2r \ln \ln \frac{1}{r}}, \quad |W_t - W_s| \leq C(\omega)\sqrt{2|t-s| \ln \frac{1}{|t-s|}}$$

for $0 \leq t, s, r \leq 1$, where $C(\omega)$ can be taken arbitrarily close to 1 on the right if $|t-s|, r$ are sufficiently small. The proofs of Theorems 1.3, 1.4 are quite complicated. It is therefore useful to keep in mind that one can easily obtain several rougher results from Lemma 1.1: if $\epsilon \in (0, \frac{1}{2})$ is fixed, then for almost every ω there exists $C(\omega) < \infty$ such that $|W_t - W_s| \leq C(\omega)|t-s|^{(1/2)-\epsilon}$ for all $0 \leq t, s \leq 1$; in particular, $|W_t| \leq C(\omega)t^{(1/2)-\epsilon}$. In fact, it is sufficient in Lemma 1.1 to take $\alpha = \frac{1}{2} - \frac{\epsilon}{2}$, $p = \frac{2}{\epsilon}$, $f(t) = W_t$ and observe that the integral on the right hand side of (1.1) is finite for almost all ω, since $\mathbf{E}|W_x - W_y|^p = N|x-y|^{p/2}$, where $N = \mathbf{E}|\xi|^p$, $\xi \sim \mathcal{N}(0,1)$, whence

$$\mathbf{E}\int_0^1\int_0^1 \frac{|W_x - W_y|^p}{|x-y|^{1+\alpha p}}\,dxdy = N\int_0^1\int_0^1 1\,dxdy < \infty.$$

Still another property exhibiting the irregularity of the trajectories of the Wiener process and playing an important role in the theory of the Itô integral, bears the name the *quadratic variation theorem*.

---

[1] The notation $\xi \sim \mathcal{N}(a,b)$ means that the random variable has a normal (Gaussian) distribution with expectation a and variance b.

Theorem 1.5. *Let $0 \leq s < t < \infty$, $s = t_{0,n} \leq t_{1,n} \leq \ldots \leq t_{k_n,n}$, $n = t$ be a sequence of subdivisions of $[s,t]$, the diameters of which approach zero as $n \to \infty$. Let W_t, b_t be two independent Wiener processes. Then as $n \to \infty$*

$$\sum_{0}^{k_n-1} (W_{t_{i+1,n}} - W_{t_{i,n}})^2 \to t - s,$$

$$\sum_{0}^{k_n-1} (W_{t_{i+1,n}} - W_{t_{i,n}})(b_{t_{i+1,n}} - b_{t_{i,n}}) \to 0,$$

where the convergence is in probability.

Proof. The proof of this theorem is completely elementary and is based on the fact that the variances of the sums under consideration are easily computed. It then turns out that they approach zero. □

It follows from the quadratic variation theorem that the usual variation of the Wiener trajectory along $[s,t]$ equals infinity with probability 1. In fact,

$$\sum_{0}^{k_n-1} (W_{t_{i+1,n}} - W_{t_{i,n}})^2 \leq \max_{i} |W_{t_{i+1,n}} - W_{t_{i,n}}| \mathrm{Var}_{[s,t]} W_r \to 0 \neq t - s$$

as $n \to \infty$ at those ω for which $\mathrm{Var}_{[s,t]} W_r(\omega) < \infty$.

From the viewpoint of the theory of functions of a real variable, almost every trajectory of W_t possesses quite exotic irregularity properties. However, they in fact allow certain functionals of a Wiener trajectory to possess unusual regularity. For example, if the function $f(x)$ is only locally Borel square-summable, then

$$\int_0^1 f(x + h_t) dt$$

will not even be a continuous function of x if h_t is a smooth function of t. At the same time, for almost every ω the function

$$\int_0^1 f(x + W_t) dt$$

is absolutely continuous with respect to x and its derivative with respect to x is square-integrable with respect to x for any finite range of variation of x. In other words, W_t "mollifies" f. This is proved by extending from smooth functions f to non-smooth ones and using the following computations obtained via the Fourier transform:

$$\mathbf{E}\int_{-\infty}^{\infty}\left(\int_0^1 f'(x+W_t)dt\right)^2 dx = \mathbf{E}\int_{-\infty}^{\infty}\left|\xi\tilde{f}(\xi)\int_0^1 e^{i\xi W_t}dt\right|^2 d\xi$$

$$= 2\int_{-\infty}^{\infty}\xi^2|\tilde{f}(\xi)|^2\mathbf{E}\int_0^1 ds\int_0^s dt\, e^{i\xi(W_s-W_t)}d\xi$$

$$= 2\int_{-\infty}^{\infty}\xi^2|\tilde{f}(\xi)|^2\int_0^1 ds\int_0^s dt\, e^{-\xi^2(s-t)/2}d\xi$$

$$= 4\int_{-\infty}^{\infty}|\tilde{f}(\xi)|^2\int_0^1\left(1-e^{-\xi^2 s/2}\right)ds\, d\xi$$

$$\leq 4\int_{-\infty}^{\infty}|\tilde{f}(\xi)|^2 d\xi$$

$$= 4\int_{-\infty}^{\infty}f^2(x)\,dx,$$

where $\tilde{f}$ is the Fourier transform of f.

1.4. Physical arguments concerning the nature of the Wiener process suggest that it must play the role of a limiting process for random polygonal lines and the same role that the Gaussian distribution plays for sums of independent identically distributed variables. To clarify this we consider the simplest model describing the process of the motion of a point particle under the impacts of others. Let η_k, $k=1,2,\ldots$, be independent identically distributed variables with $\mathbf{E}\eta_k=0$, $\mathbf{E}\eta_k^2=1$. We fix an integer $n\geq 1$ and at moments of time that are multiples of $1/n$ we let our particle undergo an action moving it by $\eta_k/\sqrt{n}$, where k is the number of the action. We suppose that the particle is initially located at zero and $S_k=\eta_1+\ldots+\eta_k$. Then at time k/n the particle will be found at the point $S_k/\sqrt{n}$ and will remain at this point during the time interval $[k/n,(k+1)/n)$. Since the motion of a real particle has a continuous trajectory, we change our piecewise-constant trajectory to a piecewise-linear one, maintaining its positions at k/n. Thus we arrive at the process

$$\xi_t^n = \frac{1}{\sqrt{n}}S_{[nt]} + (nt-[nt])\frac{1}{\sqrt{n}}\eta_{[nt]+1}.$$

This process gives an approximate representation of 1-dimensional Brownian motion, and we must obtain a more complete representation as $n\to\infty$. Incidentally, the mere desire to let n go to infinity leads to intervals between the interactions of length $\frac{1}{n}$ and a displacement at the moment of impact by $\eta_n/\sqrt{n}$, since in this case, by the central limit theorem, ξ_t^n is asymptotically normal with parameters $(0,t)$.

As in the central limit theorem, the issue is no longer one of the convergence of ξ_t^n as $n\to\infty$ to some process, but of the convergence of the *distributions* of the processes $\xi_\cdot^n$ to the Wiener process distribution. Let $C[0,1]$ be the metric space of all real continuous functions on $[0,1]$ with distance between two elements x,y of this space defined by the formula

$$\rho(x., y.) = \sup\{|x_t - y_t| : t \in [0,1]\}.$$

We take the σ-algebra of Borel sets in $C[0,1]$. It then turns out that $\xi_\cdot^n$ and $W.$ are random elements with values in $C[0,1]$. The random elements $\xi_\cdot^n$, $W.$ have distributions on $C[0,1]$, that is, the corresponding measures on the Borel subsets of $C[0,1]$ are given by

$$F_{\xi_\cdot^n}(A) = \mathbf{P}(\xi_\cdot^n \in A), \quad F_{W.}(A) = \mathbf{P}(W. \in A).$$

Let us also recall that a sequence of measures μ_n on the Borel sets of a metric space X is called *weakly convergent* to a measure μ if for all continuous bounded functions f on X

$$\lim_{n\to\infty} \int_X f(x)\mu_n(dx) = \int_X f(x)\mu(dx).$$

The next theorem, which is an analogue of the central limit theorem, is called the *functional limit theorem* and also bears the unfortunate name of the *invariance principle*. It has enormous theoretical and applied value.

Theorem 1.6. (Donsker). *The sequence of distributions $F_{\xi_\cdot^n}$ converges weakly to $F_{W.}$ on $C[0,1]$.*

This theorem has generalizations to a wide class of processes different from ξ_t^n (see Chap. 4).

As an example of the application of Donsker's Theorem, let us find the distribution of the random variable $\zeta = \max\{W_t, t \in [0,1]\}$. Since $\max\{x_t, t \in [0,1]\}$ is a continuous function of $x.$ on $C[0,1]$, by Donsker's Theorem the distributions of the random variables $\zeta^n = \max\{\xi_t^n, t \in [0,1]\}$ weakly converge to that of ζ.

Here it is absolutely essential that we are able to construct ξ_t^n from any independent variables η_k satisfying the conditions $\mathbf{E}\eta_k = 0$, $\mathbf{E}\eta_k^2 = 1$. We take η_k so that $\mathbf{P}\{\eta_k = 1\} = \mathbf{P}\{\eta_k = -1\} = \frac{1}{2}$. Then the probability of appearance of each fixed polygonal line as a realization of ξ_t^n on $[0,1]$ is the same and equals 2^{-n}. Furthermore, the number of polygonal lines favorable to the events

$$\left\{\zeta^n \geq \frac{i}{\sqrt{n}}, \xi_1^n < \frac{i}{\sqrt{n}}\right\}, \quad \left\{\zeta^n \geq \frac{i}{\sqrt{n}}, \xi_1^n > \frac{i}{\sqrt{n}}\right\},$$

where i is a positive integer, is the same. This can be shown by using the "reflection principle" (D. André), when each polygonal line suitable for the first event does not change up to the time of attaining the level $i/\sqrt{n}$, and then is symmetrically reflected with respect to this level and jumps to a polygonal line favorable to the second event. Moreover, we have a one-to-one correspondence. Consequently,

$$\mathbf{P}\left\{\zeta^n \geq \frac{i}{\sqrt{n}}, \xi_1^n < \frac{i}{\sqrt{n}}\right\} = \mathbf{P}\left\{\zeta^n \geq \frac{i}{\sqrt{n}}, \xi_1^n > \frac{i}{\sqrt{n}}\right\}.$$

Furthermore, it is clear that

$$\mathbf{P}\left\{\zeta^n \geq \frac{i}{\sqrt{n}},\ \xi_1^n > \frac{i}{\sqrt{n}}\right\} = \mathbf{P}\left\{\xi_1^n > \frac{i}{\sqrt{n}}\right\},$$

$$\mathbf{P}\left\{\zeta^n \geq \frac{i}{\sqrt{n}}\right\} = \mathbf{P}\left\{\zeta^n \geq \frac{i}{\sqrt{n}},\ \xi_1^n > \frac{i}{\sqrt{n}}\right\}$$
$$+ \mathbf{P}\left\{\zeta^n \geq \frac{i}{\sqrt{n}},\ \xi_1^n < \frac{i}{\sqrt{n}}\right\} + \mathbf{P}\left\{\xi_1^n = i/\sqrt{n}\right\}.$$

Hence

$$\mathbf{P}\left\{\zeta^n \geq \frac{i}{\sqrt{n}}\right\} = 2\mathbf{P}\left\{\xi_1^n > \frac{i}{\sqrt{n}}\right\} + \mathbf{P}\left\{\xi_1^n = i/\sqrt{n}\right\}$$

for integers $i > 0$. For $i = 0$, this equation is trivial. We then have

$$\mathbf{P}\{\zeta^n = i/\sqrt{n}\} = \mathbf{P}\{\zeta^n \geq i/\sqrt{n}\} - \mathbf{P}\{\zeta^n \geq (i+1)/\sqrt{n}\}$$

$$= 2\mathbf{P}\left\{\xi_1^n = \frac{i+1}{\sqrt{n}}\right\} + \mathbf{P}\left\{\xi_1^n = \frac{i}{\sqrt{n}}\right\} - \mathbf{P}\left\{\xi_1^n = \frac{i+1}{\sqrt{n}}\right\}$$

$$= \mathbf{P}\left\{\xi_1^n = \frac{i+1}{\sqrt{n}}\right\} + \mathbf{P}\left\{\xi_1^n = \frac{i}{\sqrt{n}}\right\},\ i \geq 0, \quad (1.5)$$

$$\mathbf{E}f(\zeta^n) = \mathbf{E}f\left(\xi_1^n - \frac{1}{\sqrt{n}}\right) + \mathbf{E}f(\xi_1^n),$$

$$\mathbf{E}f(\zeta) = 2\mathbf{E}f(W_1),$$

$$\mathbf{P}\{\max_{t \leq 1} W_t \geq x\} = 2\mathbf{P}\{W_1 \geq x\},\ x \geq 0,$$

where f is any continuous bounded function equal to zero on the negative semi-axis.

Formula (1.5), in fact, gives the desired distribution. In general,

$$\mathbf{P}\{\max_{t \leq T} W_t \geq x\} = 2\mathbf{P}\{W_T \geq x\} = \sqrt{\frac{2}{\pi T}} \int_x^\infty e^{-y^2/2T} dy, \quad (1.6)$$

where $x \geq 0$, $T > 0$. This follows, for example from (1.5) and from the elementary self-similarity property of W_t, which says that $c^{-1}W_{c^2 t}$ is a Wiener process for any constant $c \neq 0$. Not much more complicated is the following fact, which reduces the analysis of the behavior of W_t for large t to the analysis of this behavior for small t.

Theorem 1.7. *If W_t is a Wiener process, then $tW_{1/t}$ is a Wiener process for $t > 0$. Furthermore, $tW_{1/t} \to 0$ a.s. as $t \downarrow 0$.*

Turning to (1.6), let us find the distribution of the first hitting time τ_x of the level x by the process W_t. Since $\{\tau_x \leq T\} = \{\max_{t \leq T} W_t \geq x\}$, it follows that

$$\mathbf{P}\{\tau_x \leq T\} = \sqrt{\frac{2}{\pi T}} \int_x^\infty e^{-y^2/2T} dy = \sqrt{\frac{2}{\pi}} \int_{x/\sqrt{T}}^\infty e^{-y^2/2} dy,\ T > 0,\ x \geq 0. \quad (1.7)$$

This is the Wald distribution which plays an important role in probability theory, mathematical statistics and, surprisingly, in the theory of partial differential parabolic and elliptic equations. It has a density equal to

$$\frac{x}{\sqrt{2\pi T^3}} e^{-x^2/2T}.$$

It follows from the form of this density that $\mathbf{E}\tau_x^\alpha < \infty$ for $x > 0$ only when $\alpha \in (0, 1/2)$.

§2. Probabilistic Construction of the Solution of the Heat Equation. Relation Between the Wiener Process and the Laplace Operator

2.1. Brownian motion is caused by the thermal motion of molecules. As is well known, the latter is also connected with the heat equation of the form

$$u_t + \frac{1}{2} u_{xx} + f = 0. \tag{2.1}$$

It is therefore not surprising that the Wiener process is related to equation (2.1). In the simplest case of the Cauchy problem for equation (2.1), when it is considered in the strip $(t, x) \in (0, T) \times E_1$ with boundary condition

$$u(T, x) = g(x), \quad x \in E_1, \tag{2.2}$$

this relation is established in a very simple manner. Indeed,

$$\mathbf{E}\left[\int_0^{T-t} f(t+s, x+W_s)ds + g(x+W_{T-t})\right]$$
$$= \int_0^{T-t} \frac{1}{\sqrt{2\pi s}} \int f(t+s, x+y) e^{-y^2/2s} dy\, ds$$
$$+ \frac{1}{\sqrt{2\pi(T-t)}} \int g(x+y) e^{-y^2/2(T-t)} dy, \tag{2.3}$$

and this expression, as is known from the theory of differential equations, is a solution of (2.1)–(2.2) if, for example, g, f are bounded, g is continuous and f satisfies a Hölder condition in (t, x) with some exponent $\alpha \in (0, 1)$. The left hand side of (2.3) is a particular case of a general formula, which is also suitable for representing the solution u of (2.1) in an arbitrary domain $Q \subset (0, \infty) \times E_1$ that is bounded on the t-axis. It turns out that under broad assumptions we have for $(t, x) \in Q$,

$$u(t, x) = \mathbf{E}\left[\int_0^{\tau(t,x)} f(t+s, x+W_s)ds + u(t+\tau(t,x), x+W_{\tau(t,x)})\right], \tag{2.4}$$

where $\tau(t,x)$ is the first exit time of the "thermal" process $(t+s, x+W_s)$ from Q as a function of $s \in (0, \infty)$. In turn, formula (2.4) is a very particular case of the probabilistic representation of the solutions of degenerate elliptic and parabolic equations. Such formulas are most easily obtained by means of the remarkable Itô formula, which is the concern of the next section. However, for the particular case of a thermal process, one can convince oneself of the validity of (2.4) in another way, using the following arguments.

For example, let $Q \subset (0,T) \times E_1$, and suppose that one can extend the function u outside of Q to the whole of $[0,T] \times E_1$ so that it becomes a smooth function which we will also denote by u. Let us define f everywhere in $(0,T) \times E_1$ by formula (2.1). Then, as was said above, $u(t,x)$ is equal to the left hand side of (2.3) with $g(x) = u(T,x)$. Hence

$$u(t,x) = \mathbf{E}\int_0^{\tau(t,x)} f(t+s, x+W_s)ds + \mathbf{E}\bigg[\int_0^{T-\tau(t,x)-t} f(t+\tau(t,x)+s,$$
$$x + W_{\tau(t,x)} + (W_{\tau(t,x)+s} - W_{\tau(t,x)}))ds + g(x + W_{\tau(t,x)}$$
$$+ (W_{T-\tau(t,x)-t+\tau(t,x)} - W_{\tau(t,x)}))\bigg]. \tag{2.5}$$

In the second expectation $(\tau(t,x), W_{\tau(t,x)})$ and the process $W_{\tau(t,x)+s} - W_{\tau(t,x)}$ are independent and the latter process is a Wiener process. This assertion (called the *strong Markov property* of the Wiener process) is an assertion of the same type as Theorem 1.2. Therefore the second term in (2.5) equals

$$\mathbf{EE}\bigg[\int_0^{T-r} f(r+s, y+W_s)ds + g(y+W_{T-r})\bigg]\bigg|_{\substack{r=t+\tau(t,x) \\ y=x+W_{\tau(t,x)}}}$$
$$= \mathbf{E}u(t+\tau(t,x), x + W_{\tau(t,x)}),$$

where the latter is obtained by equating $u(t,x)$ with the left hand side of (2.3).

Equations of the type of formula (2.4) are interesting both from the viewpoint of probability theory and the viewpoint of the theory of differential equations. For example, suppose that we are interested in the average time required for a Wiener process to exit from the interval $(-a, b)$, where $a > 0$, $b > 0$. Let τ be the first exit time of W_t from $(-a,b)$, $Q = (0,T) \times (-a,b)$, $u(t,x) = u(x) = (b-x)(x+a)$. Then $f = 1$, $\tau(0,0) = \tau \wedge T$ and by (2.4),

$$ab = u(0,0) = \mathbf{E}[\tau \wedge T + u(W_{\tau \wedge T})]. \tag{2.6}$$

It follows from formula (1.7) that $\tau = (\tau_b \wedge \tau_{-a})$ is finite with probability 1, and since $u(W_\tau) = 0$ if $\tau < \infty$, we conclude from (2.6) by letting $T \to \infty$ that $\mathbf{E}\tau = ab$.

Formula (2.4) shows that the solution of equation (2.1) in the domain Q is uniquely determined by f and by the values of u only at those points of ∂Q that can be joined to at least one point $(t,x) \in Q$ by a continuous curve of the form $(t+s, x+x_s)$, $s \geq 0$, where $x_0 = 0$. This fact pertains also to the theory of differential equations and distinguishes the so-called parabolic boundary of Q.

Here is another simple application of (2.4) to the theory of differential equations.

Theorem 2.1. (Tikhonov). *Let $u(t,x)$ be a continuous function on $[0,T] \times E_1$ having continuous derivatives with respect to (t,x) of the form u_t, u_{xx} on $(0,T) \times E_1$ and suppose further that $u_t + \frac{1}{2} u_{xx} = 0$ on $(0,T) \times E_1$, $u(T,x) = 0$, $|u(t,x)| \le \exp(cx^2)$, where $c \in (0,\infty)$. Then $u \equiv 0$ on $[0,T] \times E_1$.*

Proof. Since we can change the origin it suffices to prove that $u(s,0) = 0$ for $s \in [0,T]$. Choosing as Q the rectangle $(0,T) \times (-x,x)$, where $x > 0$, by formula (2.4) we find that

$$u(s,0) = \mathbf{E} u(s+\tau_x, x) I(\tau_x < \tau_{-x} \wedge (T-s))$$
$$+ \mathbf{E} u(s+\tau_{-x}, -x) I(\tau_{-x} < \tau_x \wedge (T-s)).$$

Hence, from (1.7) for $s \in [T - (4c)^{-1}, T]$, $s \ge 0$,

$$|u(s,0)| \le \frac{2x}{\sqrt{2\pi}} \int_0^{T-s} r^{-3/2} e^{-x^2/2r} e^{cx^2} dr$$
$$\le \frac{2x}{\sqrt{2\pi}} \int_0^{T-s} r^{-3/2} e^{-x^2/4r} dr$$
$$= \frac{2}{\sqrt{2\pi}} \int_0^{\frac{T-s}{\sqrt{x}}} r^{-3/2} e^{-1/4r} dr,$$

since $c - 1/2r \le -1/4r$ when $r \in (0, T-s)$. This approaches zero as $x \to \infty$. Accordingly, $u(s,0) = 0$ for $s \in [T - (4c)^{-1}, T]$, $s \ge 0$. Similarly, $u(s,x) = 0$ for these same s and all x. We have proved that if $u = 0$ for $s = T$, then $u = 0$ in the strip $x \in E_1$, $s \in [T - (4c)^{-1}, T]$, $s \ge 0$. If in the above argument we take $(T - (4c)^{-1}) \vee 0$ instead of T, then we find that $u = 0$ in a strip lying still closer to the origin. After a finite number of similar steps, we see that $u = 0$ in $[0,T] \times E_1$, as required. □

Formula (2.4) has a multidimensional analogue. Indeed, let us call the d-dimensional process $W_t = (W_t^1, \ldots, W_t^d)$ a d-dimensional Wiener process if W_t^i is a Wiener (one-dimensional) process for $i = 1, \ldots, d$ and the processes $W_\cdot^1, \ldots, W_\cdot^d$ are independent. It turns out that if $u(t,x)$, $t \in [0,T]$, $x \in E_d$, is a solution of

$$u_t + \frac{1}{2} \Delta u + f = 0 \quad \left(\Delta = \frac{\partial^2}{(\partial x^1)^2} + \ldots + \frac{\partial^2}{(\partial x^d)^2}\right) \tag{2.7}$$

in $[0,T] \times E_d$ and $u(T,x) = g(x)$ on E_d, then under natural assumptions, $u(t,x)$ is equal to the left hand side of (2.3), where W_t is a d-dimensional Wiener process. This is shown with the help of a formula similar to (2.3).

Formula (2.4), along with its derivation, also naturally goes over to the multi-dimensional case.

2.2. Next, let $\mathcal{D}$ be a bounded domain in E_d, and let u be a sufficiently regular solution of the equation

$$\frac{1}{2}\Delta u + f(x) = 0 \tag{2.8}$$

in $\mathcal{D}$, let u be continuous in $\bar{\mathcal{D}}$ and $u = g$ on $\partial \mathcal{D}$. Then for $x \in \mathcal{D}$,

$$u(x) = \mathbf{E}\left[\int_0^{\tau(x)} f(x + W_t)dt + g(x + W_{\tau(x)})\right], \tag{2.9}$$

where $\tau(x) = \inf\{t : x + W_t \in \mathcal{D}\}$. This is because u does not depend on t, so that $u_t + \frac{1}{2}\Delta u + f = 0$, and by formula (2.4) for $t = T$, $Q = (0, 2T) \times \mathcal{D}$,

$$u(x) = \mathbf{E}\left[\int_0^{T \wedge \tau(x)} f(x + W_t)dt + u(x + W_{T \wedge \tau(x)})\right].$$

It remains to let $T \to \infty$.

Formula (2.9) gives a probabilistic representation of the solution of the Laplace equation. It is as important as formula (2.4). With its help, let us show, for example, that the two-dimensional and hence also the one-dimensional Wiener process are *recurrent*. Since the two-dimensional Wiener process possesses the strong Markov property, it is sufficient to establish that it reaches any circle on the plane with probability 1. Let $\tau_r(x) = \inf\{t : |x + W_t| = r\}$. It is sufficient to convince oneself that $\mathbf{P}\{\tau_r(x) < \infty\} = 1$ for all $r > 0$, $x \in E_2$, $|x| > r$.

Since $\Delta \ln |x| = 0$, it follows from (2.9) applied to $u(x) = (\ln R - \ln |x|)(\ln R - \ln r)^{-1}$ that for $R > |x| > r$ we have

$$\frac{\ln R - \ln |x|}{\ln R - \ln r} = \mathbf{E}u(x + W_{\tau_r(x) \wedge \tau_R(x)}) = \mathbf{P}\{\tau_r(x) < \tau_R(x)\}.$$

As $R \to \infty$ this gives

$$1 = \mathbf{P}\{\tau_r(x) < \lim_{R \to \infty} \tau_R(x)\} = \mathbf{P}\{\tau_r(x) < \infty\}.$$

With the help of (2.9) one can show that the d-dimensional Wiener process W_t for $d \geq 3$ is non-recurrent and, furthermore, $|W_t| \to \infty$ as $t \to \infty$ (a.s.).

§3. The Itô Integral and Differentiation Rules for Composite Stochastic Functions

3.1. Let $W_t = W_t(\omega)$, $t \geq 0$, $\omega \in \Omega$, be the standard (one-dimensional) Wiener process on some probability space $(\Omega, \mathcal{F}, \mathbf{P})$. The one-dimensional motion of a Brownian particle for some fixed temperature is quite well described by it. At another constant temperature, the variance of the Brownian particle is multiplied by a constant, and the process $fW_t(\omega)$, where $f = $ const, will be suitable for describing its motion. If, on the other hand, the temperature of the fluid changes with time, but is piecewise-constant and equals f_i on each interval $[t_i, t_{i+1})$ of some decomposition $0 = t_0 \leq t_1 \leq \ldots$ of the time semi-axis $[0, \infty)$, then the position of the one-dimensional Brownian particle at t can be represented in the form

$$\sum_i f_i(W_{t \wedge t_{i+1}} - W_{t \wedge t_i}). \tag{3.1}$$

In passing from piecewise-constant temperatures to a continuously changing one, expression (3.1) is transformed into the stochastic integral

$$\int_0^t f_s dW_s. \tag{3.2}$$

One cannot interpret (3.2) as a Stieltjes integral for any fixed ω, since $W_t(\omega)$ has unbounded variation (see Sect. 1.3). Wiener found a way of defining the integral of non-stochastic functions f_t with respect to W_t. For smooth f_t he set

$$\int_0^t f_s dW_s = f_t W_t - \int_0^t W_s f'_s ds. \tag{3.3}$$

The linear operator so obtained, which is defined on smooth functions f, turns out to possess the property (isometry) that

$$\mathbf{E}\left(\int_0^t f_s dW_s\right)^2 = \int_0^t f_s^2 ds.$$

With the help of this property the stochastic integral is extended from smooth functions to the whole of $L_2[0,t]$.

Itô (Itô 1944), (Itô 1946), (Itô 1951) observed in the 40s that the stochastic integral with respect to W_t extends to a wide class of functions f_t depending also on the "event" ω, but in a "non-anticipative" manner. He not only significantly extended the class of integrable functions, but also proved the continuity of the integral with respect to the upper limit, deduced the rules of operation for stochastic integrals, the famous "Itô formula", and thereby initiated the development of stochastic analysis. Itô considered not only integrals with respect to W_t, but also with respect to Poisson measures and centered Poisson measures.

To describe Itô's construction, we need the concept of a Wiener process with respect to a flow of σ-algebras. Suppose that for every $t \geq 0$ we are given a σ-algebra $\mathcal{F}_t$ of subsets of Ω, where $\mathcal{F}_t \subset \mathcal{F}$, $\mathcal{F}_t \subset \mathcal{F}_s$ for $t \leq s$. Then we say that there is an (expanding) flow of σ-algebras $\{\mathcal{F}_t\}$. A Wiener process $W_t(\omega)$ on $(\Omega, \mathcal{F}, \mathbf{P})$ is called *Wiener with respect to the flow* $\{\mathcal{F}_t\}$ if W_t is $\mathcal{F}_t$-measurable for each $t \geq 0$ and $W_s - W_t$ does not depend on $\mathcal{F}_t$ for any $t \leq s$. In this case we also say that $\{W_t, \mathcal{F}_t\}$ is a Wiener process. It must be said that there always exists a flow $\{\mathcal{F}_t\}$ with respect to which the process W_t is Wiener. It follows from Theorem 1.2 that one can take, for example, $\mathcal{F}_t = \mathcal{F}_t^W := \sigma(W_s, s \leq t)$.

We will call a real *stochastic process* $f_t(\omega)$, $t \geq 0$, $\omega \in \Omega$, simple if there exists a decomposition $0 = t_0 \leq t_1 \leq \ldots$ of $[0, \infty)$ such that $f_t = f_{t_i}$ for $t \in [t_i, t_{i+1})$, $i = 0, 1, \ldots$, f_t is $\mathcal{F}_t$-measurable for each t and

$$\mathbf{E} \int_0^\infty |f_t|^2 dt < \infty. \tag{3.4}$$

We denote the set of all simple processes by H_0.

For a simple process f_t we set

$$\int_0^\infty f_t dW_t = \sum_{i=0}^\infty f_{t_i}(W_{t_{i+1}} - W_{t_i}).$$

Itô takes as the basis of the construction of the stochastic integral this expression, similar to (3.1), and not the integration by parts formula (3.3). Using elementary brief calculations it is shown that for $t \in H_0$

$$\mathbf{E}\left(\int_0^\infty f_t dW_t\right)^2 = \mathbf{E} \int_0^\infty |f_t|^2 dt.$$

This clearly enables us to extend the stochastic integral to functions f for which there is a sequence $f_n \in H_0$ such that

$$\mathbf{E} \int_0^\infty |f_t - f_{nt}|^2 dt \to 0.$$

In other words, the functions $f \in \bar{H}_0 =: H$ prove to be Itô-integrable, where the closure is understood in the sense of the Hilbert space $L_2(\Omega \times (0, \infty))$. Itô proved that the set H coincides with the set of functions f_t that are $\mathcal{F}_t$-measurable for each t, measurable with respect to (ω, t) and are such that (3.4) holds (more precisely, the latter set as a subset of $L_2(\Omega \times (0, \infty))$ coincides with H).

Further, the integral with finite limits from s to t for $0 \leq s \leq t \leq \infty$ is defined by the formula

$$\int_s^t f_r dW_r = \int_0^\infty f_r I(s \leq r \leq t) dW_r.$$

Itô extended the integral with finite limits to a class of functions wider than H. We denote by S the set of all real functions $f_t = f_t(\omega)$ that are measurable with respect to (ω, t), $\mathcal{F}_t$-measurable for each t and are such that

$$\int_0^T f_t^2 dt < \infty \text{ (a.s.)}$$

for any constant $T > 0$. Using elementary facts from martingale theory Itô proved that for $f \in S$ the process

$$\int_0^t f_s dW_s \tag{3.5}$$

has a modification that is continuous with respect to t. In what follows, by the integral (3.5) we shall always mean the continuous process. The next theorem on passing to the limit under the stochastic integral sign forms the basis for extending the Itô stochastic integral from the class H to S.

Theorem 3.1. Let $f_0, f_1, f_2, \ldots \in S$, $0 \le T \le \infty$. It is asserted that

$$\sup_{t \le T} \left| \int_0^t f_{0s} dW_s - \int_0^t f_{ns} dW_s \right| \xrightarrow{\mathbf{P}} 0$$

if and only if

$$\int_0^T |f_{0s} - f_{ns}|^2 ds \xrightarrow{\mathbf{P}} 0.$$

We use this theorem to find $\int_0^t W_s dW_s$. We have

$$2\int_0^t W_s dW_s = \mathbf{P} - \lim_{n \to \infty} 2\int_0^t \sum_{i=0}^{n-1} W_{t\frac{i}{n}} I\left(t\frac{i}{n} \le s < t\frac{i+1}{n}\right) dW_s$$

$$= \mathbf{P} - \lim_{n \to \infty} \sum_{i=0}^{n-1} 2 W_{t\frac{i}{n}} (W_{t\frac{i+1}{n}} - W_{t\frac{i}{n}})$$

$$= \mathbf{P} - \lim_{n \to \infty} \sum_{i=0}^{n-1} [W_{t\frac{i+1}{n}}^2 - W_{t\frac{i}{n}}^2 - (W_{t\frac{i+1}{n}} - W_{t\frac{i}{n}})^2]$$

$$= W_t^2 - t \text{ (a.s.)},$$

where the last equality holds by virtue of Theorem 1.5. Since the first and last expressions in this chain of equations are continuous with respect to t (a.s.), it follows that almost surely we have

$$2\int_0^t W_s dW_s = W_t^2 - t \tag{3.6}$$

for all t.

3.2. This formula is a particular case of the remarkable *Itô formula*, which we now discuss. We turn straightaway to the multidimensional case. We say that a k-dimensional Wiener process $W_t = (W_t^1, \ldots, W_t^k)$ is Wiener with respect to a flow of σ-algebras $\{\mathcal{F}_t\}$ (for short, $\{W_t, \mathcal{F}_t\}$ is a Wiener process) if $\{W_t^i, \mathcal{F}_t\}$ is a Wiener process for each $i = 1, \ldots, k$. We say that the vector process $f_t = (f_t^1, \ldots, f_t^k) \in S^k$ if $f^i \in S$ for $i = 1, \ldots, k$. For $f \in S^k$ we set

$$\int_0^t f_s dW_s = \sum_{i=1}^k \int_0^t f_s^i dW_s^i.$$

If we are given a matrix process $\sigma_t = (\sigma_t^{ij}, i = 1, \ldots, d, j = 1, \ldots, k)$, then we naturally write $\sigma \in S^{dk}$, when $\sigma_t^{ij} \in S$ for all i, j. Furthermore, in this case $\int_0^t \sigma_s dW_s$ is defined as the k-dimensional process for which

$$\left(\int_0^t \sigma_s dW_s\right)^i = \sum_{j=1}^k \int_0^t \sigma_s^{ij} dW_s^j.$$

Further, it is convenient to introduce the concept of a stochastic differential. It is interesting that various authors, (Itô, Gikhman *et al.*) at different times gave an interpretation of this concept that is different from the sense in which it is described below. It appeared considerably later than the Itô stochastic integral when the Itô formula had obtained wide recognition and the necessity arose to write it in the most simple and natural way. Let $\sigma = (\sigma^{ij}) \in S^{dk}$, let $b_t = (b_t^1, \ldots, b_t^d)$ be $\mathcal{F}_t$-measurable for each t and measurable with respect to (ω, t) and let $\int_0^t |b_s| ds < \infty$ (a.s.) for any t. In this case, if for a process $\xi_t = (\xi_t^1, \ldots, \xi_t^d)$ we have almost surely

$$\xi_t = \xi_0 + \int_0^t b_s ds + \int_0^t \sigma_s dW_s$$

for all t, then we say that ξ_t has a *stochastic differential*, and write

$$d\xi_t = b_t dt + \sigma_t dW_t. \tag{3.7}$$

We call the right hand side, which is a purely formal expression, the *stochastic differential* of the process ξ_t.

Let us define the rules of operation with stochastic differentials. The addition of them and multiplication by a constant are defined in the natural manner. When multiplying stochastic differentials we organize the removal of parentheses in the usual manner, and use the following multiplication table:

$$dt\,dt = dt\,dW_t^i = dW_t^i dt = 0, \quad dW_t^i dW_t^i = dt, \quad dW_t^i dW_t^j = 0,$$

for $i \neq j$. For example,

$$\left(b_t^i dt + \sum_{j=1}^k \sigma_t^{ij} dW_t^j\right)\left(\tilde{b}_t^i dt + \sum_{j=1}^k \tilde{\sigma}_t^{ij} dW_t^j\right) = \sum_{j=1}^k \sigma_i^{ij} \tilde{\sigma}_i^{ij} dt.$$

Theorem 3.2. (Itô formula). *Let ξ_t be a d-dimensional process with a stochastic differential, and let $u(x)$ be a twice continuously differentiable function on E_d. Then the process $u(\xi_t)$ has a stochastic differential and*

$$du(\xi_t) = \sum_{i=1}^{d} u_{x^i}(\xi_t)d\xi_t^i + \frac{1}{2}\sum_{i,j=1}^{d} u_{x^i x^j}(\xi_t)d\xi_t^i d\xi_t^j. \qquad (3.8)$$

Theorem 3.2 is most simply derived from the next theorem.

Theorem 3.3. *Let the one-dimensional processes ξ_t, η_t have stochastic differentials (with a k-dimensional Wiener process W). Then the process $\xi_t \eta_t$ has a stochastic differential and*

$$d(\xi_t \eta_t) = \xi_t d\eta_t + \eta_t d\xi_t + d\xi_t d\eta_t. \qquad (3.9)$$

We immediately deduce from Theorem 3.3 that if (3.8) holds for $u(x), v(x)$, then it holds for $u(x)v(x)$. Furthermore, formula (3.8) holds trivially for linear functions. Consequently, it holds for all polynomials of $x = (x^1, \ldots, x^d)$. The proof of Theorem 3.3 is completed by passing to the limit from polynomials to smooth functions and using Theorem 3.1.

In turn, in view of the linearity of both sides of (3.9), its proof reduces separately for ξ and η to the cases when $d\xi_t = b_t dt$ or $d\xi_t = f_t dW_t$, $d\eta_t = b_t dt$ or $d\eta_t = f_t dW_t$, where b_t, b_t, f_t, f_t are simple functions. Representing a simple function as a sum of simple functions with "one step", the whole thing very quickly reduces to proving the following equalities:

$$dt^2 = 2t\,dt, \quad d(tW_t^j) - (t\,dW_t^j + W_t^t dt) = 0,$$
$$d(W_t^i)^2 = 2W_t^i dW_t^i + dt, \quad d(W_t^i W_t^j) = W_t^i dW_t^j + W_t^j dW_t^i \text{ for } i \neq j.$$

Here the first equality is well known, while the second can be proved, for example, by integrating with respect to t, squaring and computing the expectation. The third equality is another writing of (3.6), while the fourth equality is easily proved roughly in the same way as the third one.

In the case when one takes as ξ_t the sum of a non-stochastic $x \in E_d$ and a d-dimensional Wiener process W_t, the Itô formula acquires the following especially simple form

$$du(x + W_t) = \frac{1}{2}\Delta u(x + W_t)dt + u_x(x + W_t)dW_t,$$

that is,

$$u(x + W_t) = u(x) + \frac{1}{2}\int_0^t \Delta u(x + W_s)ds + \int_0^t u_x(x + W_s)dW_s. \qquad (3.10)$$

3.3. The Itô formula along with the following property of the Itô stochastic integral serves as the basis of numerous computations. Before stating this property, let us call a non-negative random variable τ taking values in $[0, \infty]$ a *Markov time* with respect to $\{\mathcal{F}_t\}$ if $\{t < \tau\} \in \mathcal{F}_t$ for any $t \geq 0$.

Theorem 3.4. (Wald identity). *Let $f \in S^k$, and τ a Markov time such that*

$$\mathbf{E} \int_0^\tau |f_t|^2 dt < \infty.$$

Then

$$\mathbf{E} \int_0^\tau f_t dW_t = 0, \quad \mathbf{E} \left| \int_0^\tau f_t dW_t \right|^2 = \mathbf{E} \int_0^\tau |f_t|^2 dt.$$

Substituting $\tau(x)$ for t in (3.10), it is again easy to obtain formula (2.9) from (3.10) with the help of Theorem 3.4. It is also easy to obtain a good deal more. For example, let u be a solution in $\mathcal{D}$ of the equation

$$\frac{1}{2} \Delta u + cu + f = 0, \tag{3.11}$$

where the functions u, c, f are sufficiently regular and $u = g$ on $\partial \mathcal{D}$. Then for $x \in \mathcal{D}$, $t \leq \tau(x)$

$$d\left(\exp\left(\int_0^t c(x+W_s)ds\right) u(x+W_t)\right) = u(x+W_t) d\left(\exp\left(\int_0^t c(x+W_s)ds\right)\right)$$

$$+ \left(\exp\left(\int_0^t c(x+W_s)ds\right)\right) du(x+W_t) + \left(d\left(\exp\left(\int_0^t c(x+W_s)ds\right)\right)\right) du(x+W_t)$$

$$= -f(x+W_t)\left(\exp\left(\int_0^t c(x+W_s)ds\right)\right) dt + \left(\exp\left(\int_0^t c(x+W_s)ds\right)\right) u_x(x+W_t) dW_t.$$

Hence, as in the preceding arguments concerning (2.9),

$$u(x) = \mathbf{E} \int_0^{\tau(x)} \left(\exp\left(\int_0^t c(x+W_s)ds\right)\right) f(x+W_t) dt$$

$$+ \mathbf{E}\left(\exp\left(\int_0^{\tau(x)} c(x+W_s)ds\right) g(x+W_{\tau(x)})\right). \tag{3.12}$$

If $f \equiv 0$, $g \equiv 1$, the above formula is called the Kac formula. It can be used, for example, to find the distribution of τ. Thus for $d = 1$, $\mathcal{D} = (-1, 1)$, $c(x) = -\lambda$, where the constant $\lambda > 0$, $f = 0, g = 1$, we have $u(x) = (\cosh\sqrt{2\lambda}x)(\cosh\sqrt{2\lambda})^{-1}$ and (3.12) gives the Laplace transform of $\tau(x)$:

$$\mathbf{E} e^{-\lambda \tau(x)} = (\cosh\sqrt{2\lambda}x)(\cosh\sqrt{2\lambda})^{-1}.$$

3.4. It is convenient to give Itô's formula (3.8) a more explicit form. Let $d\xi_t = \sigma_t dW_t + b_t dt$, then upon carrying out the necessary computations we obtain

$$du(\xi_t) = L_t u(\xi_t) dt + \sigma_t^* u_x(\xi_t) dW_t,$$

where
$$L_t u(y) = \sum_{i,j \leq d} a_t^{ij} u_{x^i x^j}(y) + \sum_{i \leq d} b_t^i u_{x^i}(y), \quad (a_t^{ij}) = \frac{1}{2} \sigma_t \sigma_t^*.$$

If $\xi_0 = x_0 \in \mathcal{D}$, x_0 is non-stochastic, $\sigma_t = \sigma(\xi_t)$, $b_t = b(\xi_t)$ for some Borel functions $\sigma(y), b(y)$ and accordingly, ξ_t is a solution of the Itô stochastic equation
$$d\xi_t = \sigma(\xi_t) dW_t + b(\xi_t) dt, \qquad (3.13)$$

then $a_t^{ij} = a^{ij}(\xi_t)$, where $(a^{ij}(y)) = \frac{1}{2}\sigma(y)\sigma^*(y)$ and, by analogy with (3.12), we obtain for the solution of the differential equation
$$\sum_{i,j \leq d} a^{ij} u_{x^i x^j} + \sum_{i \leq d} b^i u_{x^i} + cu + f = 0 \qquad (3.14)$$

in $\mathcal{D}$ with boundary data $u = g$ on $\partial \mathcal{D}$, the representation
$$u(x_0) = \mathbf{E} \int_0^\tau f(\xi_t) \exp\left(\int_0^t c(\xi_s) ds\right) dt + \mathbf{E} \exp\left(\int_0^\tau c(\xi_s) ds\right) g(\xi_\tau), \qquad (3.15)$$

where τ is the first exit time of ξ_t from $\mathcal{D}$. It is understood that (3.15) is valid only under certain conditions. However, it is not useful to formulate some general theorem on this account, since it would not encompass all possible cases, and the derivation of (3.15), noted above, can be made completely rigorously in each specific case.

Formula (3.15), which also is sometimes called the Itô formula, plays an exceptionally important role in realizing the connection between the theory of diffusion processes and the theory of partial differential equations. Let us observe, at the same time, that equation (3.14) is elliptic (generally speaking, degenerate and, accordingly, can also prove to be parabolic), since
$$\sum_{i,j} a^{ij} \lambda^i \lambda^j = \frac{1}{2} |\sigma^* \lambda|^2 \geq 0.$$

It is impossible even briefly to mention all the applications of the probabilistic representation (3.15) of the solution of (3.14). We merely show how it immediately entails the existence of a non-zero bounded solution on $(-\infty, \infty)$ of the equation
$$u'' + x^3 u' - u = 0, \qquad (3.16)$$

concerning which it is mistakenly asserted in one of the papers that it has only null bounded solutions.

Let us consider the equation
$$\xi_t = x + W_t + \int_0^t \xi_s^3 ds, \qquad (3.17)$$

where x is fixed and W_t is a one-dimensional Wiener process. Denoting $\eta_t = \xi_t - W_t$, we see that (3.17) is equivalent to the ordinary equation $\dot{\eta}_t = (\eta_t + W_t)^3$, whence it is clear that (3.17) has a solution $\xi_t = \xi_t(x)$, but only on a finite (stochastic) time interval $[0, \tau(x))$ and, what is more, $|\xi_t(x)| \to \infty$ as $t \uparrow \tau(x)$. We set
$$v(x) = \mathbf{E} e^{-\tau(x)}.$$

Since $\tau(x) < \infty$ (a.s.) (it is sufficient that $\mathbf{P}\{\tau(x) < \infty\} > 0$), it follows that $v(x) > 0$. Further, let $\tau_n(x)$ be the first exit time of $\xi_t(x)$ from $(-n, n)$. Then, clearly,
$$\tau(x) = \tau_n(x) + \tau(\xi_{\tau_n(x)}(x)),$$
which along with the strong Markov property of the Wiener process gives
$$v(x) = \mathbf{E} e^{-\tau_n(x)} v(\xi_{\tau_n(x)}(x)).$$

Hence it follows that v is a solution of (3.16). Indeed (3.16) has a solution on $(-n, n)$ equal to v at the points $\pm n$. According to formula (3.15), this solution coincides with v on $(-n, n)$. Thus v is a positive bounded solution of (3.16).

We can go slightly further in analyzing equation (3.16). It is similarly shown that
$$v_1(x) = \mathbf{E} e^{-\tau(x)} I(\lim_{t \uparrow \tau(x)} \xi_t(x) = \infty)$$
is a second bounded solution of (3.16). The solutions v and v_1 are linearly independent ($v(x) \not\to 0$, $v_1(x) \to 0$ as $x \to -\infty$), therefore any solution of (3.16) is bounded.

§4. Stochastic Differential Equations and Diffusion Processes. Girsanov's Theorem

4.1. Probably the concept of stochastic differential equation first appeared in papers of Bernshtein (Bernshtein 1934), (Bernshtein 1938), (Bernshtein 1946) approximately fifty years ago. Bernshtein understood by a stochastic differential equation a sequence of stochastic difference equations in t, and under certain hypotheses proved the convergence of one-dimensional distributions of their solutions to a limit under an unlimited decrease of the step size with respect to t. Here the question of the *existence of a limit process*, which would naturally be called the solution of the stochastic differential equation, was left aside. In the papers (Gikhman 1950), (Gikhman 1951) a more consistent viewpoint was developed. The solution of stochastic differential equation is present already in them and this equation itself is understood almost in the contemporary sense. Moreover, Gikhman was interested not only in the solvability of stochastic equations, but also in what manner one could describe analytically certain properties of their solutions. In particular, he was the first to prove the solvability of second-order linear degenerate parabolic equations, doing this in a purely probabilistic manner.

Roughly at the same time, Itô (Itô 1946), (Itô 1951) proposed a somewhat different treatment of the concept of a stochastic differential equation and proved the solvability of these equations with the help of the concept of a stochastic integral introduced by him. The approach of Itô overshadowed the papers of Gikhman, which were "discovered" again only at the start of the 60s and was a powerful stimulus to the development of the theory of Markov-type processes, thanks to the convenient, natural and well-developed apparatus of dealing with stochastic integrals. Above all, this apparatus made possible the construction and study of processes of a complex nature with the help of simpler processes, such as Wiener and Poisson process. At the present time there is a huge literature devoted to Itô stochastic equations. Part of this literature can be found in the references in the books (Gikhman and Skorokhod 1982) and (Ikeda and Watanabe 1981).

4.2. Itô's stochastic equations have the form (3.13), where σ, b can depend on t and on ω. For applications it is important that, given some process, one is able to "guess" which stochastic differential equation it satisfies. Many general results exist on this theme, mainly connected with the concepts of a Markov process, a semimartingale and their canonical representations (see Chap. 3 below). Here we shall present just one rule, which works well in many cases and which is quite close to the ideas of (Kolmogorov 1931); (Gikhman 1950), (Gikhman 1951) and (Itô 1951).

Let ξ_1 be a d-dimensional stochastic process continuous in t. We set $\mathcal{F}_t^\xi = \sigma\{\xi_s, \ s \leq t\}$, and assume that (in a suitable sense) for each $t \geq 0$ there exist the limits

$$\lim_{h \downarrow 0} h^{-1} \mathbf{E}\{\xi_{t+h} - \xi_t | \mathcal{F}_t^\xi\} =: b_t, \tag{4.1}$$

$$\lim_{h \downarrow 0} h^{-2} \mathbf{E}\{(\xi_{t+h}^i - \xi_t^i)(\xi_{t+h}^j - \xi_t^j) | \mathcal{F}_t^\xi\} =: a_t^{ij}. \tag{4.2}$$

The vector b_t is called the drift vector and the matrix $a_t = (a_t^{ij})$ the diffusion matrix. Clearly, $a_t \geq 0$. We set $\sigma_t = \sqrt{a_t}$. It turns out that under broad assumptions, either on the initial probability space or its extension, a d-dimensional Wiener process $(W_t, \mathcal{F}_t^\xi)$ exists such that $d\xi_t = b_t dt + \sigma_t dW_t$. If in (4.1), (4.2) we have $b_t = b(\xi_t), a_t = a(\xi_t)$ for some, (generally speaking, random) functions $b_t(x), a_t(x)$ on $[0, \infty) \times \Omega \times E_d$, then ξ_t satisfies the Itô stochastic equation

$$d\xi_t = b_t(\xi_t)dt + \sigma_t(\xi_t)dW_t, \tag{4.3}$$

where $\sigma_t(x) = \sqrt{2a_t(x)}$. In the case when $b_t(x), a_t(x)$ do not depend on ω, one says that the process ξ_t is a diffusion process.

As an example of an application of the rule presented above, let us consider the Ornstein-Uhlenbeck process, which is often encountered in physical and other applications, since it is the simplest continuous stationary process. A one-dimensional Gaussian process ξ_t with zero mean is called an *Ornstein-Uhlenbeck process* if $\mathbf{E}\xi_t\xi_s = e^{-|t-s|}$ for $t, s \geq 0$. One can show by arguments similar to those following Theorem 1.4 that ξ_t has a continuous modification, and therefore we will assume that the process ξ_t is continuous. Further, for $t \geq s$, $h \geq 0$

$$\mathbf{E}(\xi_{t+h} - e^{-h}\xi_t)\xi_s = e^{-(t+h-s)} - e^{-h}e^{-(t-s)},$$

from which it follows by the normal correlation theorem that $\xi_{t+h} - e^{-h}\xi_t$ does not depend on the trajectory of ξ_s for $s \leq t$, that is, it does not depend on $\mathcal{F}_t^\xi$,

$$\mathbf{E}\{\xi_{t+h} - e^{-h}\xi_t | \mathcal{F}_t^\xi\} = \mathbf{E}(\xi_{t+h} - e^{-h}\xi_t) = 0,$$
$$\mathbf{E}\{(\xi_{t+h} - e^{-h}\xi_t)^2 | \mathcal{F}_t^\xi\} = \mathbf{E}(\xi_{t+h} - e^{-h}\xi_t)^2 = 1 - e^{-2h},$$
$$\mathbf{E}\{(\xi_{t+h} - \xi_t)^2 | \mathcal{F}_t^\xi\} = \mathbf{E}\{(\xi_{t+h} - e^{-h}\xi_t + (e^{-h} - 1)\xi_t)^2 | \mathcal{F}_t^\xi\}$$
$$= 1 - e^{-2h} + (e^{-h} - 1)^2.$$

Consequently, in this case $a_t = 2$, $b_t = -\xi_t$ and (4.3) gives the *Langevin equation*

$$\xi_t = \xi_0 + \sqrt{2}W_t - \int_0^t \xi_s ds, \quad t \geq 0.$$

Another important example is the Brownian bridge or conditional Wiener process which plays an important role in statistics (even of one-dimensional random variables). The Brownian bridge is a one-dimensional Wiener process W_t considered under the condition that $W_1 = 0$. Since $\mathbf{P}\{W_1 = 0\} = 0$, it is necessary to refine this definition. Indeed, one says that a one-dimensional continuous process ξ_t, $t \in [0,1]$, is a *Brownian bridge* if for any $0 \leq t_1 \leq \ldots \leq t_n < 1$, the distribution $(\xi_{t_1}, \ldots, \xi_{t_n})$ has a density coinciding with the conditional density $(W_{t_1}, \ldots, W_{t_n})$ under the condition that $W_1 = 0$. Since this is Gaussian, ξ_t is a Gaussian process. Arguing as before, we find that the process $W_t - tW_1$, $t \in [0,1]$ does not depend on W_1. Hence for $0 \leq s \leq t \leq t + h < 1$

$$\mathbf{E}\xi_t\xi_s = \lim_{\epsilon \downarrow 0} \frac{1}{\mathbf{P}\{|W_1| \leq \epsilon\}} \mathbf{E}W_t W_s I\{|W_1| \leq \epsilon\}$$
$$= \lim_{\epsilon \downarrow 0} \frac{1}{\mathbf{P}\{|W_1| \leq \epsilon\}} \mathbf{E}(W_t - tW_1)(W_s - sW_1) I\{|W_1| \leq \epsilon\}$$
$$= \mathbf{E}(W_t - tW_1)(W_s - sW_1) = s - ts,$$

$$\mathbf{E}\left(\xi_{t+h} - \frac{1-t-h}{1-t}\xi_t\right)\xi_s = 0,$$
$$\mathbf{E}(\xi_{t+h} - \xi_t | \mathcal{F}_t^\xi) = -\xi_t \frac{h}{1-t},$$
$$b_t = -\frac{1}{1-t}\xi_t, \quad a_t = 1.$$

Accordingly, we arrive at the following relation:

$$\xi_t = \widetilde{W}_t - \int_0^t \frac{1}{1-s}\xi_s ds$$

with some Wiener process $\widetilde{W}_t$.

4.3. Up to now we have presented examples of Itô stochastic equations with a constant coefficient σ. One can often prove their solvability by appealing to the theory of ordinary equations, as was done for (3.17), and introducing the new process $\eta_t = \xi_t - \sigma W_t$. Here it is necessary that for the appropriate ordinary equation, an existence theorem for a solution be known. It is not hard to present examples showing that $\dot{x}_t = b(t, x_t)$ cannot have solutions if $b(t, x)$ is only measurable. All the more remarkable is the next result.

Theorem 4.1. (Zvonkin 1974), $(d = 1)$; (Veretennikov 1980), $(d \geq 1)$. Let $(W_t, \mathcal{F}_t)$ be a d-dimensional Wiener process, σ a real constant, $\sigma \neq 0$, and $b(t, x)$ a bounded Borel function of $(t, x) \in [0, \infty) \times E_d$ with values in E_d. Then there exists a d-dimensional continuous process ξ_t that is $\mathcal{F}_t$-measurable for each $t \geq 0$ and is such that with probability 1 we have

$$\xi_t = \sigma W_t + \int_0^t b(s, \xi_s) ds \qquad (4.4)$$

for all t. Furthermore, if the process η_t is $\mathcal{F}_t$-measurable for each $t \geq 0$ and also almost surely satisfies (4.4) for all t, then $\xi_t = \eta_t \; \forall t \geq 0$ (a.s.).

In particular, it follows from Theorem 4.1 that for almost every Wiener trajectory W_t the ordinary equation

$$\eta_t = \int_0^t b(s, \eta_s - \sigma W_s) ds \qquad (4.5)$$

has a solution. However, from the viewpoint of the theory of ordinary equations, its uniqueness is somewhat extravagant; it is not asserted that for almost every individual Wiener trajectory, a solution of (4.5) is unique, but only that two $\mathcal{F}_t$-adapted solutions of (4.5), defined for almost all trajectories of W_t, coincide almost surely. At the same time, equations (4.4), (4.5) possibly have solutions depending on the future. The question of the uniqueness of a solution of (4.5) or (4.4) for almost each trajectory of W_t remains open at present.

The simplest stochastic equation with diffusion σ depending on an "event" is the linear stochastic equation without drift

$$\xi_t = 1 + \int_0^t \xi_s \sigma_s dW_s. \qquad (4.6)$$

Using Itô's formula, it is verified by elementary means that its solution is the so-called *exponential martingale*

$$\xi_t = \exp\left\{\int_0^t \sigma_s dW_s - \frac{1}{2}\int_0^t |\sigma_s|^2 ds\right\}, \tag{4.7}$$

where σ_s, W_s can be scalar or d-dimensional; it is merely required that $\sigma \in S^d$.
In connection with the exponential martingale we have the following.

Theorem 4.2. (Girsanov 1960). *Let $(W_t, \mathcal{F}_t)$ be a d-dimensional Wiener process on a probability space $(\Omega, \mathcal{F}, \mathbf{P})$, $\sigma \in S^d$, and let $T \in (0, \infty)$ be a constant. Let*

$$\mathbf{E}\exp\left\{\int_0^T \sigma_s dW_s - \frac{1}{2}\int_0^T |\sigma_s|^2 ds\right\} = 1. \tag{4.8}$$

We define on $\mathcal{F}$ a new (probability) measure by the formula

$$\overline{\mathbf{P}}(d\omega) = \exp\left\{\int_0^T \sigma_s dW_s - \frac{1}{2}\int_0^T |\sigma_s|^2 ds\right\}\mathbf{P}(d\omega).$$

It is then claimed that the process

$$\xi_t = W_t - \int_0^t \sigma_s ds$$

is d-dimensional Wiener for $t \in [0, T]$ with respect to $\{\mathcal{F}_t\}$ on the probability space $(\Omega, \mathcal{F}, \overline{\mathbf{P}})$.

Let us note that condition (4.8) is natural in view of the connection of (4.7) with (4.6) and Theorem 3.4 on the Wald identities. The proof of Theorem 4.2 is based on the fact that equation (4.8) proves to be true when σ_s is replaced by $\sigma_s + i\lambda_s$, where λ_s is a non-random simple function. After this substitution, elementary transformations show that the left hand side of (4.8) is the expectation with respect to the measure $\overline{\mathbf{P}}$ of

$$\exp\left\{i\int_0^T \lambda_s d\xi_s + \frac{1}{2}\int_0^T |\lambda_s|^2 ds\right\},$$

whence the joint characteristic function of $\xi_{t_1}, \ldots, \xi_{t_n}$ is immediately obtained for any $t_i \in [0, T]$, $n \geq 1$.

Girsanov's Theorem 4.2 has an enormous number of uses in the theory of Itô stochastic equations and applications of them, for example, to the statistics of stochastic processes of diffusion type, control of such processes etc. In this connection it is helpful to note that condition (4.8) is satisfied, for example, if σ_t is bounded on $[0, T] \times \Omega$. There is, in fact, a more refined result.

Theorem 4.3. *For condition (4.8) to hold it is sufficient (Novikov 1972) that*

$$\mathbf{E}\exp\frac{1}{2}\int_0^T |\sigma_s|^2 ds < \infty,$$

or (Kazamaki 1976) that for any $t \leq T$

$$\mathbf{E}\exp\frac{1}{2}\int_0^t \sigma_s dW_s < \infty.$$

The following "weak" version of Theorem 4.1 is obtained in a remarkably elementary manner from Girsanov's Theorem 4.2.

Theorem 4.4. (Girsanov 1960). *Let $\sigma \neq 0$ be a constant, and $b(t,x)$ a Borel function on $[0,T] \times E_d$ with values in E_d. Then there exists a probability space $(\Omega, \mathcal{F}, \mathbf{P})$ and a d-dimensional Wiener process $(W_t, \mathcal{F}_t)$ on it such that equation (4.4) has a continuous solution on $[0,T]$ that is $\mathcal{F}_t$-measurable for each $t \in [0,T]$. Furthermore, the distributions in $C([0,T]; E_d)$ of any two solutions of (4.4) are the same.*

Indeed, to prove the existence of a solution it is sufficient to take on some probability space $(\Omega, \mathcal{F}, \overline{\mathbf{P}})$ a d-dimensional Wiener process $(\eta_t, \mathcal{F}_t)$, set $\xi_t = \sigma \eta_t$,

$$W_t = \sigma^{-1}\Big(\xi_t - \int_0^t b(s, \xi_s)ds\Big),$$

and observe that by Theorem 4.2, the process W_t is Wiener on $(\Omega, \mathcal{F}, \mathbf{P})$ for $t \in [0,T]$ with

$$\mathbf{P}(d\omega) = \exp\Big\{\int_0^T \frac{1}{\sigma}b(s,\xi_s)d\eta_s - \frac{1}{2\sigma^2}\int_0^T |b(s,\xi_s)|^2 ds\Big\}\overline{\mathbf{P}}(d\omega).$$

The second assertion of Theorem 4.3 is proved by reversing these arguments.

This theorem, by contrast with Theorem 4.1, does not allow one to assert the existence of a solution of (4.4) on a *given* probability space with a *given* Wiener process, which is the "weakness" of Theorem 4.4. However, it has considerable advantage over Theorem 4.1 since Theorem 4.3, along with the proof, goes over verbatim to the case when the coefficient b depends on the entire past process ξ_t in a non-anticipative manner.

One can use the Girsanov Theorem 4.2 in many computations. Let us find, for example, the distribution of $\zeta = \max\{t + W_t, t \leq T\}$, where $T \in (0, \infty)$ is a constant and W_t is a one-dimensional Wiener process. Setting $\overline{W}_t = t + W_t$, $\overline{\mathbf{P}}(d\omega) = \exp\big(-W_T - \frac{1}{2}T\big)\mathbf{P}(d\omega)$, we obtain

$$\mathbf{P}\{\zeta > x\} = \int_\Omega I(\zeta > x) e^{W_T + \frac{1}{2}T} e^{-W_T - \frac{1}{2}T} \mathbf{P}(d\omega)$$
$$= \int_\Omega I(\max_{t\leq T} \overline{W}_t > x) e^{\overline{W}_T - \frac{1}{2}T} \overline{\mathbf{P}}(d\omega)$$
$$= \mathbf{E} e^{W_T - \frac{1}{2}T} I(\max_{t\leq T} W_t > x),$$

where the last equality is valid because $\overline{W}_t$ is a Wiener process for $t \leq T$ with respect to the measure $\overline{\mathbf{P}}$. It is then shown similarly to (1.5) that for $x > 0$

$$\mathbf{P}(\max_{t\leq T} W_t > x,\ W_T > y) = \mathbf{P}(W_T > y) \text{ for } y \in [x,0),$$
$$\mathbf{P}(\max_{t\leq T} W_t > x,\ W_T \leq y) = \mathbf{P}(\max_{t\leq T} W_t > x,\ W_T \geq 2x - y)$$
$$= \mathbf{P}(W_T \geq 2x - y) \text{ for } y \in (-\infty, x].$$

Hence for $x > 0$

$$\mathbf{P}\{\zeta > x\} = -\int_x^\infty e^{y-\frac{1}{2}T} d_y \mathbf{P}\{W_T > y\} + \int_{-\infty}^x e^{y-\frac{1}{2}T} d_y \mathbf{P}\{W_T \geq 2x - y\}$$
$$= \int_x^\infty e^{y-\frac{1}{2}T} \frac{1}{\sqrt{2\pi T}} e^{-\frac{y^2}{2T}} dy + \int_{-\infty}^x e^{y-\frac{1}{2}T} \frac{1}{\sqrt{2\pi T}} e^{-\frac{(2x-y)^2}{2T}} dy.$$

4.4. By contrast with the refined methods of proving Theorems 4.1, 4.4, the following classical theorem is proved by the usual method of successive approximations.

Theorem 4.5 (Itô 1946), (Itô 1951). *Let $(W_t, \mathcal{F}_t)$ be a k-dimensional Wiener process, let $\sigma_t(x) = \sigma_t(\omega, x)$, $t \geq 0$, $\omega \in \Omega$, $x \in E_d$, be a matrix function of dimension $d \times k$, and let $b_t(x) = b_t(\omega, x)$ be a function with values in E_d. Suppose that $\sigma_t(x), b_t(x)$ are measurable with respect to (t, ω, x) and $\mathcal{F}_t$-measurable with respect to ω for each t, x and that for some constant K for all t, ω, x, y, i, j we have*

$$|\sigma_t^{ij}(x) - \sigma_t^{ij}(y)| + |b_t^i(x) - b_t^i(y)| \leq K|x - y|,$$
$$|\sigma_t^{ij}(x)| + |b_t^i(x)| \leq K(1 + |x|).$$

Then for each $\mathcal{F}_0$-measurable stochastic vector $\xi_0 \in E_d$ there exists a continuous process ξ_t that is $\mathcal{F}_t$-measurable for each t and is such that almost surely

$$\xi_t = \xi_0 + \int_0^t \sigma_s(\xi_s) dW_s + \int_0^t b_s(\xi_s) ds \tag{4.9}$$

for all $t \geq 0$. Furthermore, this process is unique in the same sense as in Theorem 4.1.

Let us note once more that Itô (Itô 1951) also considered equations of the form (4.9) on the right hand side of which integrals over Poisson measures and centered Poisson measures are allowed, so that their solutions proved to be locally infinitely divisible processes of a sufficiently general nature.

§5. Stochastic Differential Equations with Boundary Conditions

5.1. Up to now we have been concerned with the solutions of stochastic equations in the entire space or up to the first exit time from the domain. It may be asked what are the possibilities for extending the solution of a stochastic equation beyond the first exit time from the domain so as to preserve the continuity of its trajectories without taking it beyond the limits of the closure of this domain. This question already arises in considering Brownian motion in a glass. One of the possibilities consists in stopping the process forever at the first exit time. In this case, the process "sticks" to the boundary. The second possibility is when the process starts to move along the boundary, "forgetting" about the domain from which it exited. Thus, it "sticks" to the boundary in a more complicated manner. The third possibility consists in "reflection" from the boundary inside the domain. Finally, one can combine these possibilities at any point of the boundary with different probabilities. A result of Venttsel' (Venttsel' 1959) shows that there cannot be any other possibilities.

It is clear how the first two possibilities are realized in the language of stochastic differential equations. To study the effect of reflection, it is more natural overall to start with a one-dimensional Wiener process W_t which we want to force to be reflected "mirror-like" at $x = 0$, so that it remains all the time on $[0, \infty)$. Clearly in this case the process $|W_t|$ is obtained. Since we are interested in stochastic differential equations, we need to find $d|W_t|$, that is, a representation of $|W_t|$ in the form of a sum of stochastic and ordinary integrals. This could be easily done via the Itô formula if the function $|x|$ were twice continuously differentiable. This difficulty is avoided by approximating $|x|$ by smooth functions and it turns out that

$$|W_t| = W'_t + \phi_t, \tag{5.1}$$

where

$$W'_t = \int_0^t \operatorname{sign} W_s dW_s, \quad \phi_t = \lim_{\epsilon \downarrow 0} \frac{1}{2\epsilon} \int_0^t I(|W_s| < \epsilon) ds, \tag{5.2}$$

where this limit exists in probability. It is called the *local time* of the process W at the origin. It is apparent from (5.1) that ϕ_t has a continuous modification, and it follows from (5.2) that $\phi_t \geq 0, \phi_t$ is non-decreasing,

$$\int_0^t I(|W_s| = 0) d\phi_s = \phi_t,$$

that is, ϕ_t increases only on the set t, where $|W_t| = 0$. Furthermore, it is not hard to see, for example by using the arguments concerning (4.1), (4.2), that W'_t is a Wiener process. By the way, this implies that $\mathbf{E}\phi_t = \mathbf{E}|W_t|$ and $\phi_t \not\equiv 0$, in spite of the fact that

$$\mathbf{E}\int_0^t I(|W_s| = 0) ds = \int_0^t \mathbf{P}\{|W_s| = 0\} ds = 0,$$

that is the Lebesgue measure of the set $\{s : |W_s| = 0\}$ equals zero (a.s.). We note that the local time of W was introduced by Lévy (Lévy 1965) in terms of the second formula (5.2) and that its existence is most easily proved by using the derivation of (5.1) from the Itô formula. Considerable information regarding local times can be found in (Itô and McKean 1965).

We now pose the problem of finding an equation which must be satisfied by a 1-dimensional diffusion process with diffusion coefficient $a(x)$, "instantaneously" reflected at the origin and proceeding on the positive semi-axis. In view of what has been said above, it is natural to define it by the equation

$$\xi_t = \xi_0 + \int_0^t \sigma(\xi_s) I(\xi_s > 0) dW_s + \phi_t, \qquad (5.3)$$

where the unknowns are ξ_t, ϕ_t; $\sigma = \sqrt{2a}$, W_t is a one-dimensional Wiener process, σ corresponds to diffusion when the process ξ_t is found outside the origin, ϕ_t corresponds to reflection at the origin, and this process is sought in the class of non-decreasing non-negative processes increasing only on the set $\{t : \xi_t = 0\}$. The natural condition is also imposed that ϕ_t be such that $\xi_t \geq 0$ for all $t \geq 0$, $\phi_0 = 0$ (and, of course, the initial data $\xi_0 \geq 0$). Finally, the requirement of "instantaneousness" of reflection is expressed in the requirement that the process ξ_t spend zero time at the origin, that is, for any t

$$\int_0^t I(\xi_s = 0) ds = 0 \text{ (a.s.)}. \qquad (5.4)$$

Thus, equation (5.3) is an equation with respect to the pair of processes ξ_t, ϕ_t, where this pair is sought in a somewhat complex class of functions. It turns out that if, for example, σ is continuous and $\sigma(0) \neq 0$, then when the remaining conditions on ξ_t, ϕ_t hold, the pair of relations (5.3), (5.4) is equivalent to the single relation:

$$\xi_t = \xi_0 + \int_0^t \sigma(\xi_s) dW_s + \phi_t. \qquad (5.5)$$

Furthermore, it turns out that equation (5.5) in the above class of functions ξ_t, ϕ_t is equivalent to the two relations (Skorokhod formulas)

$$\xi_t = \xi_0 + \int_0^t \sigma(\xi_s) dW_s - \min_{0 \leq s \leq t} \left\{ \left(\xi_0 + \int_0^s \sigma(\xi_u) dW_u \right) \wedge 0 \right\}, \qquad (5.6)$$

$$\phi_t = - \min_{0 \leq s \leq t} \left\{ \left(\xi_0 + \int_0^s \sigma(\xi_u) dW_u \right) \wedge 0 \right\}. \qquad (5.7)$$

Actually, here there is only one equation (5.6) with respect to ξ_t, while (5.7) expresses ϕ in terms of ξ. The solvability of (5.6) can be proved by the method of successive approximations for Lipschitz σ, which is similar to the proof of Theorem 4.5.

5.2. In the one-dimensional case there is only one direction of reflection. Now in the two-dimensional case we can construct processes reflecting at various angles on the boundary. For example, let us solve the (one-dimensional) equation (5.3) in the class of functions described by all the conditions enumerated after it, including also (5.4), and let us take a one-dimensional Wiener process $\widetilde{W}_t$ independent of W_t and consider the pair (ξ_t, η_t), where $\eta_t = \widetilde{W}_t + a\phi_t$, $a = $ const. It turns out that it is natural to regard the process (ξ_t, η_t) as a two-dimensional process in $E_2^+ = \{(x^1, x^2) : x^1 \geq 0\}$, instantaneously reflected from the axis $x^1 = 0$ at an angle $\alpha = \tan^{-1} a$. Indeed, for $\xi_0 = 0$ we have $\mathbf{E}\xi_t = \mathbf{E}\phi_t$, $\mathbf{E}\eta_t = a\mathbf{E}\phi_t$, $\mathbf{E}\eta_t/\mathbf{E}\xi_t = a$.

The general stochastic equation for a diffusion process in a d-dimensional domain $\mathcal{D}$ with reflection in a direction $l(x)$ given on $\partial\mathcal{D}$ and directed inside $\mathcal{D}$ has the following form:

$$d\xi_t = \sigma(\xi_t)I(\xi_t \in \mathcal{D})dW_t + b(\xi_t)I(\xi_t \in \mathcal{D})dt + l(\xi_t)d\phi_t, \qquad (5.8)$$

where the pair ξ_t, ϕ_t is sought such that $\xi_t \in \mathcal{D}$, ϕ_t is non-decreasing, $\phi_0 = 0$,

$$\phi_t = \int_0^t I(\xi_s \in \partial\mathcal{D})d\phi_s, \quad \int_0^t I(\xi_s \in \partial\mathcal{D})ds = \int_0^t \rho(\xi_s)d\phi_s,$$

and where σ, b, W are respectively a matrix, a vector and a Wiener process of appropriate dimensions, and $\rho(x)$ a given non-negative function. The function $\rho(x)$ characterizes the time that the process ξ_t spends on $\partial\mathcal{D}$. If $\rho \equiv 0$, then an "instantaneous" reflection is obtained, and one says that the boundary $\partial\mathcal{D}$ is "rigid", but if $\rho > 0$, then one speaks of an "elastic" reflection, when the process ξ_t spends some (real) time also on $\partial\mathcal{D}$. Here there is already a combination of "sticking" to the boundary with "instantaneous" reflection. At the same time, we see that when $l = 0$ the process ξ_t halts on the boundary. We can describe more complex behavior of ξ_t on the boundary, including diffusion along $\partial\mathcal{D}$, if on the right hand side of (5.8) we add terms containing certain martingales (see Ikeda and Watanabe 1981).

Concerning equation (5.8) and the properties of its solutions and other approaches to the definition of diffusion processes with boundary conditions, including processes passing through the boundary, we refer the reader to (Gikhman and Skorokhod 1982); (Genis and Krylov 1973); (Anulova 1978); (Lions and Sznitman 1984); (Tanaka 1979); (Malyutov 1969); (Portenko 1982).

5.3. As with the "usual" Itô equations, equations with boundary conditions are connected with the theory of partial differential equations. One can establish this connection, for example, with the help of the Itô formula (Theorem 3.2), in which it is necessary to consider stochastic differentials of a form that is now more general than (3.7). For example, for the case of processes of type (5.8), the Itô formula has the same form as in Theorem 3.2 with the following supplement to the table of multiplication of differentials

$$(d\phi_t)^2 = dt\,d\phi_t = dW_t^i\,d\phi_t = 0.$$

It is proved in the same way as (3.15) that under certain conditions, if u is a solution of (3.14) with boundary condition $(u_x, l) = g$ on $\partial\mathcal{D}$ and ξ_t is a solution of (5.8) with initial data $\xi_t = x$, then

$$u(x) = \mathbf{E}\left[\int_0^\infty f_0(\xi_t)e^{\int_0^t c_0(\xi_s)ds}dt + \int_0^\infty g(\xi_t)e^{\int_0^t c_0(\xi_s)ds}d\phi_s\right], \tag{5.9}$$

where $(f_0, c_0) = (f, c)I(x \in \mathcal{D})$. Formula (5.9), like (3.15), is important both for probability theory and for the theory of differential equations. For example, with its help, in the paper (Malyutov 1969) the transparent probabilistic meaning of certain solvability conditions of the Poincaré directional derivative problem (which are somewhat mysterious from an analytic point of view) was completely explained.

References*

Anulova, S.V. (1978): Processes with a Lévy generating operator in a half-space. Izv. Akad. Nauk SSSR, Ser. Mat. *42*, 708–750. [English transl.: Math USSR, Izv.*13*, 9–51 (1979)] Zbl. 386.60043

Bernshtein, S.N. (1934): Principles of the theory of stochastic differential equations. Tr. Mat. Inst. Steklova *5*, 95–124. Zbl. 9,218

Bernstein, S.N. (1938): Équations différentielles stochastiques. Actual. Sci. Ind. *738*, 5–31. Zbl. 22,243

Bernshtein, S.N. (1946): Probability Theory. OGIZ, Moscow Leningrad

Freidlin, M.I. (1963): Diffusion processes with reflection and a directional derivative problem on a manifold with boundary. Teor. Veroyatn. Primen. *8*, No. 1, 80–88. [English transl.: Theory Probab. Appl. *8*, 75–83 (1963)] Zbl. 124,341

Genis, I.L., Krylov, N.V. (1973): The exact barriers in the problem with directional derivative. Sib. Mat. Zh. *14*, No. 1, 36–43. [English transl.: Sib. Math. J. *14*, 23–28 (1973)] Zbl. 256.35026

Gikhman, I.I. (1950): On the theory of differential equations of random processes. Part I: Ukr. Mat. Zh. *2*, No. 4, 37–63. Zbl. 45,405

Gikhman, I.I. (1951): On the theory of differential equations of random processes. Part II: Ukr. Mat. Zh. *3*, No. 3, 317–339. Zbl. 45,405

Gikhman, I.I., Skorokhod, A.V. (1982): Stochastic Differential Equations and their Applications. Naukova Dumka, Kiev. Zbl. 557.60041

Girsanov, I.V. (1960): The transformation of a class of random processes by means of an absolutely continuous change of measure. Teor. Veroyatn. Primen. *5*, No. 3, 314–330. [English transl.: Theory Probab. Appl. *5*, 285–301 (1962)] Zbl. 100,340

Ikeda, N., Watanabe, S. (1981): Stochastic Differential Equations and Diffusion Processes. North-Holland, Amsterdam. Zbl. 495.60005

Itô, K. (1944): The stochastic integral. Proc. Imperial Acad. Tokyo *20*, 519–524. Zbl. 60,291

Itô, K. (1946): On a stochastic integral equation. Proc. Japan Acad. *22*, 32–35

Itô, K. (1951): On stochastic differential equations. Mem. Am. Math. Soc. *4*, 1–51. Zbl. 54,58

Itô, K., McKean, H.P., Jr. (1965): Diffusion Processes and their Sample Paths. Springer, Berlin Heidelberg New York. Zbl. 127,95

---

* For the convenience of the reader, references to reviews in Zentralblatt für Mathematik (Zbl.), compiled using the MATH database, have, as far as possible, been included in this bibliography.

Kazamaki, N. (1978): The equivalence of two conditions on weighted norm inequalities for martingales. Proc. Int. Symp. Stochastic Differential Equations, Kyoto 1976, 141–152. Zbl. 429.60047

Kolmogorov, A.N. (1931): Über die analytischen Methoden in der Wahrscheinlichkeitsrechnung. Math. Ann. *104*, 415–458. Zbl. 1,149

Krylov, N.V. (1969): Diffusion on the plane with reflection II. The boundary-value problem. Sib. Mat. Zh. *10*, No. 2, 355–372. [English transl.: Sib. Math. J. *10*, 253–265 (1969)] Zbl. 177,373

Lévy, P. (1965): Processus Stochastiques et Mouvement Brownien 2nd ed. Gauthier-Villars, Paris. Zbl. 137,116

Lions, P.L., Sznitman, A.S. (1984): Stochastic differential equations with reflecting boundary conditions. Commun. Pure Appl. Math. *37*, 511–537. Zbl. 598.60060

Liptser, R.Sh., Shiryaev, A.N. (1977): Statistics of Random Processes I. General Theory. Springer, Berlin Heidelberg New York. Zbl. 364.60004

Liptser, R.Sh., Shiryaev, A.N. (1978): Statistics of Random Processes II. Applications. Springer, New York. Zbl. 369.60001

Malyutov, M.B. (1969): Poincaré's boundary-value problem. Tr. Mosk. Mat. O.-va *20*, 173–204. [English transl.: Trans. Mosc. Math. Soc. *20*, 173–204 (1971)] Zbl. 181,379

Nikol'skij, S.M. (1969): Approximation of Functions of Many Variables and Imbedding Theorems. Nauka, Moscow. [English transl.: Springer, Berlin Heidelberg New York (1975)] Zbl. 185,379

Novikov, A.A. (1972): An identity for stochastic integrals. Teor. Veroyatn. Primen. *17*, No. 4, 761–765. [English transl.: Theory Probab. Appl. *17*, 717–720 (1973)] Zbl. 284.60054

Portenko, N.I. (1982): Generalized Diffusion Processes. Naukova Dumka, Kiev. English transl.: Am. Math. Soc., Providence, 1990. Zbl. 727.60088

Tanaka, H. (1979): Stochastic differential equations with reflecting boundary conditions in convex regions. Hiroshima Math. J. *9*, No. 1, 163–178. Zbl. 423.60055

Venttsel', A.D. (1959): Boundary conditions for multidimensional diffusion processes. Teor. Veoyatn. Primen. *4*, No. 2, 172–185. [English transl.: Theory Probab. Appl. *4*, 164–177 (1960)] Zbl 89,134

Veretennikov, A.Yu. (1980): Strong solutions and explicit formulas for solutions of stochastic integral equations. Mat. Sb., Nov. Ser. *111* (*153*), No. 3, 434–452. [English transl.: Math. USSR, Sb. *39*, 387–403 (1981)] Zbl. 431.60061

Zvonkin, A.K. (1974): A transformation of the state space of a diffusion process that annihilates the drift. Mat. Sb., Nov. Ser. *93* (*135*), No. 1, 129–149.[English transl.: Math. USSR, Sb. *22*, 129–149 (1975)] Zbl. 291.60036

Chapter 2
Stochastic Differential and Evolution Equations

I. Stochastic Differential Equations (SDEs)

S.V. Anulova and A.Yu. Veretennikov

§1. Strong Solutions of Stochastic Differential Equations

1.1. The solution of an SDE with respect to a Wiener process is called strong (see Chap. 1, Sect. 4) if it is adapted to a Wiener flow of σ-field. In other words, a strong solution is a solution, the trajectory of which up to each moment t can be represented as a measurable mapping of the trajectory of a Wiener process also up to t. Apart from their independent significance in the theory of SDEs, strong solutions play a large role in the theory of control of diffusion processes and in filtration theory. It would seem that at present, the theory of strong solutions has more or less been constructed. Nevertheless, a number of highly important questions in it have not yet been resolved. One of the basic unsolved questions is to find natural sufficient conditions for the existence of the strong SDE solutions arising in filtration theory (the renewal problem). The Markov case (see below) has been studied fairly thoroughly. Here we do not touch upon equations in infinite-dimensional spaces (in particular, Part II of this chapter is devoted to them) or equations with jumps, equations with respect to semimartingales or semimartingales with multidimensional time. Further, we present theorems on strong solutions for SDEs with aftereffect and with random coefficients, Markov theory for a non-degenerate diffusion, diffusions with degeneracy, and counterexamples of Tsirel'son and Barlow.

1.2. Equations with Aftereffect. Let $(\Omega, \mathcal{F}, \mathbf{P})$ be a complete probability space, $(W_t, \mathcal{F}_t)$ a d-dimensional Wiener process on it, and $b(t,x)$ and $\sigma(t,x)$ measurable mappings, $b : \mathbb{R}_+ \times C[0,\infty; E^d] \to E^d$, $\sigma : \mathbb{R}_+ \times C[0,\infty; E^d] \to E^d \times E^d$ (the space of $d \times d$ matrices). The functions $b(t, \cdot)$ and $\sigma(t, \cdot)$ are assumed to be non-anticipative, that is, they are $\mathcal{B}[0,t; E^d]$-measurable for each $t \geq 0$, where $\mathcal{B}[0,t; E^d]$ is the Borel σ-algebra in $C[0,t; E^d]$ (coinciding with the σ-algebra generated by the cylinder sets in $C[0,t; E^d]$). We consider the stochastic differential equation (SDE)

$$dx_t = \sigma(t,x)dW_t + b(t,x)dt, \ t \geq 0, \tag{1.1}$$

with the $\mathcal{F}_0$-measurable initial condition

$$x_0 = \eta. \tag{1.2}$$

A solution x_t is said to be *strong* if for each t the random variable x_t is $\mathcal{F}_t^{W,\eta}$-measurable. Sometimes by a strong solution we also mean an $\mathcal{F}_t$-adapted solution on a given probability space; in this situation, the term "strict solution" is applied. We use it in Theorem 1. In this section by a strong solution we shall always mean an $\mathcal{F}_t^{W,\eta}$-adapted one, while if the initial condition $\eta = x$ is non-random, then an $\mathcal{F}_t^W$-adapted one. A solution is called *strongly unique* or *pathwise unique* if for any given probability space and any given Wiener process any two solutions are the same. Let us also recall that a solution is called *weakly unique* or *unique in distribution* if the distribution law is the same for all solutions (even on different probability spaces). We recall that in each case, the concept of a solution assumes that the corresponding Itô and Lebesgue integrals in (1.1) are defined for it, for which it is sufficient that

$$\mathbf{P}\Big(\int_0^T |b(t,x)|dt + \int_0^T \|\sigma(t,x)\|^2 dt < \infty, \forall T > 0\Big) = 1.$$

Theorem 1.1 (Liptser and Shiryaev 1974, Chap. 4). *Suppose that the non-anticipative functions $\sigma(t,x)$, $b(t,x)$ satisfy the Lipschitz condition*

$$|b(t,x) - b(t,x')|^2 + \|\sigma(t,x) - \sigma(t,x')\|^2$$
$$\leq L_1 \int_0^t |x_s - x'_s|^2 dK_s + L_2|x_t - x'_t|^2, \ t \geq 0,$$

and the linear growth condition

$$|b(t,x)|^2 + \|\sigma(t,x)\|^2 \leq L_1 \int_0^t (1 + |x_s|^2)dK_s + L_2(1 + |x_t|^2), \ t \geq 0,$$

where $L_1, L_2 \geq 0$, $K_s(s \geq 0)$ is a non-decreasing right-continuous function and $x, x' \in C[0, \infty; E^d]$. Let η be an $\mathcal{F}_0$-measurable random variable, $\mathbf{P}(|\eta| < \infty) = 1$. Then equation (1.1)–(1.2) has a strict solution and the solution is pathwise unique.

The proof of this and other similar theorems with Lipschitz-type conditions is usually based on the successive approximation method and the Gronwall-Bellman Inequality, see, for example, (Liptser and Shiryaev 1974). The same method enables one also to study more general equations for semimartingales.

In many linear problems it is useful to have an explicit form of a solution of an SDE.

Theorem 1.2 (Liptser and Shiryaev 1974, Chap. 4). *Let $\sigma(t,x) \equiv \sigma(t)$, and $b(t,x) = b_0(t) + b_1(t)x_t$, where $\sigma(t), b_0(t), b_1(t)$ are measurable non-random functions, $\sigma(t), b_1(t)$ are $d \times d$ matrices, $b_0(t)$ is a d-dimensional vector and*

$$\int_0^T (|b_0(t)| + \|b_1(t)\| + \|\sigma(t)\|^2)dt < \infty, \forall T > 0.$$

Then equation (1.1)–(1.2) has a unique strong solution which is representable in the form

$$x_t = \Phi_t\left[\eta + \int_0^t \Phi_s^{-1}b_0(s)ds + \int_0^t \Phi_s^{-1}\sigma(s)dw_s\right],$$

where the $d \times d$ matrix Φ_t is a solution of the fundamental equation

$$\Phi_t = E_{d\times d} + \int_0^t b_1(s)\Phi_s ds$$

($E_{d\times d}$ is the $d \times d$ identity matrix).

1.2. The concept of a weak solution of an SDE of the form (1.1)–(1.2) has already been mentioned in Chap. 1, Sect. 4. In connection with the concepts of strong and weak solutions, we note that Tsirel'son has constructed an example of a one-dimensional SDE of the form (1.1) with unit diffusion and bounded drift (depending on the entire "past") having no strong solutions, but having a weak solution.

But in the Markov case, that is, for the case (1.3) (see below), an equation with unit diffusion and a measurable bounded drift has a unique strong solution.

1.3. The Markov Case. Yamada-Watanabe Theorem. Let us consider the d-dimensional equation

$$dx_t = \sigma(t, x_t)dW_t + b(t, x_t)dt, \ t \geq 0, \tag{1.3}$$

with non-random initial condition

$$x_0 = x. \tag{1.4}$$

Here b, σ are measurable bounded functions: $b : \mathbb{R}_+ \times E^d \to E^d$, $\sigma : \mathbb{R}_+ \times E^d \to E^d \times E^d$. The following Yamada-Watanabe principle is of importance in the theory of strong solutions.

Theorem 1.3 (Yamada and Watanabe 1971). *Suppose that the equation (1.3)–(1.4) has a weak solution and the solution of (1.3)–(1.4) is pathwise unique. Then this equation has a strong solution.*

A more detailed proof of Theorem 3 than that given in (Yamada and Watanabe 1971) can be found in (Zvonkin and Krylov 1975). We note that the scheme of the proof is applicable in different situations: for SDEs with reflection, with jumps and in the non-Markov case.

Theorem 1.4 (Yamada and Watanabe 1971). *Let the coefficients σ and b be bounded and satisfy the conditions*

$$\|\sigma(t,x) - \sigma(t,x')\| \leq \rho_1(|x-x'|),$$
$$|b(t,x) - b(t,x')| \leq \rho_2(|x-x'|),$$

where $\rho_1, \rho_2 \in C(\mathbb{R}_+; \mathbb{R}_+)$, $\rho_1(0) = \rho_2(0) = 0$, ρ_1, ρ_2 are increasing and convex upward, and for any $\epsilon > 0$

$$\int_0^\epsilon \frac{du}{\rho_1(u)} = \infty, \quad \int_0^\epsilon \frac{du}{\rho_2(u)} = \infty. \tag{1.5}$$

Then equation (1.3)–(1.4) has a strong solution and the solution of the equation is pathwise unique.

In particular, if σ and b satisfy the Lipschitz condition

$$\|\sigma(t, x) - \sigma(t, x')\| + |b(t, x) - b(t, x')| \le L|x - x'|,$$

then the conditions of Theorem 1.4 are satisfied.

Essentially weaker conditions are possible in the one-dimensional case $d = 1$.

Theorem 1.5 (Yamada and Watanabe 1971). *Let $d = 1$, let the coefficients σ and b be bounded and suppose that*

$$|\sigma(t, x) - \sigma(t, x')| \le \rho_1(|x - x'|),$$
$$|b(t, x) - b(t, x')| \le \rho_2(|x - x'|),$$

where $\rho_1, \rho_2 \in C(\mathbb{R}_+; \mathbb{R}_+)$, $\rho_1(0) = \rho_2(0) = 0$, ρ_1, ρ_2 are increasing and convex upward, and for any $\epsilon > 0$

$$\int_0^\epsilon \frac{du}{\rho_1^2(u)} = \infty, \quad \int_0^\epsilon \frac{du}{\rho_2(u)} = \infty. \tag{1.6}$$

Then equation (1.3)–(1.4) has a strong solution and the solution of the equation is pathwise unique.

The condition on the modulus of continuity ρ_1 is satisfied, in particular, if $|\sigma(t, x) - \sigma(t, x')| \le L|x - x'|^{1/2}$.

Theorems 1.4 and 1.5 are proved by means of suitable approximations of $|u|$ and the Itô formula for $|x_t - x'_t|$, where x_t, x'_t are two solutions of the equation. The conditions (1.5) or (1.6) for ρ_1, as it turns out, ensure that the local time of the process $|x_t - x'_t|$ equals zero. By virtue of these same conditions (1.5) or (1.6), certain well known inequalities then ensure uniqueness (in particular, under a Lipschitz condition, this step is ensured by the Gronwall-Bellman inequality). For similar proofs see (Yamada and Watanabe 1971) or (Ikeda and Watanabe 1981). In fact, it is very close to the use of local times, see [E. Perkins, Local time and pathwise uniqueness for stochastic differential equations, Sém. de proba. XVI, Lect. Notes Math. 920, 201-208, Springer-Verlag, Berlin (1982)],[J.-F. Le Gall, Sém. de Proba. XVII, Proc. 1981/82, Lect. Notes Math, 986, 15-31, Springer-Verlag, Berlin (1983)].

1.4. Equations with Measurable Drift. Nakao (Nakao 1972) established the existence of a strong solution (more precisely, strong uniqueness, from which the existence of a strong solution also follows) of the one-dimensional SDE

$$dx_t = \sigma(x_t)dW_t + b(x_t)dt, \quad x_0 = x,$$

with bounded measurable σ and b under the condition $\sigma(x) \geq \epsilon > 0$ and var $\sigma < \infty$ using beautiful martingale methods.

Zvonkin (Zvonkin 1974) proved a similar result for the one-dimensional equation

$$dx_t = \sigma(t, x_t)dW_t + b(t, x_t)dt, \quad x_0 = x,$$

with bounded (or linearly increasing) measurable coefficients under the condition

$$\sigma^2(t,x) \geq \epsilon > 0 \text{ and } |\sigma(t,x) - \sigma(t,x')| \leq L|x - x'|^{1/2}$$

by means of a special change of coordinates, annihilating the drift: it turned out that the given transformation (see below) preserves the Hölder modulus of continuity of diffusion with exponent $\frac{1}{2}$, that is, it reduces the problem to Theorem 1.4 of Yamada and Watanabe. In the multidimensional case, Zvonkin (Zvonkin 1974) proved a similar result regarding a strong solution under the condition

$$\lambda^* \sigma \sigma^*(t,x) \lambda \geq \epsilon |\lambda|^2 \ (\epsilon > 0), \quad |\sigma(t,x) - \sigma(t,x')| \leq L|x - x'|,$$

and a Dini condition on the drift coefficient:

$$|b(t,x) - b(t',x')| \leq \rho(|x - x'| + |t - t'|^{1/2}),$$

where $\rho(u) \geq 0$, $u \geq 0$, the function ρ is increasing and convex upwards, $\rho(0) = 0$, for any $\epsilon > 0$

$$\int_0^\epsilon \frac{\rho(u)}{u} du = \infty,$$

and $u^\alpha / \rho(u) \to 0$, $u \to 0$, for each $\alpha > 0$. Under precisely such conditions, the same coordinate change leaves the diffusion Lipschitz, that is, the problem reduces to the classical Itô theorem (an equation with coefficients satisfying a Lipschitz condition).

In (Veretennikov 1979) strong uniqueness was established for the one-dimensional SDE under conditions combining the Nakao and Zvonkin conditions. We do not present the formulation here, since in the next section a very similar result is formulated for a more general case with degeneracy. Nevertheless, we shall say a few words regarding the proof. The idea of it was proposed by Krylov. The proof itself relies on an estimate of the distribution of the stochastic integral and the technique applied by Yamada and Watanabe. It turns out that there exist maximal and minimal solutions. If they do not coincide, then this contradicts the weak uniqueness of the solution of the equation. Similar constructions were used by Torondzhadze and Chitashvili (1980) (1981). In the multidimensional case, there is the following result.

Theorem 1.6. *Suppose that the conditions*

$$\|\sigma(t,x) - \sigma(t,x')\| \leq L|x - x'|,$$

and
$$\lambda^* \sigma \sigma^*(t,x)\lambda \geq \epsilon |\lambda|^2, \ \epsilon > 0$$
hold. Then equation (1.3)–(1.4) has a strong solution and the solution of the equation is pathwise unique.

The idea of the proof is to use Zvonkin's change of variables annihilating drift.[1] Here, it has to be said, the diffusion coefficient is spoilt; in the new variables it no longer satisfies a Lipschitz condition in general. Nevertheless, it turns out that it has a derivative that is locally summable to a high degree, and this can be used in the case of a non-degenerate diffusion. Here Krylov's estimate of the distribution of the stochastic integral is of help. A detailed proof (under somewhat more stringent conditions on the diffusion) can be found in (Veretennikov 1980). The change of variables itself relies on results on the solvability of the Cauchy problem for systems of parabolic equations, see (Ladyzhenskaya, Solonnikov, Ural'tseva 1967) (where the formulations and proofs pertain to boundary-value problems) and (Zvonkin 1974). In both papers an additional condition of uniform continuity with respect to (t,x) is imposed on the matrix $\sigma\sigma^*$. A necessary result without this condition is contained in the paper (Veretennikov 1982). Similar results (not for systems, but for a single parabolic equation) were established in (Stroock and Varadhan 1979, Appendix). A convergence of approximations to a strong solution on a given probability space was established in [I. Gyongy, N. Krylov, Existence of strong solutions for Itô's stochastic equations via approximations, Probab. Theory Relat. Fields, 1996, 105, 143-158].

1.5. Equations with Degenerate Diffusion. How does one deal with strong uniqueness and the existence of a strong solution if the drift coefficient is non-smooth (measurable), and the diffusion can degenerate? We present two versions of a (partial) answer to this question. The first result is a generalization of Theorem 1.6.

Let $d = d_1 + d_2$, where d_1, d_2 are natural numbers, $\sigma_1 := (\sigma_{ij} : 1 \leq i \leq d_1, \ 1 \leq j \leq d)$, $\sigma_2 := (\sigma_{ij} : d_1 < i \leq d, \ 1 \leq j \leq d)$, $b_1 := (b^i : 1 \leq i \leq d_1)$, $b_2 := (b^i : d_1 < i \leq d)$, $x_1 = (x^1, \ldots, x^{d_1})$, $x_2 = (x^{d_1+1}, \ldots, x^d)$. We assume that the matrix $\sigma_1 \sigma_1^*$ is non-singular: for all $t \geq 0$, $x \in E^d$, $\lambda \in E^{d_1}$

$$\lambda^* \sigma_1 \sigma_1^*(t,x)\lambda \geq \nu |\lambda|^2 \ (\nu > 0). \tag{1.7}$$

Lemma. *Let condition (1.7) hold, and suppose that all the coefficients are continuous in $x_2 = (x^{d_1+1}, \ldots, x^d)$. Then equation (1.3)–(1.4) has a solution on some probability space.*

This lemma (Veretennikov 1983, Theorem 1), which was proved in the homogeneous case by Nisio (Nisio 1973), allows one, by the Yamada-Watanabe

---

[1] Editor's Note. The change of variables annihilating drift was, in fact, known earlier (see Gikhman, Skorokhod :"Stochastic Differential Equations", Naukova Dumka, Kiev, 1968). Zvonkin first applied this change to prove the existence of strong solutions.

theorem, to prove only strong uniqueness, from which the existence of a strong solution follows.

Theorem 1.7. *Suppose that condition (1.7) holds, and all the coefficients satisfy a Lipschitz condition in x_2 with constant C not depending on (t, x_1), the functions σ_1, σ_2 and b satisfy a Lipschitz condition in x_1 with the same constant C, and the functions $a_1(t, x_1, \cdot) = \sigma_1 \sigma_1^*(t, x_1, \cdot)$ and $b(t, x_1, \cdot)$ have two uniformly bounded equicontinuous derivatives for all t, x_1. Then equation (1.3)–(1.4) has a strong solution and the solution is pathwise unique.*

The idea of the proof is the same as in the non-degenerate case, but now it is necessary to transform only some of the coordinates. Moreover, a Cauchy problem arises for the parabolic system of equations, where the remaining variables emerge as parameters. To apply the Itô formula, two continuous derivatives of the coefficients a_1 and b_1 with respect to x_2 are required. For a detailed proof see (Veretennikov 1983). Note that in (Veretennikov 1983) the condition of equicontinuity of $\nabla^2_{x_1} a_1(t, x_1, \cdot)$ is omitted; furthermore, differentiability conditions are imposed on σ_1, and not on a_1 in this paper. If σ_1 is a symmetric positive definite root of $2a_1$, then for non-degenerate a_1, these versions are equivalent.

Thus one of the possibilities for strong uniqueness consists in assuming additional smoothness along those directions for which the diffusion degenerates. It has to be said that these assumptions of additional smoothness stem from the method, and possibly they can be weakened.

Another possibility is allowed by Theorem 1.8 below, in which results are presented only for the one-dimensional case, since up to now the multidimensional results (see Kleptsyna and Veretennikov 1981) have considerably less generality. Thus, let $d = 1$, and suppose that σ is representable in the form

$$\sigma(t, x) = \sigma_1(t, x)\sigma_2(t, \sigma_3(x)), \tag{1.8}$$

where $\sigma_1, \sigma_2, \sigma_3$ are Borel bounded functions, σ_1 and σ_2 satisfy the Hölder condition

$$|\sigma_1(t, x) - \sigma_1(t, x')| + |\sigma_2(t, x) - \sigma_2(t, x')|$$
$$\leq C|x - x'|^{1/2}, \quad t \geq 0, \quad x, x' \in E^1, \tag{1.9}$$

σ_2 is isolated from zero:

$$\inf_{t,x} \sigma_2(t, x) > 0, \tag{1.10}$$

and σ_3 is a function of locally bounded variation: for any $N > 0$

$$\mathrm{var}_{[-N,N]} \sigma_3 < \infty. \tag{1.11}$$

Finally, it is assumed that the drift $b(t, x)$ is representable in the form

$$b(t, x) = \beta(t, x) + \sigma(t, x)B(t, x), \tag{1.12}$$

where β and B are bounded measurable functions, while β satisfies a Lipschitz condition
$$|\beta(t,x) - \beta(t,x')| \leq C|x - x'|, \qquad (1.13)$$
$t \geq 0$, $x, x' \in E^1$. No kind of smoothness is assumed for B.

Theorem 1.8 (Kleptsyna 1984). *Let $d = 1$, and assume that conditions (1.8)–(1.13) hold. Then equation (1.3)–(1.4) has a strong solution and the solution of the equation is pathwise unique.*

See the paper (Veretennikov and Kleptsyna 1983) for other theorems on strong uniqueness.

1.6. Comparison Theorems. Let $d = 1$. Consider the two simultaneous equations
$$dx_t^i = \sigma(t, x_t^i)dW_t + b^i(t, x_t^i)dt, \quad x_0^i = x^i, \qquad (1.14)$$
where $i = 1, 2$, and b^1, b^2 are bounded Borel functions. It is assumed that for $i = 1, 2$ the conditions of Theorem 1.8 are satisfied, in particular, each equation has a strong solution and the solution is pathwise unique.

Theorem 1.9. *Let $x^1 \geq x^2$, and let $b^1(t,x) \geq b^2(t,x)$ for all t, x. Then $\mathbf{P}(x_t^1 \geq x_t^2, \ t \geq 0) = 1$.*

This result was established by (Kleptsyna 1985), announced in [Kleptsina M.L., Veretennikov A. Yu. (1982), Theorems of comparison, existence and uniqueness for stochastic Itô equations, IV USSR-Japan Symp. on Probab. Th. and Math. Stat., Tbilisi 1982, Abstr. Comm., vol. 2, 21-22]. Earlier in other situations similar results were proved in (Skorokhod 1987); (Yamada 1973); (Manabe and Shiga 1979). The basic idea of the proof consists in first reducing the problem to the situation $b^1(t,x) > b^2(t,x)$, $x^1 > x^2$, and then establishing the equality $\mathbf{P}(\tau := \inf(t \geq 0: \ x_t^1 < x_t^2) = \infty) = 1$ in this situation. If the b^i are continuous, then this is obtained from the following considerations. Let $\tau' := \inf(t \geq 0: \ x_t^1 = x_t^2) < \infty$. Then by the condition $b^1 - b^2 > 0$ and the continuity of b^i in some neighborhood of τ', we obtain $x_t^1 > x_t^2$ ($t \neq \tau'$), from which it follows that $\tau = \infty$. For discontinuous b^i, one succeeds by using an estimate of Krylov.

1.7. Theorems on the Existence of a Strong Solution. The comparison theorems, in particular Theorem 1.9, gives an interesting possibility for establishing the existence of a strong solution by a monotonic passage to the limit without reference to uniqueness of trajectories, which in fact, may not hold.

Theorem 1.10. *Let $d = 1$, suppose that conditions (1.8)–(1.11) hold, and the function $b(t, \cdot)$ is continuous for any $t \geq 0$. Then equation (1.3)–(1.4) has at least one strong solution.*

We re-emphasize that pathwise uniqueness for a fixed function b is not asserted. However, if one considers a family of functions (b^α, $\alpha \in E^1$) strictly monotonically dependent on the parameter α (for example, $b^\alpha(t,x) := b(t,x) + \alpha$), then that same comparison theorem also guarantees strong uniqueness for all values of α except, perhaps, countably many.

Theorem 1.11. *Let $d = 1$, suppose that conditions (1.8)–(1.11) hold, $b^\alpha(t,x) = b(t,x) + \alpha$, and the function $b(t,\cdot)$ is continuous for all $t \geq 0$. Then for all $\alpha \in E^1$, excluding perhaps a countable number of values, the solution of*
$$dx_t = \sigma(t,x_t)dW_t + b^\alpha(t,x_t)dt, \quad x_0 = x,$$
is pathwise unique.

The idea of the proof is to use monotonicity with respect to α: a monotonic function cannot have more than a countable number of discontinuities. Theorem 1.11 was proved in (Kleptsyna 1985). The method of monotonic approximations itself was first applied by Mel'nikov (Mel'nikov 1979), who proved strong uniqueness for a piecewise-continuous drift. Existence theorems without uniqueness were then obtained by this method in (Fedorenko 1982); (Mel'nikov 1982); (Kleptsyna, Fedorenko 1984) and (Kleptsyna 1985).

1.8. Pathwise Solution of Stochastic Differential Equations. The very concept of a solution of a stochastic differential equation is not a pathwise one; a solution must be given directly for all $\omega \in \Omega$, this being the specific character of stochastic integration. However, if $\sigma \equiv 1$, and the coefficient b is sufficiently smooth, then one can solve equation (1.3)–(1.4) for each ω, that is, for each trajectory $W_t(\omega)$, $t \geq 0$, knowing nothing about the trajectories of W for other values of ω. The natural question arises: in which situations can one understand (and solve) a stochastic differential equation as a pathwise equation for each ω and in this case can one understand the operation of stochastic integration in some pathwise sense, that is, determine a value of the variable $\int \sigma(s, x_s(\omega))dW_s(\omega)$ only along the trajectories $x_t(\omega)$ and $W_t(\omega)$ for fixed ω. To date there is no satisfactory answer to this question in the general situation: all the known results on this theme require the satisfaction of a complete integrability condition or a Frobenius condition with respect to the diffusion matrix and a certain smoothness of the coefficients which, as a rule, is sufficiently restrictive. The condition of complete integrability of σ is usually formulated using derivatives as well. However, the reason for including this condition is not in fact connected with the smoothness of the coefficients. This condition itself arises from a method in which a Wiener trajectory is replaced by some smooth approximation, and the possibility is established of continuously extending the corresponding operator with smooth ones on all the continuous trajectories. The continuity condition of this operator is also the complete continuity condition. Nevertheless, in the case $d = 1$ a complete integrability condition, as a rule, is satisfied almost automatically (see below).

Complete Integrability Condition. Let the function $\sigma(t,\cdot)$ be continuously differentiable for each $t \geq 0$. Then for each $t \geq 0$ $\sigma(t,\cdot)$ is completely integrable if for all $\xi, \eta \in E^d$

$$\left(\left(\frac{\partial}{\partial x}\sigma(t,x)\right)\sigma(t,x)\xi\right)\eta = \left(\left(\frac{\partial}{\partial x}\sigma(t,x)\right)\sigma(x,t)\eta\right)\xi.$$

The sense of this condition which, in fact, is not connected with the differentiability of σ, consists in the unique solvability in C^1 of the problem

$$\frac{dy}{dx} = \sigma(t,y),$$

(for fixed t) with arbitrary initial condition $y(x_0) = y_0$; see, for example, (Cartan 1971).

The next theorem was established in (Krasnosel'skij and Pokrovskij 1978) (in this paper the more general situation was considered, when the diffusion matrix depends also on the value of W_t : $\sigma(t, x_t, W_t)$). It is more convenient to state it for a stochastic equation in Stratonovich form.

Theorem 1.12. *Suppose that the complete integrability condition for $\sigma(t, \cdot)$ holds for each $t \geq 0$, the function σ is twice continuously differentiable with bounded first derivatives and the function b is continuously differentiable. Then the solution of the stochastic differential equation in the Stratonovich form*

$$dx_t = \sigma(t, x_t) \circ dW_t + b(t, x_t)dt, \quad t \geq 0, \ x_0 = x,$$

is representable in the form $x_t = V_t[0, x]W$, where $V_t[0, x]u$ is a continuous extension to $C[0, \infty; E^d]$ of the operator converting the function u in $C^1[0, \infty; E^d]$ to a solution of the equation

$$dX_t = \sigma(t, X_t)u_t dt + b(t, X_t)dt, \ t \geq 0, \ X_0 = x. \tag{1.15}$$

A continuous extension of this operator exists.

Similar results (with $\sigma = \sigma(t,x)$) were established in (Doss 1976, 1977) and (Sussman 1978). New theorems on continuous dependence for equations with respect to semimartingales have been obtained by Matskyavichus (Matskyavichus 1985, etc.) and by Gyöngy [I. Gyöngy (1988), on the approximation of stochastic differential equations, Stochastics, 23, 331-352].

In general, the "extra" smoothness requirements on σ and b cannot be weakened, even in the one-dimensional case. Here there are only partial results, one of which is presented below.

Theorem 1.13. *Let $d = 1$, $\sigma \equiv 1$ and suppose that the function $b(x)$ is bounded and continuous ($x \in E^1$). Then the solution of the equation*

$$x_t(\omega) = x + W_t(\omega) + \int_0^t b(x_s(\omega))ds \tag{1.16}$$

is unique with probability 1.

Let us note that in view of the condition of continuity of b, equation (1.16) has at least one solution for any continuous trajectory W_t, $t \geq 0$.

Uniqueness is understood here in the sense of ordinary differential equations, that is, for a fixed $\omega \in \Omega$. The proof (see Veretennikov, Kleptsyna 1989) is based on the decomposition of a Wiener process on a fixed interval of time into a series in sines and cosines (see, for example, Ikeda and Watanabe 1981) and on a comparison theorem for ordinary differential equations.

The result of Theorem 1.13 is extended also to the case $\sigma \not\equiv 1$ under the condition $\inf_x \sigma(x) > 0$ and $\sigma \in C^1(E^1)$ (Veretennikov and Kleptsyna 1989). In this case, however, the question arises *what is a pathwise solution of a stochastic differential equation?* One possible method of reasoning is as follows. We make a change of coordinates, after which the diffusion becomes identically equal to unity. We solve the equation in these new coordinates and then perform the reverse change. In this case there arises the natural question: is the "pathwise" stochastic integral

$$\int_0^t \sigma(x_s)dW_s := x_t - x_0 - \int_0^t b(x_s)ds$$

defined in some sense precisely as an integral (and not simply as the right hand side of the given equation)? It turns out that the answer to this question is affirmative: this stochastic integral actually can be understood as the limit of Riemann-Itô integral sums under an unlimited refinement of the intervals of the partition, if the filter of partitions is fixed in advance, for example, if the binary rational points are taken as the partition points. The proof is based on the "pathwise" Itô formula, which in turn is based on the definition of the quadratic variation of a "typical" Wiener trajectory. A similar approach was used in (Krasnosel'skij and Pokrovskij 1978).

The almost sure uniqueness in the sense of ordinary differential equations for equation (1.16) actually implies uniqueness in the sense of strong solutions (which, it is true, is already known; this is just another way of establishing it).

The above method not only yields results in the cases $d > 1$, for $d = 1$ and measurable function b, but even for $d = 1$ and continuous b depending on time.

Yet another approach to pathwise solutions, connected with the description of all non-anticipative and anticipative solutions was proposed in (Torondzhadze and Chitashvili 1981).

1.9. Counterexamples: Theorems of Tsirel'son and Barlow. The existence of weak SDE solutions, of Markov and non-Markov type, has been established under very broad restrictions. The first simple counterexamples to the existence of a strong solution for the equation $dx_t = \sigma(x_t)dW_t$, $x_0 = x$, with non-degenerate bounded diffusion were constructed by Tanaka and Krylov. In the Tanaka example, the dimension $d = 1$, $\sigma(x) = I(x \geq 0) - I(x < 0)$; in the Krylov example, $d = 2$, and the diffusion matrix has the form

$$\sigma(x) = \begin{pmatrix} \dfrac{x_1}{\sqrt{x_1^2 + x_2^2}} & \dfrac{-x_2}{\sqrt{x_1^2 + x_2^2}} \\ \dfrac{x_2}{\sqrt{x_1^2 + x_2^2}} & \dfrac{x_1}{\sqrt{x_1^2 + x_2^2}} \end{pmatrix},$$

see the discussion of these and other examples in the article (Zvonkin and Krylov 1975). Let us note one interesting property of both examples: if x_t is a (weak) solution, then its modulus $|x_t|$ is $\mathcal{F}_t^W$-adapted, although, as was said, x_t cannot possess this property.

In the paper (Zvonkin and Krylov 1975), the three most important unsolved problems at that time of the theory of strong solutions were formulated. Now they have all been solved. The existence of a strong solution of the SDE $dx_t = b(t, x_t)dt + dW_t$, $x_0 = x$, with bounded measurable drift b ($d \geq 1$) follows from Theorem 1.6 given above. The other two problems obtained negative solutions in the examples of Tsirel'son and Barlow.

The example of (Tsirel'son 1975) pertains to the one-dimensional equation

$$dx_t = \alpha(t, x_0^t)dt + dW_t, \quad x_0 = x, \tag{1.17}$$

where $\alpha(t, x_0^t)$ is a measurable bounded non-anticipative functional. Let $\{a\}$ be the fractional part of a, $(t_k, k = 0, -1, -2, \ldots)$ a sequence of positive numbers $\lim_{k \to -\infty} t_k = 0$, $0 < t_{k-1} < t_k$. Let

$$\alpha(t, x_0^t) = \begin{cases} \left\{ \dfrac{x_{t_k} - x_{t_{k-1}}}{t_k - t_{k-1}} \right\}, & t \in [t_k, t_{k+1}), \ k = -1, -2, \ldots, \\ 0, & t = 0 \text{ or } t \geq t_0. \end{cases}$$

Theorem 1.14 (Tsirel'son 1975). *Equation* (1.17) *does not have a strong solution.*

One can read the completed proof of the theorem of Tsirel'son, proposed by Krylov in the book (Ikeda and Watanabe 1981).

The result of Barlow pertains to the one-dimensional SDE

$$dx_t = \sigma(x_t)dW_t, \quad x_0 = x, \tag{1.18}$$

with non-degenerate continuous diffusion.

Theorem 1.15 (Barlow 1982). *Let $\sigma(x)$ be a continuous function and suppose that there exist positive constants c_1, c_2, c_3, c_4, α, β such that*
  a) $0 < \alpha < 1/2$;
  b) $\alpha \leq \beta < a + \min(\alpha/(\alpha+2), (\alpha - 2\alpha^2)/(1+2\alpha))$;
  c) $0 < c_1 < \sigma^2(x) < c_2$, $x \in [x_1, x_2]$;
  d) $|\sigma(x) - \sigma(y)| \leq c_3|x-y|^\alpha$, $x, y \in [x_1, x_2]$;
  e) *for any x, y, $x_1 \leq x < y \leq x_2$, there exist x', y', $x \leq x' < y' \leq y$ such that $|x' - y'| > c_4|x - y|$ and $|\sigma(x') - \sigma(y')| > c_4|x' - y'|^\beta$.*
Then for $x_0 \in (x_1, x_2)$ the solution of (1.18) *cannot be pathwise unique.*

The proof, which is extremely cunning, can be found in (Barlow 1982). A simple example was also constructed there of a function $\sigma(x)$ satisfying the conditions of the theorem, and it was thereby shown that the set of such functions is non-empty.

§2. Weak Solutions of Stochastic Differential Equations with Non-Smooth Coefficients in E^d

2.1. In its development the theory of stochastic differential equations recapitulated the theory of ordinary differential equations: from the theorem of the existence of a solution of an equation with Lipschitz coefficients, the step was taken to continuous coefficients (Skorokhod 1961). The idea of the proof was taken from ordinary differential equations, namely, Euler approximations and the Arzelà-Ascoli compactness condition. Of course, the specific character of stochastic objects was bound to bring in new problems; it turned out that for limiting processes, it is impossible to guarantee relative compactness in the sense of convergence in probability. Skorokhod overcame this difficulty, with the help of the method of a single probability space: a probability space was constructed on which the Euler approximations converged in probability. For some time no one observed that the limiting process, although it is in fact a solution, proved to be not "strong", but "weak"; it was not adapted to the flow generated by a Wiener process. The interpretation of this fact led to the conclusion that for irregular diffusion and drift coefficients, it is necessary to describe the corresponding processes not as solutions of stochastic differential equations, but as solutions of a certain martingale problem.

2.2. Let us fix a natural number d. Let us denote by $C = \{X\}$ the space of continuous E^d-valued functions on $[0,\infty)$; $\mathcal{C}$ is the σ-algebra of Borel sets of C, **C** is the natural flow of σ-algebras on C.

Suppose that on $[0,\infty) \times C$ we are given functions $\sigma = \sigma_t(X)$ and $b = b_t(X)$ with values in the space of $d \times d$-dimensional matrices and in E^d, respectively, which are measurable **C**-non-anticipative stochastic processes.

We consider the equation

$$dX_t = \sigma_t(X)dW_t + b_t(X)dt. \tag{1.19}$$

Definition 1.1. By a weak solution of equation (1.19) with initial condition $x \in E^d$ is meant a process X with values in E^d and continuous trajectories defined on some stochastic basis $(\Omega, \mathcal{F}, \mathbf{F}, \mathbf{P})$ with an **F**-Wiener process W such that $X_0 = x$ **P**-a.s., X is **F**-adapted and (1.19) is satisfied.

We set $a = \frac{1}{2}\sigma\sigma^*$.

Definition 1.2 (Stroock and Varadhan 1979, Chap. 6, Sect. 6.0). By a solution (a,b) of the *martingale problem* is meant a probability **P** on $(C,\mathcal{C})$ such that for any $f \in C_0^2(E^d)$ the process

$$f(X_t) - \int_0^t \Big(\sum_{i,j=1}^d a_s^{ij}(X)\frac{\partial^2 f}{\partial x_i \partial x_j}(X_s) + \sum_{i=1}^d b_s^i(X)\frac{\partial f}{\partial x_i}(X_s)\Big)ds,$$

is a **C**-martingale.

The next theorem shows that in these two definitions we are in fact dealing with the same object.

Theorem 1.16 (Ikeda and Watanabe 1981, Chap. IV, §2, Proposition 2.1). *The distribution of every weak solution is a solution of the martingale problem. Every solution of the martingale problem is a distribution of some weak solution.*

Definition 1.3. We say that a solution of (1.19) is weakly unique if for any $x \in E^d$ all the weak solutions of (1.19) with initial condition $X_0 = x \in E^d$ are identically distributed in $(C, \mathcal{C})$.

Theorem 1.17 (A corollary of Lemma 1.2, Chap. IV of (Ikeda and Watanabe 1981)). *Strong uniqueness of equation (1.10) entails weak uniqueness.*

The next theorem shows that in questions of the uniqueness and existence of a solution of (1.19) the coefficient σ plays a determining role.

Theorem 1.18 (Ikeda and Watanabe 1981, Chap. IV, §4, Theorem 4.2). *Let c be a bounded function on $[0, \infty) \times C$ with values in E^d and defining a measurable **C**-adapted process. The existence (or uniqueness) of the solution of equation (1.19) is equivalent to the existence (or uniqueness) of the solution of the equation*

$$dX_t = \sigma_t(X)dW_t + [b_t(X) + \sigma_t(X)c_t(X)]dt.$$

Corollary. *If the function $\sigma^{-1}b$ is defined and bounded, then the existence (or uniqueness) of a solution of equation (1.19) is equivalent to the existence (or uniqueness) of a solution of the equation without drift*

$$dX_t = \sigma_t(X)dW_t.$$

2.3. We will say that a function f on $[0, \infty) \times E^d$ is Markov if f depends only on (t, X_t).

Theorem 1.19. *Suppose that for some $c > 0$ we have*

$$(\|\sigma\|^2 + |b|^2)_t(X) \leq c(1 + \sup_{s \leq t}|X_s|^2)$$

for all t, X. Then a weak solution of equation (1.19) exists in each of the following cases:
1) the functions σ and b are continuous in X for each t (Skorokhod 1961a); (Conway 1971); (Lebedev 1983);

2) *the function σ is Markov and uniformly non-degenerate (Krylov 1977, Theorem 1, Chap. II, §6)*;

3) *for $d \geq 2$ and some $k \in \{1, \ldots, d-1\}$, the functions σ, b are continuous with respect to $(x_1, \ldots, x_k)$ and the matrix $(\sigma^{i,j})_{i=k+1, j=1}^{i,j=d}$ is Markov and uniformly non-degenerate (Nisio 1973); (Veretennikov 1983);*

Proof. Let us first note that since the coefficients σ and b have linear growth, we can assume without loss of generality that they are bounded (see Stroock and Varadhan 1979, Chap. 10).

In each of the cases 1), 2), 3) the idea of the proof is the same; to approximate σ and b by "good" functions σ^n and b^n, for which there exists a strong solution X^n of the equation

$$dX_t^n = \sigma_t^n(X^n)dW_t + b_t^n(X^n)dt, \tag{1.20}$$

and then pass to the weak limit of its distribution with respect to n.

$1°$. Suppose that 1) holds. For $X \in C$ and $t \geq 0$ we denote by $X_{\wedge t}$ the function taking the value $X_{s \wedge t}$ at time s. We take a natural number n, set $t_n = \frac{1}{n}[n, t]$ (the brackets [ ] denote the integral part) and set $\sigma_t^n(X) = \sigma_t(X_{\wedge t_n})$, $b_t^n(X) = b_t(X_{\wedge t_n})$, $t \geq 0$. Equation 1.20 has a strong solution X^n since its coefficients "almost do not depend" on X. Let us verify that the "characteristics"

$$\int_0^t b_s^n(X^n)ds, \quad \int_0^t a_s^n(X^n)ds$$

of the process X^n converge to the limits

$$\int_0^t b_s(X)ds, \quad \int_0^t a_s(X)ds.$$

Let $K \subseteq C$ be a compact set. For fixed s the restriction of b_s to K gives a bounded uniformly continuous function. We denote by V_s the modulus of continuity of this function. Then for $X \in K$ we have

$$\int_0^t |b_s - b_s^n|(X)ds = \int_0^t |b_s(X) - b_s(X_{\wedge s_n})|ds$$

$$\leq \int_0^t V_s(\sup_{u \in [s_n, s]} |X_u - X_{s_n}|)ds.$$

By the Arzelà-Ascoli Theorem (Billingsley 1968, Appendix 1)

$$\sup_{u \in [s_n, s]} |X_u - X_{s_n}| \to 0 \text{ as } n \to \infty$$

uniformly for $X \in K$. Hence by the Lebesgue dominated convergence theorem we deduce that the right hand side of the inequality tends to zero uniformly on K as $n \to \infty$; hence, so does the left hand side. Similarly

$$\int_0^t |a_s - a_s^n|(X)ds \to 0 \text{ as } n \to \infty$$

uniformly on K. In accordance with the version of the limit theorem for semimartingales (see Chap. 4), this means that the distribution of the solution X^n has a limit point and it is a solution (a, b) of the martingale problem.

2°. Let 2) be satisfied. By the corollary to Theorem 1.18 we can assume that $b \equiv 0$. Since σ is Markov, there exists a Borel function σ_M on $[0, \infty) \times E^d$ such that $\sigma_t(X) = \sigma_M(t, X_t)$. We fix $t > 0$ and construct a sequence of continuous functions σ^n converging to σ_M in $L_{d+1}([0, t] \times E^d)$. Let X^n be a solution of (1.20). By Krylov's estimate (Krylov 1977, Chap. II, §3, Theorem 4)

$$\mathbf{E} \int_0^t \|\sigma_M - \sigma^n\|(s, X_s^n)ds \leq \text{const} \cdot \|\sigma_M - \sigma^n\|_{L_{d+1}},$$

where the constant does not depend on n. The proof is concluded as in case 1°, with the help of the limit theorem for semimartingales (see Chap. 4).

3°. In case 3) the proof is obtained by a combination of the proofs for cases 1) and 2). □

Theorem 1.20. *Let the functions σ and b be locally bounded and satisfy one of the following conditions on bounded subsets of C:*

1) for some $L > 0$ and non-increasing continuous function $K : [0, \infty) \to [0, \infty)$ we have

$$\|\sigma_t(X) - \sigma_t(Y)\|^2 + |b_t(X) - b_t(Y)|^2 \leq \int_0^t |X_s - Y_s|dK_s + L|X_t - Y_t|$$

for all t, X, Y; (Liptser and Shiryaev 1974, Chap. 4, §4, Theorem 4.6).

2) σ is Markov, the matrix $\frac{1}{2}\sigma\sigma^*$ is uniformly non-singular and for all $x \in E^d$ and $T \geq 0$

$$\lim_{y \to x} \sup_{s \in [0,T]} \|a^M(s, x) - a^M(s, y)\| = 0;$$

(Stroock and Varadhan 1979, Chap. 7).

3) $d = 1$, σ is Markov and the matrix $\frac{1}{2}\sigma\sigma^*$ is uniformly non-singular (Stroock and Varadhan 1979, Exercise 7.3.3);

4) $d = 2$, σ is Markov and the matrix $\frac{1}{2}\sigma\sigma^*$ does not depend on time and is uniformly non-singular (Stroock and Varadhan 1979, Exercise 7.3.4);

5) σ is Markov, uniformly non-singular and piecewise-constant with respect to some partition of E^d into a finite number of polygons (Krasnosel'skij and Pokrovskij 1983).

Then the solution of equation (1.19) is weakly unique.

Proof. The proof of uniqueness in cases 2)–4) is based in broad outline on the following ideas. It is sufficient to establish that for any smooth finite f the expectation

$$\mathbf{E}\int_0^\infty e^{-\lambda t} f(X_t) dt$$

has the same value for all weak solutions of (1.19) with initial condition $X_0 = x$. We denote it by $u(x)$, $x \in E^d$. Using the Itô formula we can reduce this problem to the question of the solvability of

$$\sum_{i,j=1}^d a^{ij} \frac{\partial^2 u}{\partial x_i x_j} + \sum_{i=1}^d b^i \frac{\partial u}{\partial x_i} - \lambda u = f$$

in the class C_b^2. Its solvability is established by perturbation theory methods. □

§3. Differentiation of Solutions of SDEs with Respect to the Initial Data

3.1. Differentiation with respect to the initial data is a particular case of differentiation with respect to a parameter, but it is placed in the title of the section in view of its importance. Two types of theorems are known here: *differentiation in the L_p norms* and *pointwise differentiation*. The first is simpler to derive, and it requires less smoothness of coefficients. The second requires somewhat greater smoothness; this is because the derivation is based on imbedding theorems. Our account basically follows (Krylov 1977) and (Blagoveshchenskij and Freidlin 1961).

Definition 1.4. We denote by $\mathcal{L}[0,T]$ (or $\mathcal{L}[0,T;\ E^d]$) the space of real-valued (or d-dimensional) stochastic processes x_t on $[0,T]$ that are measurable with respect to (t,ω) and have finite norm

$$\left(\mathbf{E}\int_0^T |x_t|^q dt\right)^{1/q}$$

for all $q \geq 1$. We denote by $\mathcal{LB}[0,T]$ (or $\mathcal{LB}[0,T;\ E^d]$) the space of all separable processes x_t with finite norm

$$(\mathbf{E}\sup_{[0,T]} |x_t|^q)^{1/q}$$

for all $q \geq 1$. If $T > 0$ is fixed, then instead of $\mathcal{L}[0,T]$, $\mathcal{LB}[0,T]$, we will usually write $\mathcal{L}$, $\mathcal{LB}$ and we will also retain the notation for $\mathcal{L}[0,T;\ E^d]$ and $\mathcal{LB}[0,T;\ E^d]$ if this does not lead to confusion. Convergence, continuity and differentiability in $\mathcal{L}$ and $\mathcal{LB}$ are defined in the natural manner (see Krylov 1977, Chap. 2, §7). For example, the sequence of processes $x_t^n \in \mathcal{L}(n \geq 1)$ converges to $x_t \in \mathcal{L}$ if for all $q \geq 1$

$$\lim_{n\to\infty} \mathbf{E}\int_0^T |x_t^n - x_t|^q dt = 0.$$

The limit x_t is denoted by $(\mathcal{L})\lim$ (or $(\mathcal{L}B)\lim$ in $\mathcal{L}B$).

Suppose that the process $x_t \in \mathcal{L}$ (or $\mathcal{L}B$) depends on the parameter p in E^d: $x_t = x_t^p$, $p \in E^d$ and that $l \in E^d$ is a given unit vector. The process $y \in \mathcal{L}$ (or $\mathcal{L}B$) is called the $\mathcal{L}$-derivative (or $\mathcal{L}B$-derivative) of x_t^p at the point p_0 in the direction l,

$$y_t = (\mathcal{L})\frac{\partial}{\partial l} x_t^p \Big|_{p=p_0} \quad (\text{or } y_t = (\mathcal{L}B)\frac{\partial}{\partial l} x_t^p \Big|_{p=p_0}),$$

if

$$y_t = (\mathcal{L})\lim_{r\downarrow 0} \frac{1}{r}(x_t^{p_0+rl} - x_t^{p_0}) \quad (\text{or } y_t = (\mathcal{L}B)\lim_{r\downarrow 0} \frac{1}{r}(x_t^{p_0+rl} - x_t^{p_0})).$$

The process x_t^p is once $\mathcal{L}$-differentiable (or $\mathcal{L}B$-differentiable) at p_0 if it has $\mathcal{L}$-derivatives (or $\mathcal{L}B$-derivatives) at p_0 in all directions. The process x_t^p is called i times ($i \geq 2$) $\mathcal{L}$-differentiable (or $\mathcal{L}B$-differentiable) at p_0 if it is once $\mathcal{L}$-differentiable (or $\mathcal{L}B$-differentiable) in some neighborhood of p_0 and each $\mathcal{L}$-derivative (or $\mathcal{L}B$-derivative) of it in any direction is $i-1$ times $\mathcal{L}$-differentiable (or $\mathcal{L}B$-differentiable) at p_0.

If $(\mathcal{F}_t)$ is some flow of σ-fields and the process x_t^p is adapted to this flow, then all its $\mathcal{L}$- and $\mathcal{L}B$-derivatives that are defined can be chosen to be $(\mathcal{F}_t)$-adapted (see Krylov 1977). Let us note that the theorem on the continuity and differentiability of a composite function holds (Krylov 1977, Theorem 2.7.9).

2.2. We now state the theorem on the continuity and differentiability with respect to a parameter of solutions of an SDE. Let E be some Euclidean space, $\mathcal{D} \subset E$ be a domain (of variation of the parameters), $T, L \geq 0$ fixed constants, and suppose that for $t \in [0, T]$, $x \in E^d$, $p \in \mathcal{D}$ the $d \times d$ stochastic matrices $\sigma_t(p, x)$ are defined, $b_t(p, x)$ are stochastic d-dimensional vectors, they are all $(\mathcal{F}_t)$-adapted, and for all p, t, ω, x, y

$$\|\sigma_t(p, x) - \sigma_t(p, y)\| + |b_t(p, x) - b_t(p, y)| \leq L|x - y|.$$

Let $\xi_t(p)$ ($t \in [0, T]$, $p \in \mathcal{D}$) be an $(\mathcal{F}_t)$-adapted stochastic process, $\xi_t(p) \in \mathcal{L}$. We consider the solution x_t^p of the SDE

$$x_t^p = \xi_t(p) + \int_0^t \sigma_s(p, x_s^p) dW_s + \int_0^t b_s(p, x_s^p) ds, \tag{1.21}$$

where W_t is a d-dimensional Wiener process. Equation (1.21) has a strong solution belonging to the space $\mathcal{L}$ (Krylov 1977, Corollary 2.5.6 and Theorem 2.5.7). If $\xi_t(p) \in \mathcal{L}B$ for all p, then $x_t^p \in \mathcal{L}B$ as well (Krylov 1977, Corollary 2.5.10).

Theorem 1.21 (Krylov 1977, Theorem 2.8.1). *Let $\sigma_t(p, x) \to \sigma_t(p_0, x)$, $b_t(p, x) \to b_t(p_0, x)$ in $\mathcal{L}$ as $p \to p_0 \in \mathcal{D}$ for each $x \in E^d$ and $\xi_t(p) \to \xi_t(p_0)$ in $\mathcal{L}$ as $p \to p_0$. Then $x_t^p \to x_t^{p_0}$ in $\mathcal{L}$ as $p \to p_0$.*

If, in addition, $\xi_t(p) \to \xi_t(p_0)$ in $\mathcal{L}B$ as $p \to p_0$, then $x_t^p \to x_t^{p_0}$ in $\mathcal{L}B$ as $p \to p_0$.

Theorem 1.22 (Krylov 1977, Theorem 2.8.4). *Let $\xi_t(p)$ be an i times ($\mathcal{L}$-continuously) $\mathcal{L}$-differentiable process at a point $p_0 \in \mathcal{D}$. Suppose that for all s, ω the functions $\sigma_s(p, x)$, $b_s(p, x)$ are i times continuously differentiable with respect to p, x for $p \in \mathcal{D}$, $x \in E^d$, and that all these derivatives up to order i inclusively do not exceed $L(1 + |x|)^m$ in norm ($m \geq 0$) for any $p \in \mathcal{D}$, s, ω, x. Then the process x_t^p is i times ($\mathcal{L}$-continuously) $\mathcal{L}$-differentiable at p_0. If, in addition, the process $\xi_t(p)$ is i times ($\mathcal{L}B$-continuously) $\mathcal{L}B$-differentiable at p_0, then the process x_t^p also possesses this same property.*

A particular case of Theorem 1.22 is the theorem on the differentiability of a solution of an SDE with respect to the initial data: here the (non-random) value x_0 plays the role of the parameter p, and the coefficients do not depend on it.

We now state a result of Blagoveshchenskij and Freidlin on almost sure differentiability with respect to initial data, that is, we assume that $\xi_t(p) \equiv p$.

Theorem 1.23 (Blagoveshchenskij and Freidlin 1961). *Let $\xi_t(p) \equiv p$, and suppose that for all s, ω the functions $\sigma_s(p, x), b_s(p, x)$ are i times continuously differentiable with respect to p, x for $p \in \mathcal{D}$, $x \in E^d$, and that all these derivatives up to order i inclusively, do not exceed $L(1 + |x|)^m$ in norm ($m \geq 0$) for any $p \in \mathcal{D}$, s, ω, x. Then the process x_t^p is $(i-1)$ times continuously differentiable with respect to p.*

Proof. The idea of the proof consists in using the Kolmogorov criterion for the continuity of a process. Let $i = 1$ (for $i > 1$ the proof is carried out by induction). We form the difference

$$\zeta_l(p, t, y) := \frac{1}{y}(x_t^{p+yl} - x_t^p),$$

where $y > 0$, $l \in E^d$, $|l| = 1$. It can be proved that

$$\mathbf{E}|\zeta_l(p, t, y) - \zeta_l(p', t', y')|^{2p} \leq C(|y - y'|^{2p} + |p - p'|^{2p} + |t - t'|^p). \quad (1.22)$$

By the Kolmogorov criterion, this estimate enables us to derive (for sufficiently large p) the continuous extendability of the field $\zeta_l(p, t, y)$ as $y \downarrow 0$. Indeed let $y_n = 2^{-n}$, and

$$\zeta_l(p, t, 0) := \zeta_l(p, t, 1) + \sum_{n=1}^{\infty}(\zeta_l(p, t, y_n) - \zeta_l(p, t, y_{n-1})).$$

By (1.22) the series converges in any L_q norm, $q \geq 1$ (by the Cauchy criterion), and the limit value $\zeta_l(p, t, 0)$ can be substituted in the left hand side of (1.22), that is, we can set $y' = 0$ in it. Then we obtain from the Kolmogorov criterion the continuity of $\zeta_l(p, t, y)$ with respect to (p, t, y) for $y \geq 0$. Hence it follows that the process x_t^p is differentiable in any direction continuously with respect to (p,t). □

As was already noted above, one can also deduce almost differentiability by applying the Sobolev imbedding theorem.

§4. Invariant Measure of a Diffusion Process

4.1. Let us consider the homogeneous d-dimensional stochastic differential equation (SDE)
$$dx_t = b(x_t)dt + \sigma(x_t)dW_t, \ t \geq 0, \tag{1.23}$$
with bounded measurable coefficients σ and b. For the moment we do not fix the initial value x_0. We will assume that any solution of (1.23) with any non-random initial condition x_0 is weakly unique (for this it suffices, for example, that the diffusion be non-degenerate and continuous, or that σ and b satisfy a Lipschitz condition). In this case, the solution (x_t^x) (x is the initial condition) is a strong Markov process (see Krylov 1973). We denote by (x_t) the Markov family of processes $(x_t^x, \ x \in E^d, \ t \geq 0)$.

General theorems on the existence of an invariant measure are applicable to (x_t) (as they are to every strong Markov process) under specific recurrence conditions.

Theorem 1.24. *For some $x \in E^d$ let the distributions of the temporal means $T^{-1}\int_0^T x_s^x ds$ be weakly compact as $T \to \infty$. Then the process (x_t) has at least one stationary distribution.*

The *uniqueness* of a stationary measure holds when the process does not have different (essential) ergodic classes. One can give the following sufficient condition of uniqueness. Denote the column vector of the matrix σ by $\sigma^i (1 \leq i \leq d)$. Lie$(\sigma)(x)$ is the linear space spanned by the vectors $\sigma^i(x)$, $1 \leq i \leq d$, and all possible Lie brackets $[\sigma^i, \sigma^j](x)$, $[[\sigma^i, \sigma^j], \sigma^k](x), \ldots$, where $1 \leq i, j, k, \ldots, \leq d$.

Theorem 1.25. *Suppose that at least one of the following two conditions holds:*
  A. $\inf_x \lambda^* \sigma \sigma^*(x) \lambda \geq \gamma |\lambda|^2, \ \gamma > 0$.
  B. $b, \sigma \in C_0^\infty(E^d)$ *and* $\inf_x \dim \mathrm{Lie}(\sigma) = d$.
Then there exists at most one stationary measure of the process x_t.

For *general Markov processes*, uniqueness conditions of a stationary measure can be found in (Skorokhod 1987, Theorem 1.23). The conditions of this theorem use: a) the concepts of irreducibility and boundedness in probability, which are discussed in (Skorokhod 1987, Theorem 1.24) for solutions of SDEs (in this theorem, the conditions are more general than in those in Theorem 1.25 of this section, however, the coefficients have to be continuous); b)

Theorem 1.25 in (Skorokhod 1987) (in which the boundedness of the process in probability is deduced from the existence of a certain Lyapunov function).

4.2. The *rate* of convergence to an invariant measure under some kind of non-degeneracy condition, for example, under conditions (A) or (B) of Theorem 1.25 depends on how good the recurrence properties of the process are. We present several results on convergence in variation, as well as results on estimates of the rate of convergence for general Markov processes. We then give simple sufficient conditions for verifying these estimates in terms of the coefficients of the SDE. We denote by $\mathbf{P}_t$ the distribution of x_t and by $\mathbf{P}(x, t, \cdot)$ the transition probability of the process x_t, $t \geq 0$.

Theorem 1.26 (Sevast'yanov 1957). *A homogeneous Markov process on a state space $(X, \mathcal{B})$ has a unique stationary distribution μ, which is ergodic (that is $\mathrm{var}(\mathbf{P}_t - \mu) \to 0$ as $t \to \infty$ for any $\mathbf{P}_0$) if for any $\epsilon > 0$ there is a set $C \in \mathcal{B}$, a probability measure π on X, numbers $t_1 > 0$, $m > 0$, $M > 0$, and for any initial distribution $\mathbf{P}_0$ there exists t_0 such that*
1) $m\pi(A) \leq \mathbf{P}(x, t_1, A)$ $\forall x \in C$, $A \subset C$, $A \in \mathcal{B}$ for any $t \geq t_0$;
2) $\mathbf{P}_t(C) \geq 1 - \epsilon$;
3) $\mathbf{P}_t(A) \leq M\pi(A) + \epsilon$ $\forall A \subset C$, $A \in \mathcal{B}$.

The next two theorems are proved by a method similar to that applied in (Veretennikov 1987) for estimating the rate of mixing.

Theorem 1.27. *Suppose that at least one of the conditions (A) or (B) of Theorem 1.25 holds, x_0 is non-random, and for any $\delta \in (0, \delta_0]$ there exists a bounded set $\mathcal{K}_\delta \in \mathcal{B}^d$ and a function $0 \leq h_\delta(x)$, $x \in E^d$, identically equal to unity on $\mathcal{K}_\delta$ such that $\lim_{|x| \to \infty} h_\delta(x) = \infty$. Suppose further that the following two inequalities hold:*
1) *for each $x \notin \mathcal{K}_\delta$*
$$\mathbf{E} \exp(\delta \tau(x)) \leq h_\delta(x),$$
where $\tau(x) := \inf(t \geq 0 : x_t^x \in \mathcal{K}_\delta)$;
2) *for all $x \in E^d$ and $t \geq 0$*
$$\mathbf{E}_x h_\delta(x_t) \leq C_\delta h_\delta(x), \quad C_\delta \geq 1,$$
where $C_\delta \downarrow 1$ as $\delta \downarrow 0$.
Then for some $C, \lambda > 0, \delta > 0$ we have the bound
$$\mathrm{var}(\mathbf{P}_t - \mu) \leq C \exp(-\lambda t) h_\delta^2(x_0)$$
for all $t \geq 0$, $x_0 \in E^d$.

Theorem 1.28. *Suppose that at least one of the conditions (A) or (B) of Theorem 1.25 holds, x_0 is non-random and there exists $k \geq 1$, a bounded set $\mathcal{K} \in \mathcal{B}^d$ and a function $h(x)$, $x \in E^d$, such that $\lim_{|x| \to \infty} h(x) = \infty$ and $h(x) I(x \in \mathcal{K}) \equiv 1$. Suppose further that the following two inequalities hold:*

1) $\mathbf{E}\tau(x)^k \leq h(x)$;
2) for all $x_0 \in E^d$, $t \geq 0$

$$\mathbf{E}h(x_t) \leq C_k h(x) \ (C_k > 0).$$

Then for some $C > 0$ we have

$$\text{var}(P_t - \mu) \leq C(1+t)^{-k}(\ln t)^{k-1} h^2(x_0).$$

For general strong Markov processes, one can replace the non-degeneracy conditions (A) or (B) by the following condition of "local Doeblin condition" type:

$$\inf_{t \geq 0} \inf_{x,x' \in \mathcal{K}_\delta} \int_{\mathcal{K}_\delta} 1 \wedge \frac{\mathbf{P}_{t,x'}(x_{t+1} \in dy)}{\mathbf{P}_{t,x}(x_{t+1} \in dy)} \mathbf{P}_{t,x}(x_{t+1} \in dy) > 0 \quad (1.24)$$

($d\mathbf{P}_1/d\mathbf{P}_2$ denotes the derivative of the absolutely continuous part). If condition (B) holds, then (1.24) also holds. If condition (A) holds, then (1.24) also holds; this can be deduced from Harnack's inequality, proved in (Krylov and Safonov 1980).

Conditions (1) and (2) of Theorem 1.27 are satisfied for a SDE, in particular, if the following equality holds:

$$\limsup_{|x| \to \infty} (b(x), x/|x|) < 0.$$

If one assumes linear growth of σ and b, then these same conditions are ensured by the relation

$$\lim_{|x| \to \infty} \frac{b(x), x/|x|)}{\|\sigma\sigma^*(x)\|} = -\infty.$$

§5. Support of a Diffusion

5.1. How does one construct the support of the solution of the stochastic differential equation

$$dX_t = \sigma(X_t)dW_t + b(X_t)dt \,? \quad (1.25)$$

It turns out that it is much easier to answer this question if one rephrases it slightly: how does one construct the support of the solution of the stochastic differential equation

$$dX_t = \sigma(X_t) \circ dW_t + b(X_t)dt, \quad (1.26)$$

where the sign ∘ indicates that the differential is understood in the Stratonovich sense (Ikeda and Watanabe 1981, Chap. III, §1). The answer has the following form. We set

$$\mathcal{S}(x) = \left\{ X : X = x + \int_0^\cdot \sigma(X_t)\dot{W}_t dt + \int_0^\cdot b(X_t) dt, \right.$$

$$\left. W \text{ is an arbitrary smooth function} \right\}.$$

Then the support of the distribution of the solution of (1.26) with initial condition $X_0 = x$ is $\overline{\mathcal{S}(x)}$. This fact is based on the following two circumstances:

1) for smooth trajectories of W, equation (1.26) (as well as (1.25)) can be solved as an ordinary differential equation;

2) the mapping $W \to X$ defined by (1.26) is "continuous" (see Krasnosel'skij and Pokrovskij 1978, Sect. 8, §1 and Sect. 32.5).

It remains only to understand why the equation in the Stratonovich form (1.26) appears in these problems. It turns out that the Stratonovich form is needed in order to ensure the continuity property of the solution as a function of a Wiener trajectory. Let us take the simplest one-dimensional equation:

$$dX_t = \sigma(X_t) dW_t, \qquad (1.27)$$

where the function σ is smooth and uniformly positive. We define a mapping on the set of smooth trajectories of W which establishes a correspondence between a smooth function of W and a solution X of the ordinary differential equation

$$dX_t = \sigma(X_t)\dot{W}_t dt$$

with initial condition $X_0 = x$. Let us convince ourselves that this mapping admits a continuous extension to all the continuous trajectories of W. We set

$$F(u) = \int_x^u \frac{dv}{\sigma(v)}, \quad u \in (-\infty, \infty),$$

where G is the inverse of F. Then $X_t = G(W_t)$, $t \in [0, \infty)$, will be the desired extension. But $X_t = G(W_t)$, $t \in [0, \infty)$ is not a solution of (1.27): by the Itô formula,

$$dX_t = \frac{dG}{dx}(W_t) dW_t + \frac{1}{2} \frac{d^2 G}{dx^2}(W_t) dt,$$

or, going from G to F,

$$dX_t = \sigma(X_t) dW_t + \frac{1}{2} \sigma \frac{d\sigma}{dx}(X_t) dt \equiv \sigma(X_t) \circ dW_t\,!$$

Finally, we show what the answer to the original question looks like in the one-dimensional case. Equation (1.25) can be rewritten in the form

$$dX_t = \sigma(X_t) \circ dW_t + \left(b - \frac{1}{2} \sigma \frac{d\sigma}{dx} \right)(X_t) dt,$$

(see Ikeda and Watanabe 1981, Chap. III, §1, formula (1.10)). Thus in order to obtain the support of the distribution of the solution of (1.25) with initial condition $X_0 = x$ it is necessary to close the set

$$\Big\{X : X = x + \int_0^{\cdot} \sigma(X_t)\dot{W}_t dt + \int_0^{\cdot} \Big(b - \frac{1}{2}\sigma\frac{d\sigma}{dx}\Big)(x_t)dt,$$

W is an arbitrary smooth function$\Big\}$.

We now turn to the precise formulation. Let d be a fixed natural number. Let C be the space of continuous functions on $[0,\infty)$ with values in E^d with the topology of uniform convergence on finite intervals, $\mathcal{C}$ its Borel σ-algebra, and $\mathbf{C}$ the natural flow of σ-algebras on C.

Let $\sigma = \sigma(x)$ and $b = b(x)$ be functions defined on E^d satisfying a Lipschitz condition with values in the space of $d \times d$ matrices and in E^d, respectively.

Theorem 1.29 ((Stroock, 1982, Chap. II, §4, Theorem 4.20) or (Ikeda and Watanabe 1981, Chap. VI, Theorem 8.1)). *Let $x \in E^d$, X a solution of equation (1.26) with initial condition $X_0 = x$, and $\mathbf{P}_x$ its distribution. Then* $\operatorname{supp} \mathbf{P}_x = \bar{S}(x)$.

The proof is carried out in the following manner. Let us fix a probability space $(\Omega, \mathcal{F}, \mathbf{P})$ and a Wiener process W defined on it. We introduce a piecewise-linear process W^ϵ, approximating the Wiener process W as $\epsilon \to 0$, and construct in terms of it the process X^ϵ satisfying the equation

$$X^\epsilon = x + \int_0^{\cdot} \sigma(X_t^\epsilon)dW_t^\epsilon + \int_0^{\cdot} b(X_t^\epsilon)dt.$$

The distribution X^ϵ is shown to weakly converge to the measure $\mathbf{P}_x$ as $\epsilon \to 0$. Hence we have the inclusion $\operatorname{supp} \mathbf{P}_x \subseteq \bar{S}_p(x)$, where

$$S_\mathbf{P}(x) = \Big\{X \in C : X = x + \int_0^{\cdot} \sigma(X_t)dW_t + \int_0^{\cdot} b(X_t)dt,$$

W is an arbitrary piecewise-smooth function from C$\}$,

It is not hard to see that $\overline{S_\mathbf{P}(x)} = \overline{S(x)}$, and hence $\operatorname{supp} \mathbf{P}_x \subseteq \overline{S(x)}$. In fact, here we have equality. This is a consequence of the following theorem.

Theorem 1.30 ((Stroock 1982, Chap. II, §4, Theorem 4.20), or (Ikeda and Watanabe 1981, Chap. VI, Theorem 8.2)). *Let $x \in E^d$, $V \in C$ a smooth function, X^V a solution of the equation*

$$X^V = x + \int_0^{\cdot} \sigma(X_t^V)\dot{V}_t dt + \int_0^{\cdot} b(X_t^V)dt,$$

and X a solution of (1.26) with initial condition $X_0 = x$. Then for any $\epsilon > 0$ and $T > 0$

$$\lim_{\delta \downarrow 0} \mathbf{P}\Big(\sup_{t \in [0,T]} |X_t - X_t^V| > \epsilon \,\Big|\, \sup_{t \in [0,T]} |W_t - V_t| \leq \delta\Big) = 0.$$

The Itô and Stratonovich stochastic differentials are connected by the relation

$$\sigma(X_t) \circ dW_t = \sigma(X_t)dW_t + \frac{1}{2}\hat{\sigma}(W_t)dt,$$

where $\hat{\sigma}$ is a d-dimensional vector, the ith coordinate of which is equal to $\mathrm{tr}\left(\frac{\partial \sigma_i}{\partial x}\sigma\right)$ and σ_i is the ith row of σ (see Ikeda and Watanabe 1981, Chap. III, §1). We therefore modify Theorem 1.29 for the Itô equation (1.25) in the following manner.

Theorem 1.31 Let $x \in E^d$, and X a solution of equation (1.25) with initial condition $X_0 = x$. Then the support of the distribution X is the closure of the set

$$\left\{X : X = x + \int_0^{\cdot} \sigma(X_t)\dot{W}_t dt + \int_0^{\cdot} (b - \hat{\sigma})(X_t)dt, \right.$$

$$\left. W \text{ is an arbitrary smooth function}\right\}.$$

The well known deep connection between the theory of diffusion processes and the theory of partial differential equations, as it applies to Theorem 1.30, appears in the maximum principle, which can be deduced as a corollary of this theorem. We state it as a separate theorem (see Theorem 1.32 below).

Let G be an open subset of $E^1 \times E^d$, $u \in C^{1,2}(G)$,

$$\frac{\partial u}{\partial t} + \frac{1}{2}\sum_{i,j=1}^{d}(\sigma\sigma^*)^{ij}\frac{\partial^2 u}{\partial x_i \partial x_j} + \sum_{i=1}^{d}\left(b^i + \frac{1}{2}\hat{\sigma}^i\right)\frac{\partial u}{\partial x_i} \geq 0$$

in G, $(t_0, x_0) \in G$, and let $G(t_0, x_0)$ be the closure in G of the points $(t_1, Y_{t_1-t_0})$, where $t_1 \geq t_0$, $Y \in \mathcal{S}(x_0)$ and $(t, Y_{t-t_0}) \in G$ for all $t \in [t_0, t_1]$.

Theorem 1.32 (Stroock 1982, Chap. II, §4, Corollary (4.21)). If $u(t_0, x_0) = \max_G u$, then $u|_{G(t_0,x_0)} \equiv u(t_0, x_0)$.

The proof is based on computing the expectation $u(t, X_t)$ by means of the Itô formula. In this case, the relation presented above between the Itô and Stratonovich stochastic differentials is used.

4.2. Let X be a diffusion process. Sometimes the following questions arise: which of two given smooth curves starting at the same point is more probable as the trajectory of X; and which among all the possible smooth curves joining two given points is most probable as the trajectory of X? The computation of the measure of a tube domain around a smooth curve provides the answer to these questions.

Let $T \geq 0$, $x \in E^d$, and let $b : E^d \to E^d$ be a C^∞ function with bounded derivatives,

$$X = x + W + \int_0^{\cdot} b(X_t)dt,$$

$$Y \in C^2([0,\infty), E^d), \quad Y_0 = 0.$$

Theorem 1.33 (Stroock 1982, Chap. II, §5, Theorem 5.5). *There exist C and $\lambda > 0$ such that*

$$\mathbf{P}\left(\sup_{t\in[0,T]} |X_t - Y_t| < \epsilon\right) \sim Ce^{-\lambda T/\epsilon^2}$$
$$\times \exp\left(-\frac{1}{2}\int_0^T |\dot{Y}_t - b(Y_t)|^2 dt - \frac{1}{2}\int_0^T \operatorname{div} b(Y_t) dt\right)$$

as $\epsilon \downarrow 0$.

Example. (Stroock 1982, Chap. II, §5, Remark 5.15). Let X be a process that passes from the origin to $x \in E^d$ during time T. Then the most probable trajectory Y must minimize the functional

$$\frac{1}{2}\int_0^T |\dot{Y}_t - b(Y_t)|^2 dt + \frac{1}{2}\int_0^T \operatorname{div} b(Y_t) dt.$$

The usual technique of variational calculus allows one to obtain the Euler equation for this problem. In particular, for $b \equiv 0$ we clearly have $Y_t = (t/T)x$, $t \in [0,T]$.

In conclusion, we note that the results set forth in this section are due to Stroock and Varadhan (Stroock and Varadhan 1972a, 1972b).

§6. Stochastic Differential Equations in Domains

6.1. Informal Description and Basic Notation. Suppose that we are given a closed domain in Euclidean space. We shall attempt to explain how to construct a process assuming values in this closed domain and behaving like a solution of a stochastic differential equation. Evidently such a process must experience some effect on the boundary ensuring its retention in the closed domain. It is natural to represent this effect as a special drift in the stochastic differential equation and assign to each point of the boundary a drift direction. Thus, we consider the equation

$$dX_t = \sigma(X_t)dW_t + b(X_t)dt + d\phi_t,$$

where ϕ, the "drift on the boundary", is a continuous process of bounded variation, increasing only on the boundary, the direction of which at a point x of the boundary is $\gamma(x)$, $|\gamma(x)| = 1$:

$$d\phi_t = I_{\partial O}(X_t)\gamma(X_t)d|\phi|_t,$$

(∂O is the boundary of the domain, $|\phi|$ denotes the variation of ϕ). By contrast with the usual drift, this drift ϕ, as a function of time is *singular* as a rule. Here is the simplest example of a Wiener process in $[0,\infty)$ reflected at the point 0:

$$dX_t = dW_t + d\phi_t,$$

where ϕ_t is a continuous increasing process

$$d\phi_t = I_{\{0\}}(X_t)d\phi_t, \quad \phi_0 = 0, \quad W_0 = 0.$$

Since $W_t + \phi_t = X_t \geq 0$, it follows that $\phi_t \geq -W_t$, but $\phi_t = \max_{s \in [0,t]} \phi_s \geq \max_{s \in [0,t]}(-W_s) = \min_{s \in [0,t]} W_s$. It turns out, however, (as was established by Skorokhod (McKean 1969, Sect. 3.9)) that $\phi_t = -\min_{s \in [0,t]} W_s$. Since, by the law of the iterated logarithm (see Itô and McKean 1965, Sect. 1.8), with probability 1 the function $\min_{s \in [0,t]} W_s$ is a singular function of t of "Cantor staircase" type, it follows that ϕ too is a singular function with respect to t.

How does one seek a solution of a stochastic differential equation in the domain ? As in the case of the entire space, it is natural to consider successive approximations (cf. the proof of Theorem 4.6 in (Liptser and Shiryaev 1974, Chap. 4)):

$$dX_t^n = \sigma(X_t^{n-1})dW_t + b(X_t^{n-1})dt + d\phi_t^n,$$
$$d\phi_t^n = I_{\partial O}(X_t^n)\gamma(X_t^n)d|\phi_t^n|.$$

Thus, for each elementary event we must solve the following deterministic problem: for a given trajectory X^{n-1} construct in the given case the reflecting process (ϕ^n) and the reflected trajectory (X^n) for the function

$$\int_0^\cdot (\sigma(X_t^{n-1})dW_t + b(X_t^{n-1})dt).$$

As it turns out, all problems of stochastic differential equations with reflection reduce, in principle, to the solution of this problem. It is called the Skorokhod problem. He first considered it in the one-dimensional case (Skorokhod 1961), (Skorokhod 1962) (see the account in (Ikeda and Watanabe 1981, Chap. III, Sect. 4.2)).

Before turning to the precise formulation we introduce our notation. Here, for the sake of greater generality, we allow an entire set of directions of reflection at each point of the boundary, rather than just one direction. Such a situation naturally arises in the case of a discontinuous field γ; at the points of discontinuity, it is necessary to admit the convex closure of the limit values of γ. For example, all the normals at a given point are admitted as possible reflection directions for a normal reflection in a convex domain.

Let d be a fixed natural number. We denote by C the space of continuous functions on $[0,\infty)$ with values in E^d with the topology of uniform convergence on finite intervals; $\mathcal{C}$ denotes its Borel σ-algebra, and $\mathbf{C}$ the natural flow of σ-algebras on C.

O is a domain in E^d, ∂O is its boundary $\bar{O} = O \cup \partial O$, $\gamma : \partial O \to E^d$ is a multivalued function, associating with $x \in \partial O$ some set $\gamma(x)$ of vectors in E^d with unit length.

6.2. Skorokhod Problem. For a function $W \in C$ with $W_0 \in \bar{O}$ we consider the "Skorokhod problem": to find a function $\Phi(W)$ such that

1) the function $\Phi(W)$ assumes values in $\bar{O}$;
2) the function $\phi = \Phi(W) - W$ equals zero at the origin, has locally bounded variation $|\phi|$ and for almost all t (with respect to the measure $|\phi|$), $d\phi_t/d|\phi|_t \in I_{\partial O}\gamma(X_t)$ (that is, the function ϕ changes only on the set $\{X_t \in \partial O\}$ and the increments of ϕ at t "have a direction" in $\gamma(X_t)$).

Theorem 1.34. *Assume that one of the following conditions holds:*
1) *O is a bounded convex domain, γ is the field of normals (Tanaka 1979);*
2) *$O = \{x \in E^d : x^i > 0, i = 1, \ldots, d\}$, γ on the face $\{x^i = 0\}$ is given by the ith row $(p^{i1}, \ldots, p^{id})$ of a matrix P with ones on the diagonal, with non-positive elements elsewhere and such that the spectral radius of $(I - P)$ is strictly less than 1 (on the intersection of the faces, γ is the normalized convex closure of its values on the corresponding faces) (Harrison and Reiman 1981).*

Then for any $W \in C$ with $W_0 \in \bar{O}$ there exists a unique solution $\Phi(W)$ of the Skorokhod problem. The mapping $\Phi : C \to C$ is continuous.

Corollary 1. *Under the conditions of Theorem 1.34 Φ is a measurable $\mathbf{C}$-non-anticipative process on C.*

Corollary 2. *Let $O = \{x \in E^d : x^1 > 0\}$, and γ a normal to the bounding hyperplane. Then Φ is determined by the formula*

$$\Phi_t^1(W) = - \inf_{s \in [0,t]} W_s^1 \wedge 0, \quad \Phi_t^i = W_t^i, \quad i = 2, \ldots, d.$$

6.3. Stochastic Differential Equation with Reflection. Suppose that on $[0, \infty) \times C$ we are given functions $\sigma = \sigma_t(X)$ and $b = b_t(X)$ with values in the space of $d \times d$ matrices and in E^d, respectively, which are $\mathbf{C}$-adapted stochastic processes. We denote $\frac{1}{2}\operatorname{tr} \sigma\sigma^*$ by $\|\sigma\|$.

We consider in $\bar{O}$ the stochastic differential equation with reflection on the boundary:

$$dX_t = \sigma_t(X)dW_t + b_t(X)dt + d\phi_t. \tag{1.28}$$

Definition 1.5. By a *weak solution* of (1.28) is meant a process X with values in $\bar{O}$ and continuous trajectories on some basis $(\Omega, \mathcal{F}, \mathbf{F}, \mathbf{P})$ with an $\mathbf{F}$-Wiener process W such that X is $\mathbf{F}$-adapted, the process

$$\phi = \left(\phi_t \stackrel{\text{def}}{=} X_t - X_0 - \int_0^t \sigma_s(X)dW_s - \int_0^t b_s(X)ds, \ t \in [0, \infty)\right)$$

has locally bounded variation $|\phi|$ and

$$\frac{d\phi_t}{d|\phi|_t} \in I_{\partial O}\gamma(X_t) d\mathbf{P} \, d|\phi|\text{-almost everywhere.}$$

Definition 1.6. A weak *solution* of equation (1.28) is called *strong* if it is adapted to the flow generated by X_0 and W.

Theorem 1.35. *Let σ and k be bounded and continuous with respect to x for each t. If the Skorokhod problem has a unique solution Φ and the mapping $\Phi : C \to C$ is continuous, then equation (1.28) has a weak solution for any initial distribution in $\bar{O}$.*

Proof. The proof is based on the following observation: if X is a solution, then by the uniqueness of the solution of the Skorokhod problem we have $X = \Phi(Y)$ with

$$Y_t = X_0 + \int_0^t \sigma_s(X)dW_s + \int_0^t b_s(X)ds, \ t \in [0, \infty)$$

and hence,
$$dY_t = \sigma_t(\Phi(Y))dW_t + b_t(\Phi(Y))dt. \tag{1.29}$$

We now use the argument in the "reverse direction". By a variant of Theorem 1.19, 1) in Sect. 2, equation (1.29) with a given initial distribution has a weak solution. But then $X = \Phi(Y)$ satisfies (1.28). □

Remark. We have not encountered this simple theorem anywhere. In all the works on stochastic differential equations with reflection (Ikeda and Watanabe 1981); (Harrison and Reiman 1981); (Lions and Sznitman 1984); (Tanaka 1979) the authors first establish the existence, uniqueness and continuity of the solution of the Skorokhod problem and then apply special constructions — contracting mappings and time change, for the proof of the existence of a solution of (1.28).

With the help of Theorems 1.34 and 1.35 one can obtain the next theorem.

Theorem 1.36. *Let σ and b be bounded and continuous with respect to X for each t. Equation (1.28) has a weak solution in each of the following cases:*
1) O is a convex bounded domain, γ is its field of normals;
2) $O = \{x \in E^d : x^i > 0, \ i = 1, \ldots, d\}$, $\gamma|_{\{x^i=0\}}$ is a constant vector, $i = 1, \ldots, d$, with assumption 2) of Theorem 1.34 being satisfied (on the intersection of the faces, γ is the normalized convex closure of its values on the corresponding faces);
3) O is bounded, O and γ are C^2-smooth, the projection of γ on the normal to the boundary is positive.

Proof. Under conditions 1) and 2) the assertion of the theorem is a direct consequence of Theorems 1.34 and 1.35. Under condition 3) the solution exists locally: one needs to take a chart in which the boundary has the form of a hyperplane, and the field γ is the set of normals to it. In view of the boundedness of the coefficients and the form of Φ (Corollary 2 of Theorem 1.34), the process does not change the charts of the atlas too often and hence it can be defined on $[0, \infty)$. □

Applying the technique of localization (see Stroock and Varadhan 1979, Chap. 10), one can obtain the following results in unbounded domains with increasing coefficients.

Chapter 2.I. Stochastic Differential Equations

Theorem 1.37. *Let σ and b be continuous functions of X for each t and suppose that for some $c > 0$ and for all t, X*

$$(\|\sigma\|^2 + |b|^2)_t(X) \leq c(1 + \sup_{s \leq t} |X_s|^2).$$

Weak solutions of equation (1.28) exist in each of the following cases:
  1) *O is a convex domain, γ is the field of normals (Tanaka 1979);*
  2) *assumption 2) of Theorem 1.36 is satisfied;*
  3) *O is a half-space, γ is a C^2-smooth field, the projection of γ on the normal is uniformly positive,*

$$\left(\sum_{i=1}^{d}(\sigma^{1i})^2 + (b^1)^2\right)_t(X) \leq c(1 + \sup_{s \leq t} |X_s^1|^2).$$

Proof. For simplicity, let $X_0 = 0 \in O$. By $C(c,t)$ we will understand different constants depending only on c and t.

1°. For any bounded domain we can construct a local solution — up to the time of exit from this domain. To prove the existence of a global solution, it is sufficient to show that all the local solutions are uniformly bounded in probability on finite time intervals, see (Stroock and Varadhan 1979, Chap. 10).

2°. Suppose that we are given on the stochastic basis $(\Omega, \mathcal{F}, \mathbf{F}, \mathbf{P})$ a d-dimensional $\mathbf{F}$-Wiener process W and measurable $\mathbf{F}$-adapted processes: Z in $[0, \infty)$ with continuous trajectories, τ in E^d, f in E^1. If for each t **P**-a.s.

$$Z_t \leq Z_0 + \int_0^t \tau_s dW_s + \int_0^t f_s ds$$

and

$$(|\tau|^2 + f^2)_t \leq c(1 + \sup_{s \in [0,t]} Z_s^2),$$

then

$$\mathbf{E} \sup_{s \in [0,t]} Z_s^2 \leq C(c,t)(1 + \mathbf{E}Z_0^2).$$

In fact, $\sup_{s \in [0,t]} Z_s^2$ does not exceed the sum of the same suprema of each term of the right hand side. Upon taking the expectation of these suprema and estimating $\sup_{s \leq t}\left(\int_0^t \tau_s dW_s\right)^2$ by means of Doob's Inequality (Liptser and Shiryaev 1974, Chap. 1, §9, Theorem 2), we obtain the desired bound from the Gronwall-Bellman Lemma (Liptser and Shiryaev 1974, Chap. 4, §4, Lemma 4.13).

3°. Let the domain O be bounded. Then for any $t > 0$ there exists $C(c,t)$ such that $\mathbf{E}\sup_{s \in [0,t]} |X_s|^2 \leq C(c,t)$ for any weak solution X of equation (1.28). In fact,

$$d|X_t|^2 = 2X_t dX_t + \|\sigma_t(X)\| dt$$
$$\leq 2X_t \sigma_t(X) dW_t + 2X_t b_t(X) dt + \|\sigma_t(X)\| dt,$$

since $X_s d\phi_s \leq 0$. The inequality follows from the fact that for any $x \in \partial O$ and any normal n at x we have $(x, n) = -(0 - x, n) \leq 0$ in view of the convexity of O. Applying the estimate in 2° to the process $|X_t|^2$, we obtain

$$\mathbf{E} \sup_{s \in [0,t]} |X_s|^4 \leq C(c, t).$$

4°. The assertion of the theorem under condition 1) is a consequence of 1° and 3°.

5°. Under condition 2) the proof is constructed in exactly the same way, since $X_s d\phi_s$ is again non-positive.

6°. We set $\sigma^1 = (\sigma^{11}, \ldots, \sigma^{1d})$. Suppose that condition 3) holds and the coefficients σ, b are bounded. Consider the process X^1 in $[0, \infty)$:

$$d(X_t^1)^2 = 2X_t^1 dX_t^1 + \|\sigma_t(X)\| dt$$
$$= 2X_t^1 \sigma_t^1(X) dW_t + 2X_t^1 b_i^1(X) dt + \|\sigma_t(X)\| dt.$$

Applying the estimate in 2° to the process $(X_t^1)^2$, we obtain:

$$\mathbf{E} \sup_{s \in [0,t]} (X_s^1)^4 \leq C(c, t).$$

But, by Corollary 1 of Theorem 1.34,

$$\phi_t^1 = -\inf_{s \in [0,t]} \left(X_0^1 + \int_0^s \sigma_u^1(X) dW_u + \int_0^s b_u^1(X) du \right) \wedge 0,$$

therefore

$$\phi_t^1 \leq \left(-\inf_{s \in [0,t]} \int_0^s \sigma_u^1(X) dW_u - \inf_{s \in [0,t]} \int_0^s b_u^1(X) du \right) \vee 0.$$

Hence by Doob's Inequality (Liptser and Shiryaev 1974, §9, Theorem 2) and the estimate for $\mathbf{E} \sup_{s \in [0,t]} (X_s^1)^4$, we obtain $\mathbf{E}(\phi_t^1)^2 \leq C(c, t)$. By Itô's formula we then have

$$d|X_t|^2 = 2X_t dX_t + \|\sigma_t(X)\| dt$$
$$= 2X_t \sigma_t(X) dW_t + 2X_t b_t(X) dt + \|\sigma_t(X)\| dt + \gamma(X_t) d\phi_t,$$

whence, by 2° and the above estimate for $\mathbf{E}(\phi_t^1)^2$, we obtain $\mathbf{E}|X_t|^4 \leq C(c, t)$.

7°. The assertion of the theorem under condition 3) now follows from 1° and 6°. □

By analogy with the definition of the pathwise uniqueness of the solution of equation (1.1) in Sect. 1 we can give a definition of the pathwise uniqueness of the solution of equation (1.28) in the entire space.

Definition 1.7. We say that for equation (1.28) there is strong or pathwise uniqueness of a solution if for any two of its weak solutions X and X' defined on one stochastic basis $(\Omega, \mathcal{F}, \mathbf{F}, \mathbf{P})$ with the same $\mathbf{F}$-Wiener process W, the equality $X_0 = X_0'$ $\mathbf{P}$-a.s. implies that $X_t = X_t'$ for all $t \geq 0$ $\mathbf{P}$-a.s.

In the case of a "domain" we have the following analog of Theorem 4.6 in (Liptser and Shiryaev 1974, Chap. 4).

Theorem 1.38. *Suppose that for some $L > 0$ and non-decreasing continuous function $K : [0, \infty) \to [0, \infty)$ we have*

$$\|\sigma_t(X) - \sigma_t(Y)\|^2 + |b_t(X) - b_t(Y)|^2 \leq \int_0^t |X_s - Y_s|^2 dK_s + L|X_t - Y_t|^2$$

for all $t \in [0, \infty)$ and $X, Y \in C$. Then strong uniqueness holds in the following cases:

1) O *is a convex domain, γ is the field of normals (Tanaka 1979).*
2) $O = \{x \in E^d : x^1 > 0\}$, γ *is C^2-smooth and the projection of γ onto the normal is uniformly positive (apparently, this result is due to (Lions and Sznitman 1984)).*

Proof. 1°. Suppose that condition 1) holds, X and X' are two solutions of equation (1.28), and ϕ and ϕ' are the corresponding processes of bounded variation. Then

$$|X_t - X_t'|^2 = \int_0^t 2(X_s - X_s') d(X_s - X_s') + \int_0^t \|\sigma_s(X) - \sigma_s(X')\| ds$$

$$\leq \int_0^t 2(X_s - X_s')(\sigma_s(X) - \sigma_s(X'))dW_s$$

$$+ \int_0^t 2(X_s - X_s')(b_s(X) - b_s(X'))ds$$

$$+ \int_0^t \|\sigma_s(X) - \sigma_s(X')\| ds,$$

since $(X_s - X_s')(d\phi_s - d\phi_s') \geq 0$ (the inequality follows from the fact that for any $x \in \partial O$, any normal n at a point x and any $x' \in \bar{O}$ we have $(x' - x, n) \geq 0$ by the convexity of $\bar{O}$). The proof is completed as in the case of the entire space (see Liptser and Shiryaev 1974, Chap. 4, §4, Theorem 4.6).

2°. Case 2) reduces to case 1) with the help of local coordinates sending γ to the field of normals. $\square$

Applying Theorem 1.3 (Yamada-Watanabe) of Sect. 1, we obtain:

Theorem 1.39. *Suppose that the conditions of Theorems 1.37 and 1.38 hold. Then there exists a strong solution of equation (1.28).*

Equation (1.28) was created to describe processes instantaneously reflected from the boundary. The following theorem shows this.

Theorem 1.40. *Let $O = \{x \in e^d : x^1 > 0\}$ and suppose that the sum $\sum_{i=1}^{d}(\sigma^{1i})^2|_{\{X_s^1=0\}}$ is non-zero. Then for any solution X of equation (1.28)*

$$\int_0^\infty I_{\partial O}(X_s)ds = 0 \quad \mathbf{P}\text{-}a.s.$$

Proof. Consider the semimartingale $\int_0^t I_{\partial O}(X_s)dX_s$. Approximating this integral by Riemann sums (see Liptser and Shiryaev 1974d, Chap. 4, §2, Lemma 4.4), we see that it is non-decreasing with respect to t, which means that it has zero quadratic variation. On the other hand, its quadratic variation equals

$$\frac{1}{2}\int_0^t I_{\partial O}(X_s) \sum_{i=1}^{d}(\sigma_s^{1i}(X))^2 ds,$$

whence the assertion of the theorem follows. □

6.4. Stochastic Differential Equations with Reflection and with Delay on the Boundary. Apart from the processes considered with instantaneous reflection, there are also processes with reflection spending non-zero time on the boundary. But inside the domain they behave as solutions of stochastic differential equations, as previously.

Suppose that in addition to O, γ, σ, b we are given a function

$$\rho : [0, \infty) \times C \to [0, \infty), \quad \rho = \rho I_{\{X_s \in \partial O\}}$$

(ρ is the course of time on the boundary). We consider the equation

$$dX_t = I_O(X_t)(\sigma_t(X)dW_t + b_t(X)dt) + d\phi_t. \tag{1.30}$$

Definition 1.8. By a weak solution of equation (1.30) is meant a process X defined on some stochastic basis $(\Omega, \mathcal{F}, \mathbf{F}, \mathbf{P})$ with an **F**-Wiener process W, such that X is **F**-adapted with values in $\bar{O}$, the process

$$\phi = \left(\phi_t \stackrel{\text{def}}{=} X_t - X_0 - \int_0^t I_O(X_s)(\sigma_s(X)dW_s + b_s(X)ds), \ t \in [0, \infty)\right)$$

has bounded variation $|\phi|$, $d\phi_t \in I_{\partial O}\gamma(X_t)d|\phi|_t$ $d\mathbf{P}dt$ almost everywhere and $I_{\partial O}(X_t)dt = \rho_t(X)d|\phi|_t$.

If $\rho|_{\{X_s=0\}} \equiv 0$, then we obtain instantaneous reflection.

If $\rho|_{\{X_s=0\}} \equiv 1$, then the solution of (1.30) spends positive time on the boundary; the drift γ on the boundary is just the usual drift of a diffusion process in the entire space.

Theorem 1.41 (Stroock and Varadhan 1971, Theorem 3.1; see also Ikeda and Watanabe 1981, Chap. IV, §7). *Let O be C_b^2-smooth, and γ continuous. Suppose that on the set $\{X_s \in \partial O\}$ $(\rho + (\gamma, n))_s(X)$ is uniformly positive and* bounded, σ is continuous, uniformly positive definite and bounded, and b is measurable and bounded. Then a solution of (1.30) exists.

Sketch of Proof. First, let $\rho = I_{\{X_s \in \partial O\}}$. We "release" a solution of a stochastic differential equation with coefficients σ, b up to the time of hitting the boundary of the domain. At the point of hitting, which we denote by x, it sits for a random time s, $\mathbf{P}(s > t) = e^{-\lambda t}$, then jumps to the point $\frac{1}{\lambda}\gamma(x)$, and so on. As $\lambda \to \infty$, the distribution X weakly converges, the limit being a solution of (1.30) (cf. Theorem 1.16 of Sect. 2.2 on the equivalence of solutions of the martingale problem and weak solutions). In order to obtain an arbitrary ρ in $\rho = I_{\{X_s \in \partial O\}}$, it is necessary to make the time change

$$\tau_t = \int_0^t (\rho_s(X) + X_{\{X_s \in O\}}) ds.$$

$\square$

Theorem 1.42 (Stroock and Varadhan 1971, Theorem 5.7). *Let σ and ρ be Markov functions, that is, there exist Borel functions on σ_M and ρ_M on $[0, \infty) \times E^d$, taking values in the space of $d \times d$-matrices and in $[0, \infty)$, respectively, such that $\sigma_t(X) = \sigma_M(t, X_t), \rho_t(X) = \rho_M(t, X_t)$, and:*

1) $\sigma_M \sigma_M^*$ *is uniformly continuous in x with respect to t and positive definite;*

2) γ *and ρ satisfy a Lipschitz condition, where ρ is either identically zero or non-vanishing;*

3) b *is bounded.*

Then the solution of (1.30) is unique in the sense of a distribution on $(C, \mathcal{C})$ (that is, it is weakly unique).

The proof of weak uniqueness is carried out in the same way as in the entire space, see the end of Sect. 1c.

One could also define strong solutions of equation (1.30), but apparently, in general, there are none of them; in his report at the conference "Probabilistic models of control and reliability processes", 24–27 May 1988 in Donetsk, R.Ya. Chitashvili disproved the existence of a strong solution even for a Wiener process in $[0, \infty)$ for $\rho = I_{\{X_s = 0\}}$.

References*

Barlow, M.T. (1982): One-dimensional stochastic differential equations with no strong solution. J. Lond. Math. Soc., II. Ser. *26*, No. 2, 335–347. Zbl. 456.60062

Bass, R.F., Pardoux, E. (1987): Uniqueness for diffusions with piecewise constant coefficients. Probab. Theory Relat. Fields *76*, 557–572. Zbl. 617.60075

Billingsley, P. (1968): Convergence of Probability Measures. Wiley, New York. Zbl. 172,212

Blagoveshchenskij, Yu.N., Freidlin, M.I. (1961): Some properties of diffusion processes depending on a parameter. Dokl. Akad. Nauk SSSR *138*, No. 3, 508–511. [English transl.: Sov. Math. Dokl. *2*, 633–636 (1961)] Zbl. 114,78

Cartan, H.P. (1967): Differential Calculus. Differential Forms.Hermann, Paris. Zbl. 156,361; Zbl. 184,127

Chitashvili, R.J., Toronzhadze, T.A. (1981): On one-dimensional stochastic differential equations with unit diffusion coefficient. Structure of solutions. Stochastics *4*, No. 4, 281–315. Zbl. 454.60056

Conway, E.D. (1971): Stochastic equations with discontinuous drift. Trans. Am. Math. Soc. *157*, 235–245. Zbl. 276.60058

Doss, H. (1976): Liens entre équations différentielles stochastiques et ordinaires. C. R. Acad. Sci., Paris, Ser. A *283*, No. 13, 939–942. Zbl. 352.60044

Doss, H. (1977): Liens entre équations différentielles stochastiques et ordinaires. Ann. Inst. Henri Poincaré, Nouv. Ser., Sect. B *13*, No. 2, 99–125. Zbl. 359.60087

Fedorenko, I.V. (1982): The existence of a strong solution of a stochastic differential equation with continuous drift, in Third Republic Symposium on Differential and Integral Equations Odessa 1982, Odessa State University, 266–267

Harrison, J.M., Reiman, M. (1981): Reflected Brownian motion on an orthant. Ann. Probab. *9*, No. 2, 302–308. Zbl. 462.60073

Ikeda, N., Watanabe, S. (1981): Stochastic Differential Equations and Diffusion Processes. North Holland, Amsterdam. Zbl. 495.60005

Itô, K., McKean, H.P., Jr. (1965): Diffusion Processes and their Sample Paths. Springer, Berlin Heidelberg New York. Zbl 127,95

Kleptsyna, M.L. (1984): Strong solutions of stochastic equations with degenerate coefficients. Teor. Veroyatn. Primen. *29*, No. 2, 392–396. [English transl.: Theory Probab. Appl. *29*, 403–407 (1985)] Zbl. 551.60061

Kleptsyna, M.L. (1985): Comparison, existence and uniqueness theorems for stochastic differential equations. Teor. Veroyatn. Primen. *30*, No. 1, 147–152. [English transl.: Theory Probab. Appl. *30*, 164–169 (1986)] Zbl. 567.60060

Kleptsyna, M.L., Fedorenko, I.V. (1984): The existence of a strong solution of a stochastic differential equation with degenerate diffusion, in: Control in Complex Nonlinear Systems, Moscow 1984, 39–41. Zbl. 563.60052

Kleptsyna, M.L., Veretennikov, A.Yu. (1981): Strong solutions of stochastic equations with degenerate diffusion, in: XV All-Union Winter Mathematics School — Colloquium on Probability Theory and Mathematical Statistics Bakuriani 1981 Metsniereba Press, Tbilisi, pp. 10–11

Krasnosel'skij, M.A., Pokrovskij, A.V. (1978): Natural solutions of stochastic differential equations. Dokl. Akad. Nauk SSSR *240*, No. 2, 264–267. [English transl.: Sov. Math., Dokl. *19*, 578–582 (1978)] Zbl. 432.60075

---

* For the convenience of the reader, references to reviews in Zentralblatt für Mathematik (Zbl.), compiled using the MATH database, have, as far as possible, been included in this bibliography.

Krasnosel'skij, M.A., Pokrovskij, A.V. (1983): Systems with hysteresis. Nauka, Moscow. [English transl.: Springer, Berlin Heidelberg New York, 1989] Zbl. 665.47038

Krylov, N.V. (1969): Itô stochastic integral equations. Teor. Veroyatn. Primen. *14*, No. 2, 340–348. [English transl.: Theory Probab. Appl. *14*, 330–336 (1969)] Zbl. 214,443; Zbl. 309.60038

Krylov, N.V. (1973): The selection of a Markov process from a system of processes and the construction of quasidiffusion processes. Izv. Akad. Nauk SSSR, Ser. Mat. *37*, No. 3, 691–708. [English transl.: Mat. USSR, Izv. *7*, 691–709 (1974)] Zbl. 295.60057

Krylov, N.V. (1977): Controlled Diffusion Processes. Nauka, Moscow. [English transl.: Springer, Berlin Heidelberg New York, 1980] Zbl. 513.60043

Krylov, N.V., Safonov, M.V. (1980): A certain property of solutions of parabolic equations with measurable coefficients. Izv. Akad. Nauk SSSR, Ser. Mat. *44*, No. 1, 161–175. [English transl.: Mat. USSR, Izv. *16*, 151–164 (1981)] Zbl. 439.35023

Ladyzhenskaya, O.A., Solonnikov, V.A., Ural'tseva, N.N. (1967): Linear and Quasilinear Equations of Parabolic Type. Nauka, Moscow. [English transl.: Am. Math. Soc., Providence (1968)] Zbl. 164,123

Lebedev, V.A. (1983): On the existence of weak solutions to stochastic differential equations with driving martingales and random measures. Stochastics *9*, 37–76. Zbl. 513.60059

Lions, P.L., Sznitman, A.S. (1984): Stochastic differential equations with reflecting boundary conditions. Commun. Pure Appl. Math. *37*, 511–537. Zbl. 598.60060

Liptser, R.Sh., Shiryaev, A.N. (1974): Statistics of Random Processes (Nonlinear Filtering and Related Questions). Nauka, Moscow. Zbl. 279.60021 [English transl.: Springer, Berlin Heidelberg New York, 1977, 1978. Zbl. 364.60004; Zbl. 369.60001

Manabe, S., Shiga, T. (1973): On one-dimensional stochastic differential equations with non-sticky boundary conditions. J. Math. Kyoto Univ. *13*, No. 3, 595–603. Zbl. 276.60061

Matskyavichus, V. (1985): S^p-stability of solutions of symmetric stochastic differential equations. Litov. Mat. Sb. *25*, No. 4, 72–84. [English transl.: Lith. Math. J. *25*, 343–352 (1985)] Zbl. 588.60050

McKean, H.P., Jr. (1969): Stochastic Integrals. Academic Press, New York. Zbl. 191.466

Mel'nikov, A.V. (1979): Strong solutions of stochastic differential equations with non-smooth coefficients. Teor. Veroyatn. Primen. *24*, No. 1, 146–149. [English transl.: Theory Probab. Appl. *24*, 147–150 (1979)] Zbl. 396.60055

Mel'nikov, A.V. (1982): On properties of strong solutions of stochastic equations with respect to semimartingales. Stochastics *8*, No. 2, 103–119. Zbl. 497.60056

Nakao, S. (1972): On the pathwise uniqueness of solutions of one-dimensional stochastic differential equations. Osaka J. Math. *9*, No. 3, 513–518. Zbl. 255.60039

Nisio, M. (1973): On the existence of solutions of stochastic differential equations. Osaka J. Math. *10*, No. 1, 185–208. Zbl. 268.60057

Sevast'yanov, B.A.: (1957): An ergodic theorem for Markov processes and its applications to telephone systems with refusals. Teor. Veroyatn. Primen. *2*, No. 1, 106–116. Zbl. 82,131

Skorokhod, A.V. (1961a): Studies in the Theory of Random Processes. Kiev University, Kiev. [English transl.: Reading, Mass. (1965)] Zbl. 108,148

Skorokhod, A.V. (1961b): Stochastic equations for diffusion processes in a bounded domain. Teor. Veroyatn. Primen. *6*, 287–298. [English transl.: Theory Probab. Appl. *6*, 264–274 (1961)] Zbl. 215,535

Skorokhod, A.V. (1962): Stochastic equations for diffusion processes in a bounded domain. II. Teor. Veroyatn. Primen. *7*, 5–25. [English transl.: Theory Probab. Appl. *7*, 3–23 (1962)] Zbl. 201,493

Skorokhod, A.V. (1987): Asymptotic Methods in the Theory of Stochastic Differential Equations. Naukova Dumka, Kiev. [English transl.: Am. Math. Soc., Providence (1989)] Zbl. 659.60055

Stroock, D.W. (1982): Lectures on Topics in Stochastic Differential Equations. Springer, Berlin Heidelberg New York. Zbl. 516.60065

Stroock, D.W., Varadhan, S.R.S. (1971): Diffusion processes with boundary conditions. Commun. Pure Appl. Math. *24*, No. 2, 147–225. Zbl. 227.76131

Stroock, D.W., Varadhan, S.R.S. (1972a): On the support of diffusion processes with applications to the strong maximum principle, in: Proceedings of the 6th Berkeley Symp. Math., Stat., Probab., Univ. California, 1976, No. 3, 333–359. Zbl. 255.60056

Stroock, D.W., Varadhan, S.R.S. (1972b): On degenerate elliptic-parabolic operators of second order and their associated diffusions. Commun. Pure Appl. Math. *25*, No. 2, 651–714. Zbl. 344.35041

Stroock, D.W., Varadhan, S.R.S. (1979): Multidimensional Diffusion Processes. Springer, Berlin Heidelberg New York. Zbl. 426.60069

Sussmann, H.J. (1978): On the gap between deterministic and stochastic ordinary differential equations. Ann. Probab. *6*, No. 1, 19–41. Zbl. 391.60056

Tanaka, H. (1979): Stochastic differential equations with reflecting boundary conditions in convex regions. Hiroshima Math. J. *9*, No. 1, 163–177. Zbl. 423.60055

Toronzhadze, T.A., Chitashvili, R.Ya. (1980): The equivalence of strong and weak regular stochastic differential equations. Soobshch. Akad. Nauk Gruz. SSR *98*, No. 1, 37–40. Zbl 428.60072

Tsirel'son, B.S. (1975): An example of a stochastic equation having no strong solution. Teor. Veroyatn. Primen. *20*, No. 2, 427–430. [English transl.: Theory Probab. Appl. *20*, 416–418 (1975)] Zbl. 353.60061

Veretennikov, A.Yu. (1979): On strong solutions of stochastic differential equations. Teor. Veroyatn. Primen. *24*, No. 2, 348–360. [English transl.: Theory Probab. Appl. *24*, 354–366 (1980)] Zbl. 418.60061

Veretennikov, A.Yu. (1980): On strong solutions and explicit formulas for solutions of stochastic integral equations. Mat. Sb., Nov. Ser. *111(153)*, No. 3, 434–452. [English transl.: Math. USSR, Sb. *39*, 387–403 (1981)] Zbl. 431.60061

Veretennikov, A.Yu. (1982): Parabolic equations and Itô's stochastic equations with coefficients that are discontinuous with respect to time. Mat. Zametki *31*, No. 4, 549–557. [English transl.: Math. Notes *31*, 278–283 (1982)] Zbl. 563.60056

Veretennikov, A.Yu. (1983): On stochastic equations with partially degenerate diffusion variables. Izv. Akad. Nauk SSSR, Ser. Mat. *47*, No. 1, 189–196. [English transl.: Math. USSR, Izv. *22*, 173–180 (1984)] Zbl. 514.60056

Veretennikov, A.Yu. (1987): On estimates for the mixing rate for stochastic equations. Teor. Veroyatn. Primen. *32*, No. 2, 299–308. [English transl.: Theory Probab. Appl. *32*, 273–281 (1987)] Zbl. 663.60046

Veretennikov, A.Yu., Kleptsyna, M.L. (1983): On strong solutions of non-Markov stochastic equations, in: Methods of Investigation of Non-Linear Systems of Equations, Moscow, 3–8

Veretennikov, A.Yu., Kleptsyna, M.L. (1989): On the trajectory approach to stochastic differential equations, in: Statistics and Control of Stochastic Processes, Collect. Artic., Moscow, 22–23. Zbl. 712.60064

Yamada, T. (1973): On a comparison theorem for solutions of stochastic differential equations and its applications. J. Math. Kyoto Univ. *13*, No. 3, 497–512

Yamada, T., Watanabe, S. (1971): On the uniqueness of solutions of stochastic differential equations, I, II. J. Math. Kyoto Univ. *11*, No. 1, 155–167; ibid. *11*, No. 3, 553–563. Zbl. 236.60037; Zbl. 229.60039

Zvonkin, A.K. (1974): A transformation of the state space of a diffusion process which annihilates the drift. Mat. Sb., Nov. Ser. *93(135)*, No. 1, 129–149. [English transl.: Math. USSR, Sb. *22*, 129–149 (1975)] Zbl. 291.60036

Zvonkin, A.K., Krylov, N.V. (1975): On strong solutions of stochastic differential equations, in: Proc. Sem. Theory of Random Processes, Vilnius 1974, Part II, 9–88. [English transl.: Sel. Math. Sov. *1*, 19–61 (1991)] Zbl. 481.60062

II. Stochastic Evolution Equations

A.Yu. Veretennikov

§1. Introduction

1.1. In the first ten years of development of the theory of Itô stochastic differential equations, the investigations in this area concentrated on finite-dimensional equations with bounded (or locally bounded) coefficients. However, with time it became necessary to broaden substantially the class of stochastic differential equations under study. In the filtration theory of diffusion processes, and also in a whole series of areas of physics and technology, stochastic equations appeared with partial derivatives which, as a rule, can be treated as stochastic differential equations in Hilbert or Banach space and with unbounded coefficients. The theory of such equations has now been created and is successfully being developed, although the number of uninvestigated problems is great.

In this part we give an introduction to the theory of stochastic evolution equations in Banach space of the form

$$du(t,\omega) = A(u(t,\omega),t,\omega)dt + B(u(t,\omega),t,\omega)dW(t), \qquad (2.1)$$

where $A(\,\cdot\,,t,\omega)$ and $B(\,\cdot\,,t,\omega)$ are families of non-anticipative operators in Banach spaces, $W(t)$ is a process with independent increments and with values in some Hilbert space; ω is an "event".

The stochastic evolution equations (2.1) with bounded operators B were first considered by Daletskij (Daletskij 1967) and Baklan (Baklan 1963, 1964). A short summary of the papers in this area up to 1979 can be found in the review (Krylov and Rozovskij 1979), where previous results of (Pardoux 1975) are generalized. In the main, this exposition follows the papers (Krylov and Rozovskij 1979) and (Rozovskij 1983).

1.2. It should be noted that, despite the essential complication of the situation by comparison with finite-dimensional "ordinary" stochastic differential equations (SDEs), many basic ideas and methods continue to work also in Banach space (clearly, with appropriate modifications). Among them are, for example, the method of successive approximations. Conditions of monotonicity and coercivity and methods of the theory of monotone operators play an important role in this theory.

Regarding the convergence of discretized equations in Banach spaces, see the paper (Alyushina and Krylov 1988). For equations in Banach spaces with respect to semimartingales relating to filtration problems, see the papers (Gyöngy and Krylov 1980, 1981).

§2. Martingales and Stochastic Integrals in Hilbert Space

2.1. Let (S, Σ, μ) be a complete measure space with a measure μ, $(X, \mathcal{X})$ a Banach space with the Borel σ-algebra, and X^* the dual of X.

Definition 1. The *mapping* $x : S \to X$ is called *measurable* if for each $\Gamma \in \mathcal{X}$ $\{s : x(s) \in \Gamma\} \in \Sigma$.

The *mapping* $x : S \to X$ is called *weakly measurable* if for each $x^* \in X^*$ the mapping $xx^* : S \to E^1$ is measurable.

The *mapping* $x : S \to X$ is called *strongly measurable* if there exists a sequence of simple measurable mappings converging μ-almost surely to x.

If X is separable, then strong measurability is equivalent to measurability.

Theorem 2.1 (Pettis; see, for example, Dunford and Schwartz 1958). *The mapping $x : S \to X$ is strongly measurable if and only if it is weakly measurable and there exists a set $B \in \Sigma$ such that $\mu(B) = 0$ and the set $\{y : y = x(s), s \in S\backslash B\}$ is separable.*

In particular, if S is separable, then strong measurability is equivalent to weak measurability.

Let $(\Omega, \mathcal{F}, \mathbf{P})$ be a probability space with an increasing system of σ-fields $(\mathcal{F}_t)$, $\mathcal{F}_t \subseteq \mathcal{F}$, $t \geq 0$.

Definition 2. By a random variable in X is meant a measurable mapping of $(\Omega, \mathcal{F}, \mathbf{P})$ to X.

Definition 3. A process $x(t, \omega)$, $t \geq 0$, with values in X is called $(\mathcal{F}_t)$-adapted if for each t $x(t, \cdot)$ is a measurable mapping from $(\Omega, \mathcal{F}_t, \mathbf{P})$ to X.

Definition 4. Subsets in $[0, \infty) \times \Omega$ are called completely measurable if they belong to the smallest σ-field with respect to which all the real $(\mathcal{F}_t)$-adapted processes that are right continuous and have limits to the left are measurable in (t, ω). A process $x : [0, \infty) \times \Omega \to X$ is called completely measurable if for any $\Gamma \in \mathcal{X}$ the set $\{(t, \omega) : x(t, \omega) \in \Gamma\}$ is completely measurable.

2.2. Let V, H be a separable Banach space, $V \subset H$, and suppose that the imbedding operator $V \to H$ is continuous (that is, $\forall v \in V$, $\|v\|_H \leq N|v|_V$, $N > 0$).

Lemma 2.1. a) *If x is a random variable with values in V (with respect to the Borel σ-algebra of (V)), then x is a random variable with values in H.*
 b) *If x is a random variable with values in H, then $\{\omega : x(\omega) \in V\} \in \mathcal{F}$.*

We now give the definition of a martingale with values in a real Hilbert space H and of a stochastic integral with respect to such a martingale. Here we will consider only martingales $m = m(t, \omega)$ that are strongly (that is, in norm) continuous in t and whose range of values is separable.

We denote by H_1 the closed linear span of the set of values $m(t, \omega)$ for $t \geq 0$, $\omega \in \Omega$. If $h(t, \omega) \perp H_1$, then it is natural to set $\int_0^t h(s, \omega) dm(s, \omega) = 0$.

Therefore, it is sufficient to define the integrals of functions with values in H_1. In accordance with this principle, the next assumption is natural, which is in force to the end of this section: H is a separable Hilbert space and H^* is identified with H.

For $h_1, h_2 \in H$ we denote by $h_1 h_2$ the scalar product of h_1 and h_2, and by $|h|$ the norm of $h \in H$.

Let G be a sub-σ-algebra of $\mathcal{F}$, x a random variable in H, and $\mathbf{E}|x| < \infty$.

Definition 5. By the conditional expectation of x with respect to G is meant a random variable $\mathbf{E}(x|G)$ with values in H such that for any $y \in H$

$$y\mathbf{E}(x|G) = \mathbf{E}(yx|G) \ (a.s.).$$

The random variable defined in this manner exists and is unique (a.s).

Definition 6. A stochastic process $x(t)$, $t \geq 0$, with values in H is called a martingale with respect to the flow $(\mathcal{F}_t)$ if
 a) x is $(\mathcal{F}_t)$-adapted in H,
 b) $\mathbf{E}|x(t)| < \infty$, $t \geq 0$,
 c) $\mathbf{E}(x(t)|\mathcal{F}_s) = x(s)$ a.s. for all $0 \leq s \leq t$.

The next result stems from the equivalence of the three concepts of measurability in a separable space.

Theorem 2.2. *A stochastic process $x(t)$, $t \geq 0$, with values in H and with finite expectation $\mathbf{E}|x(t)| < \infty$, $t \geq 0$, is a martingale with respect to a flow $(\mathcal{F}_t)$ if and only if the process $yx(t)$ is a one-dimensional martingale with respect to $(\mathcal{F}_t)$ for any $y \in H$.*

Definition 7. A process $x(t)$, $t \geq 0$, with values in H is called a local martingale if there exists a sequence of Markov times $\tau_n \uparrow \infty$ (a.s.) such that $x(t \wedge \tau_n)$ is a martingale for each n.

The class of local martingales is denoted by $\mathcal{M}_{\text{loc}}(\mathbb{R}_+, H)$, and the sequence (τ_n) is called localizing. In what follows we will consider only local martingales and martingales that are strongly continuous with respect to t. We denote by $\mathcal{M}^c_{\text{loc}}(\mathbb{R}_+, H)$ the class of continuous local martingales emanating from the origin.

Theorem 2.3. *If $x \in \mathcal{M}^c_{\text{loc}}(\mathbb{R}_+, H)$ then there exists a localizing sequence of Markov times (τ'_n) such that $\mathbf{E} \sup_{t \geq 0} |x(t \wedge \tau'_n)|^2 < \infty$. If, furthermore, $\mathbf{E}|x(t)|^2 < \infty \ \forall t \geq 0$, then $\mathbf{E}\sup_{s \leq t} |x(s)|^2 \leq 4\mathbf{E}|x(t)|^2 \forall t \geq 0$.*

We assume throughout that $(\tau'_n) \equiv (\tau_n)$.

Theorem 2.4. *If $x \in \mathcal{M}^c_{\text{loc}}(\mathbb{R}_+, H)$, then $|x(t)|^2$ is a local submartingale.*

Definition 8. We denote by $\langle x \rangle_t$ for $x \in \mathcal{M}^c_{\text{loc}}(\mathbb{R}_+, H)$ the increasing process (compensator) for $|x(t)|^2$ in the Doob-Meyer decomposition.

By the Doob-Meyer Theorem the "bracket" $\langle x \rangle_t$ is defined uniquely (a.s.) and is continuous with respect to t.

If $x, y \in \mathcal{M}_{\text{loc}}^c(\mathbb{R}_+, H)$, then we set $\langle x, y \rangle_t = \frac{1}{4}(\langle x + y \rangle_t - \langle x - y \rangle_t)$. If (τ_n) is a localizing sequence for $x(t)$ and for $y(t)$, then for $0 \leq s \leq t$

$$\mathbf{E}[(x(t \wedge \tau_n) - x(s \wedge \tau_n))(y(t \wedge \tau_n) - y(s \wedge \tau_n))|\mathcal{F}_s]$$
$$= \mathbf{E}[\langle x, y \rangle_{t \wedge \tau_n} - \langle x, y \rangle_{s \wedge \tau_n}|\mathcal{F}_s] \text{ a.s.}$$

Theorem 2.5 (Davis Inequality). *Let $x \in \mathcal{M}_{\text{loc}}^c(\mathbb{R}_+, H)$, and τ a Markov time, $\mathbf{P}(\tau < \infty) = 1$. Then*

$$\mathbf{E}\sup_{s \leq \tau} |x(t)| \leq 3\mathbf{E}[\langle x \rangle_\tau^{1/2}].$$

We fix an orthonormal basis $\{h_i, \, i \geq 1\}$ in H and set $x^i(t) = h^i x(t)$. It is well known that almost surely $d\langle x^i, x^j \rangle_t \ll d\langle x \rangle_t$ for any i, j.

Let E be some separable Hilbert space. We identify E^* with E. Let $\{e_i, \, i \geq 1\}$ be an orthonormal basis in E, $\mathcal{L}(H, E)$ the space of continuous linear operators from H to E, and $\mathcal{L}_2(H, E)$ the subspace of $\mathcal{L}(H, E)$ consisting of all Hilbert-Schmidt operators. The space $\mathcal{L}_2(H, E)$ is a separable Hilbert space with norm

$$\|B\| = \left(\sum_i |Bh^i|^2\right)^{1/2} = \left(\sum_{ij} (e_j, Bh^i)^2\right)^{1/2},$$

and $\|B\|$ does not depend on the choice of bases in H, E.

Let Q be a symmetric, non-negative nuclear operator in $\mathcal{L}(H, H)$. We denote by $\mathcal{L}_Q(H, E)$ the set of all linear, generally speaking, unbounded operators B defined on $Q^{1/2}H$, sending $Q^{1/2}H$ to E and such that $B^{1/2}Q \in \mathcal{L}_2(H, E)$.

For $B \in \mathcal{L}_Q(H, E)$ we set

$$|B|_Q = \|BQ^{1/2}\|.$$

It is well known that if $B \in \mathcal{L}_2(H, E)$, then

$$|B| \equiv |B|_{\mathcal{L}(H,E)} \leq \|B\|, \; B \in \mathcal{L}_Q(H, E),$$

and

$$|B|_Q \leq |B|(\operatorname{tr} Q)^{1/2}.$$

Let us return to $x \in \mathcal{M}_{\text{loc}}^c(\mathbb{R}_+, H)$. There exists a completely measurable process $Q_x(t)$ with values in $\mathcal{L}_2(H, H)$ such that for all (t, ω) the operator $Q_x(t, \omega)$ is a symmetric, non-negative nuclear operator with $\operatorname{tr} Q = 1$ and

$$h^i Q_x(t) h^j = \frac{d\langle x^i, x^j \rangle_t}{d\langle x \rangle_t} \quad (d\mathbf{P} \times d\langle x \rangle\text{-a.s.})$$

for all i, j for each basis $\{h^i\}$. The operator Q_x is called the correlation operator of x.

If $B(t)$ is a completely measurable process with values in $\mathcal{L}_2(H, E)$ and

$$\mathbf{E}\int_0^t \|B(s)\|^2 d\langle x\rangle_s < \infty$$

for each $t \geq 0$, then there exists a square-integrable martingale y_t that is continuous with respect to t, takes values in E and is such that for any orthonormal basis $\{h^i\}$ and any $y \in E$, $T \geq 0$

$$\lim_{n\to\infty} \mathbf{E}\sup_{t\leq T}\left|yy(t) - \sum_{i=1}^n \int_0^t yB(s)h^i d(h^i x(s))\right|^2 = 0. \qquad (2.1)$$

Two processes possessing this property are almost surely identical for all t. Therefore the expression

$$y(t) = \int_0^t B(s)dx(s) \qquad (2.2)$$

is well defined and we have the representation

$$\langle y\rangle_t = \int_0^t |B(s)|^2_{Q_x(s)} d\langle x\rangle_s, \qquad (2.3)$$

which along with the inequality $|B|_{Q_x} \leq \|B\|$ (let us recall that $\operatorname{tr} Q_x \equiv 1$) enables us to extend the integral (2.2) (preserving property (2.3)) in the usual fashion to the space of all completely measurable functions $B(s)$ for which for all $t \geq 0$

$$\int_0^t \|B(s)\|^2 d\langle x\rangle_s < \infty \text{ a.s.}$$

The stochastic integral so defined is continuous with respect to t and is a local martingale. In the next theorem the concept of a stochastic integral is extended to a still wider class of processes $B(s)$.

Theorem 2.6. *For each (s,ω) let there be defined an operator $B(s) = B(s,\omega) \in \mathcal{L}_{Q_x(s,\omega)}(H, E)$ such that the process $B(s)Q_x^{1/2}(s)$ is completely measurable (in $\mathcal{L}_2(H,E)$) and for each t the right hand side of (2.3) is finite a.s. Then the sequence*

$$y_n(t) \equiv \int_0^t B(s)Q_x^{1/2}(s)\left(\frac{1}{n} + Q_x^{1/2}(s)\right)^{-1} dx(s)$$

converges uniformly with respect to $t \leq T$ ($T > 0$) in probability to a limit (say, $y(t)$). Furthermore, $y(\,\cdot\,) \in \mathcal{M}^c_{\mathrm{loc}}(\mathbb{R}_+, E)$ and representation (2.3) is valid.

Remark. The above construction follows the paper (Krylov and Rozovskij 1979). A somewhat different construction was given in (Pardoux 1975).

The process $y(t)$ constructed in Theorem 2.6 is by definition taken to be equal to the right hand side of (2.2). If X is a separable Hilbert space and $A \in \mathcal{L}(E, X)$, then almost surely we have

$$Ay(t) = \int_0^t AB(s)dx(s) \tag{2.4}$$

for all $t \geq 0$. Let $e \in E$. We define the operator $\hat{e} \in \mathcal{L}(E, R^1)$ by the formula $\hat{e}y = ey$ (the scalar product is in E). Then by (2.4) we have

$$ey(t) = \int_0^t \hat{e}B(s)dx(s).$$

If $h(s) \in H$, $h(\,\cdot\,)$ is completely measurable, and for each $t \geq 0$

$$\int_0^t |\hat{h}(s)|^2_{Q_x(s)} d\langle x \rangle_s \equiv \int_0^t |Q_x^{1/2}(s)h(s)|^2 d\langle x \rangle_s < \infty \text{ (a.s.)},$$

then we have

$$\int_0^t \hat{h}(s)dx(s) := \int_0^t h(s)dx(s).$$

2.3. We now introduce the concept of a Wiener process in H.

Definition 9. Let Q be a symmetric non-negative nuclear operator on H, and $\operatorname{tr} Q < \infty$. By a Wiener process (with respect to $(\mathcal{F}_t)$) in H with covariance operator Q is meant a *continuous martingale* $W(t)$ *with values in* H and correlation operator $(\operatorname{tr} Q)^{-1}Q$ such that $W(0) = 0$, $\langle W \rangle_t = \operatorname{tr} Q \cdot t$.

For any symmetric non-negative nuclear Q with $\operatorname{tr} Q > 0$ one can construct a Wiener process corresponding to it on some probability space. Here $EW^2(t) = \operatorname{tr} Q \cdot t$. One can define a stochastic integral with respect to a Wiener process, not only for completely measurable $B(s)$, but also for $B(s)$ measurable with respect to (s, ω) and $(\mathcal{F}_s)$-adapted with $\int_0^t |B(s)|_{Q^2} ds < \infty$ a.s. $(t > 0)$.

§3. The Itô Formula for the Square of the Norm

3.1. Let V be a Banach space, V^* its dual, and H a Hilbert space (all spaces are real). If $v \in V$ ($h \in H$, $v^* \in V^*$), then we denote by $|v|$ ($|h|, |v^*|$) the norm of v (respectively, h, v^*) in V (in H, in V^*); if $h_1, h_2 \in H$, then $h_1 h_2$ denotes the scalar product of h_1 and h_2 and the result of the action of $v^* \in V^*$ on $v \in V$ is denoted by $v^*v = vv^*$. Let $\Lambda : V \to H$ be a bounded linear operator, where ΛV is everywhere dense in H.

Let $v(t, \omega) \in V$, $h(t, \omega) \in H$, $v^*(t, \omega) \in V^*$, $t \geq 0$, be three processes given on a complete probability space $(\Omega, \mathcal{F}, \mathbf{P})$ with an increasing flow of σ-algebras $(\mathcal{F}_t), \mathcal{F}_t \subseteq \mathcal{F}$, $t \geq 0$. Let $v(t, \omega)$ be (Lebesgue) strongly measurable with respect to (t, ω), weakly measurable with respect to ω relative to $\mathcal{F}_t$ for almost all t, and for each $v \in V$ suppose that the variable $vv^*(t, \omega)$ is $\mathcal{F}_t$-measurable with respect to ω for almost all t and measurable with respect to (t, ω). It is assumed that $h(t, \omega)$ is strongly continuous with respect to t, strongly measurable relative to $\mathcal{F}_t$ for each t and is a local semimartingale, that is, there exist in H strongly $\mathcal{F}_t$-measurable processes $A(t)$, $m(t)$ that are continuous in t and are such that $m(t)$ is a local martingale, the trajectories of $A(t)$ (for each ω) have finite variation on any finite interval, and $h(t) = A(t) + m(t)$.

We fix $p \in (1, \infty)$ and set $q = p/(p-1)$. We assume that $|v(t)| \in L_p[0,T]$ (a.s.) for each $T \geq 0$ and there exists a measurable function $f(t, \omega)$ with respect to (t, ω) such that $f(t) \in L_q[0,T]$ (a.s.) for each $T \geq 0$, $|v^*(t)| \leq f(t)$ for all (t, ω) (note that the function $|v^*(t)|$ is non-measurable in general).

3.2. The next result (the "Itô formula for the square of the norm") is fundamental for this section.

Theorem 2.7. *Let τ be a Markov time and suppose that for every $v \in V$ we have*

$$\Lambda v \Lambda v(t) = \int_0^t vv^*(s)ds + \Lambda v h(t) \tag{2.5}$$

almost everywhere on $\{t, \omega\} : t < \tau(\omega)\}$. Then there exists a set $\Omega' \subseteq \Omega$ and a function $\tilde{h}(t)$ with values in H such that:

a) $\mathbf{P}(\Omega') = 1$, $\tilde{h}(t)$ *is strongly $\mathcal{F}_t$-measurable on the set $\{\omega : t < \tau(\omega)\}$ for each t, $\tilde{h}(t)$ is continuous with respect to t on $[0, \tau(\omega))$ for each ω, and $\Lambda v(t) = \tilde{h}(t)$ (a.e. $\{(t, \omega) : t < \tau(\omega)\}$);*

b) *for $\omega \in \Omega'$, $t < \tau(\omega)$*

$$\tilde{h}^2(t) = h^2(0) + 2\int_0^t v(s)v^*(s)ds + 2\int_0^t \tilde{h}(s)dh(s) + \langle m \rangle_t;$$

c) *if V is separable, then for $\omega \in \Omega'$, $t < \tau(\omega)$, $v \in V$*

$$\Lambda v \tilde{h}(t) = \int_0^t vv^*(s)ds + \Lambda v h(t);$$

d) *if V is separable and (2.5) is satisfied for some $t \geq 0$ for each $v \in V$ (a.s.) on $\{\omega : t < \tau(\omega)\}$, then $\Lambda v(t) = h(t)$ (a.s.) on $\{\omega : t < \tau(\omega)\}$.*

For a discussion of Theorem 3.1, its proof and some corollaries, see the papers (Krylov and Rozovskij 1979) and (Rozovskij 1983).

§4. Stochastic Differential Equations of Monotone Type in Banach Spaces

4.1. We shall be considering the stochastic evolution equation

$$v(t, \omega) = u_0(\omega) + \int_0^t A(v(s, \omega), s, \omega)ds$$

$$+ \int_0^t B(v(s, \omega), s, \omega)dW(s, \omega) + z(t, \omega) \tag{2.6}$$

Chapter 2.II. Stochastic Evolution Equations

on some complete probability space $(\Omega, \mathcal{F}, \mathbf{P})$ with an increasing flow of complete σ-algebras $(\mathcal{F}_t)$, $t \in [0, T]$, $0 < T < \infty$. Here we assume that we are given real separable Hilbert spaces H and E, identified with their duals, $W(t) = W(t, \omega)$ is a Wiener process in E with nuclear covariance operator Q (see above), and $z(t) = z(t, \omega)$ is a square-integrable continuous martingale in H. We also assume that we are given a real separable reflexive Banach space V and its dual V^* such that:
 a) $V \subseteq H \equiv H^* \subseteq V^*$,
 b) V is dense in H (in the norm of H),
 c) for all $v \in V$, $|v|_H \leq c|v|_V$ ($c > 0$),
 d) if $v \in V$, $v^* \in H(\subseteq V^*)$, then $vv^* = (v, v^*)_H$.
(The most important examples of such spaces H and V are the spaces $V = W_p^m(G)$, $p \in (1, \infty)$, and $H = L_2(G)$, where G is a bounded domain in $\mathbb{R}^d$ and $dp \geq 2(d - mp)$.) It is assumed that for each $(v, t, \omega) \in V \times [0, T] \times \Omega$ the variables $A(v, t, \omega)$ and $B(v, t, \omega)$ in (2.6) are operators:

$$A(v, t, \omega) \in V^*, \quad B(v, t, \omega) \in \mathcal{L}_Q(E, H),$$

and the numbers $p \in (1, \infty)$ and $q = p/(p-1)$ are fixed.

We will also assume that for each $v \in V$ the functions $A(v, t, \omega)$, $B(v, t, \omega)$ are Lebesgue measurable with respect to (t, ω) relative to the measure $dt \times d\mathbf{P}$ and $(\mathcal{F}_t)$-adapted, that is, for each $v \in V$, $t \in [0, T]$ they are $\mathcal{F}_t$-measurable with respect to ω. (We recall that, in view of the separability of V^* and $\mathcal{L}_Q(E, H)$, the concepts of weak and strong measurability coincide here, which is why we merely talk about measurability.) The variable u_0 in (2.6) is an $\mathcal{F}_0$-measurable function on Ω with values in H.

Definition 1. By a solution, or V-solution of equation (2.6) is meant a function $v(t, \omega)$ with values in V defined on $[0, T] \times \Omega$, measurable with respect to (t, ω), $(\mathcal{F}_t)$-adapted and satisfying the inequality

$$\mathbf{E} \int_0^T (|v(t)|_V^p + |v(t)|_H^2) dt < \infty \tag{2.7}$$

and equality (2.6), understood as equality of elements of V^* for almost all $(t, \omega) \in [0, T] \times \Omega$.

Thus the solution is understood in the strong sense. As regards the theory of weak solutions, see the papers (Viot 1976) and (Métivier and Viot 1988).

Equation (2.7) is considered under the following conditions: there exist constants K, $\alpha > 0$ and a non-negative $(\mathcal{F}_t)$-adapted measurable function $f(t, \omega)$ (with respect to (t, ω)) on $[0, T] \times \Omega$ such that for all $v, v_1, v_2 \in V$ and $(t, \omega) \in [0, T] \times \Omega$ the following properties hold:

A_1) (semicontinuity of A): the function $vA(v_1 + \lambda v_2)$ is continuous with respect to λ on E^1;

A_2) (monotonicity of (A, B)):

$$2(v_1 - v_2)(A(v_1) - A(v_2)) + |B(v_1) - B(v_2)|_Q^2 \leq K|v_1 - v_2|_H^2;$$

A_3) (coercivity of (A, B)):

$$2vA(v) + |B(v)|_Q^2 + \alpha|v|^p \leq f + K|v|_H^2;$$

A_4) (boundedness of the growth of A):

$$|A(v)|_{V^*} \leq f^{1/q} + K|v|_V^{p-1};$$

A_5) $\quad \mathbf{E}|\mu_0|_H^2 < \infty, \quad \mathbf{E}\int_0^T f(t)dt < \infty.$

It turns out that under such conditions the functions $A(v(t), t)$ and $B(v(t), t)$ are measurable with respect to (t, ω) and $(\mathcal{F}_t)$-adapted, and under the condition that inequality (2.7) holds, the integrals in (2.6) are defined and finite (see (Krylov and Rozovskij 1979); (Rozovskij 1983)).

Definition 2. By an H-solution of equation (2.6) is meant a function $u(t, \omega)$ with values in H, defined on $[0, T] \times \Omega$, strongly continuous in H with respect to t, $(\mathcal{F}_t)$-adapted and such that

1) $u \in V$ (a.e. (t, ω))[1] and

$$\mathbf{E}\int_0^T (|u(t)|_V^p + |u(t)|_H^2)dt < \infty;$$

2) there exists a set of full measure $\Omega' \subseteq \Omega$ on which

$$u(t) = u_0 + \int_0^t A(u(s), s)ds + \int_0^t B(u(s), s)dW(s) + z(t) \qquad (2.8)$$

for all $t \in [0, T]$, where the equality is understood as an equality in V^* and the integrals are understood as

$$\int_0^t A(u(s), s)ds := \int_0^t I(u(s) \in V)A(u(s), s)ds,$$

$$\int_0^t B(u(s), s)dW(s) := \int_0^t I(u(s) \in V)B(u(s), s)dW(s),$$

where the function $I(u(s) \in V)$ ($= 1$ a.e. (t, ω)) is measurable with respect to (t, ω) and $(\mathcal{F}_t)$-adapted by Lemma 2.1.

Definition 3. An H-solution u is said to be a continuous modification in H of a solution v if $u(t, \omega) = v(t, \omega)$ a.e. (t, ω).

Theorem 2.8. *Suppose that all the above conditions hold. Then there exists a V-solution v of equation (2.6) which has a continuous modification in H. This modification is unique and*

---

[1] *Translator's note*: A property is said to hold a.e. (t, ω) if for a.e. t the property holds for almost all ω.

$$\mathbf{E}\sup_{t\leq T}|u(t)|_H^2 + \mathbf{E}\int_0^T |v(t)|_V^p dt \leq c\Big(\mathbf{E}|u_0|_H^2 + \mathbf{E}\int_0^T f(t)dt + \mathbf{E}|z(T)|_H^2\Big),$$

where c depends only on K, p, T, α.

Theorem 2.9. *Suppose that the assumptions of Theorem 2.8 hold, A and B do not depend on ω, $z(t) \equiv 0$, $v(t)$ is a solution of equation (2.6) and $u(t)$ is its continuous modification in H. Then $u(t)$ is a Markov random function.*

For the proof of Theorems 2.8 and 2.9, see (Krylov and Rozovskij (1979) and (Rozovskij 1983). The existence of a V-solution is proved by passing to the limit by the monotonicity method from the finite-dimensional situation, for which dimension-independent estimates of the solution are preliminarily established. Let us observe that in the finite-dimensional case also, monotonicity and coercivity conditions give a result that is new by comparison with the "usual" theory (cf. Sect. 1). The existence of a continuous modification in H of the solution stems from Theorem 2.7. A theorem on the stability of the solution with respect to the initial data is also valid.

§5. Stochastic Partial Differential Equations I.

The First Boundary-Value Problem for Parabolic-Type Non-Linear Equations

5.1. The entire discussion in this and the next section is carried out in Sobolev spaces. Let E^d be a d-dimensional Euclidean space, G a domain in E^d, Γ the boundary of G, m an integer, $m \geq 1$, $p \in (1, \infty)$. Let us recall that by a Sobolev space $W_p^m(G)$ is meant the space of real functions on G with finite norm

$$\|u\|_{m,p} := \Big(\sum_{\alpha_1,\ldots,\alpha_m} \|\mathcal{D}^{\alpha_1}\ldots\mathcal{D}^{\alpha_m}u\|_p^p\Big)^{1/p},$$

where $\mathcal{D}^{\alpha_1}\ldots\mathcal{D}^{\alpha_m}$ are generalized derivatives; by the *Sobolev space* $\overset{\circ}{W}_p(G)$ is meant the closure of $C_0^\infty(G)$ in the norm $\|\cdot\|_{m,p}$.

Let $(\Omega, \mathcal{F}, \mathbf{P})$ be a complete probability space with an increasing flow of complete σ-algebras $(\mathcal{F}_t)$, $\mathcal{F}_t \subseteq \mathcal{F}$, G a bounded domain with a regular boundary (concerning regularity conditions, see, for example, (Nikol'skij 1969)) (or $G = E^d$), $p = 2$, $2(d - mp) \leq dp$. Let $z(t)$ be a square-integrable martingale (with respect to $(\mathcal{F}_t)$) taking values in $L_2(G)$ and continuous with respect to t in $L_2(G)$, and let $W(t)$ be a Wiener process with values in some separable Hilbert space E and covariation operator Q. The spaces $V = \overset{\circ}{W}_p^m(G)$, $H = L_2(G)$, $V^* = W_q^{-m}(G)$, $q = p/(p-1)$, satisfy the assumptions a)–d) of Sect. 4.

For fixed $T > 0$ we consider in the cylinder $[0,T] \times G$ the non-linear problem

$$du(t,x,\omega) = -(-1)^{|\alpha_1|+\ldots+|\alpha_m|}\mathcal{D}^{\alpha_1}\ldots$$
$$\ldots \mathcal{D}^{\alpha_m} A_{\alpha_1\ldots\alpha_m}(\mathcal{D}^{\beta_1}\ldots\mathcal{D}^{\beta_m}u(t,x,\omega),t,x,\omega)dt$$
$$+ B(\mathcal{D}^{\beta_1}\ldots\mathcal{D}^{\beta_m}u(t,x,\omega),t,x,\omega)dW(t) + dz(t,x,\omega), \quad (2.9)$$
$$u(0,x,\omega) = u_0(x,\omega), \quad x \in G, \quad (2.10)$$
$$\mathcal{D}^{\beta_0}\ldots\mathcal{D}^{\beta_{m-1}}u|_S = 0 \quad (2.11)$$

for all $\beta_0,\ldots,\beta_{m-1}$ such that

$$|\beta_0| + \ldots + |\beta_{m-1}| \le m-1,$$

where S is the lateral surface of the cylinder. Here, as usual, summation over repeating indices α_i is assumed, the functions A, B depend on t, x, ω and all derivatives of u with respect to x of order up to m; A, u are real functions, B is a function with values in E, in the second term BdW in (2.9), we have the scalar product in E in mind.

We suppose that for any set of real numbers $\xi = (\xi^{\beta_1,\ldots,\beta_m})$ and any $e \in E$ the functions $A(\xi,t,x,\omega)$, $B(\xi,t,x,\omega)e = (B(\xi,t,x,\omega),e)_E$ are measurable with respect to (t,x,ω), for each $t \in [0,T]$ they are measurable with respect to (x,ω) relative to the product of $\mathcal{B}^d$ (the Borel σ-field in E^d) and $\mathcal{F}_t$, and that for each t, x, ω they are continuous with respect to ξ. We suppose further that there exist a constant $K > 0$ and a non-negative function $f(t,x,\omega)$ possessing these same measurability properties as $A(0,t,x,\omega)$ such that for all $t, x, \omega, \xi, \alpha_1, \ldots, \alpha_m$

$$|A_{\alpha_1\ldots\alpha_m}(\xi,t,x,\omega)| \le f^{1/q}(t,x,\omega) + K\sum_{\beta_1,\ldots,\beta_m} |\xi^{\beta_1,\ldots,\beta_m}|^{p-1}, \quad (2.12)$$

$$|B(\xi,t,x,\omega)|_E^2 \le f(t,x,\omega) + K\sum_{\beta_1,\ldots,\beta_m} |\xi^{\beta_1,\ldots,\beta_m}|^p + K|\xi^{0,\ldots,0}|^2. \quad (2.13)$$

Finally, it is assumed that $u_0(x,\omega)$ is measurable with respect to $\mathcal{B}^d \times \mathcal{F}_0$ and for all t, ω, $\|u_0\|_2 < \infty$, $\|f(t)\|_1 < \infty$

$$\mathbf{E}\|u_0\|_2^2 < \infty, \quad \mathbf{E}\int_0^T \|f(t)\|_1 dt < \infty.$$

These assumptions allow one to give a precise meaning to equation (2.9) and conditions (2.10), (2.11).

Definition 1. By a V-solution (or H-solution) of problem (2.9)–(2.11) is meant a V-solution (or H-solution) of equation (2.6). A continuous modification in H of the solution (2.9)–(2.11) is similarly defined. One can also formulate an equivalent definition with the help of the integral identity $((\cdot,\cdot)_0$ is the scalar product in $L_2(G)$):

$$(v(t),\eta)_0 = (u_0,\eta)_0 - \int_0^t (A_{\alpha_1\ldots\alpha_m}(\mathcal{D}^{\beta_1}\ldots\mathcal{D}^{\beta_m}v,s), \mathcal{D}^{\alpha_1}\ldots\mathcal{D}^{\alpha_m}\eta)_0 ds$$
$$+ \int_0^t \int_{R^d} B(\mathcal{D}^{\beta_1}\ldots\mathcal{D}^{\beta_m}v,s,x)\eta(x)dxdW(s) + (z(t),\eta)_0 \quad (2.14)$$

a.e. (t,ω) for any $\eta \in V$.

5.2.

Definition 2. Equation (2.9) is said to satisfy a strong parabolicity condition if the operators A, B satisfy conditions $(A_2), (A_3)$ in Sect. 4. The next theorem automatically follows from the results of Sect. 4.

Theorem 2.10. *If equation (2.9) is strongly parabolic, then the assertions of Theorems 2.8 and 2.9 hold.*

Strong parabolicity conditions are difficult to verify in general form. All the same, generalizing results of Vishik (Vishik 1963), one can give the following simple sufficient condition for the strong parabolicity of equation (2.9), called the algebraic strong-parabolicity condition:

A) there exist constants N and $\epsilon > 0$ such that for all $\xi^{\alpha_1,\ldots,\alpha_m}$, $\eta^{\alpha_1,\ldots,\alpha_m}$, t, x, ω

$$-2A^{\beta_1\ldots\beta_m}_{\alpha_1\ldots\alpha_m}(\xi^{\beta_1,\ldots,\beta_m},t,x,)\eta^{\alpha_1,\ldots,\alpha_m}\eta^{\beta_1,\ldots,\beta_m}$$
$$+ |B^{\beta_1\ldots\beta_m}(\xi^{\beta_1,\ldots,\beta_m},t,x)\eta^{\beta_1,\ldots,\beta_m}|^2_Q$$
$$+ \epsilon \sum_{|\alpha_1|+\ldots+|\alpha_m|=m} |\xi^{\alpha_1,\ldots,\alpha_m}|^{p-2}|\eta^{\alpha_1,\ldots,\alpha_m}|^2$$
$$\leq N(\eta^{0,\ldots,0})^2, \quad (2.15)$$

where

$$A^{\beta_1\ldots\beta_m}_{\alpha_1\ldots\alpha_m} := \frac{\partial}{\partial \xi^{\beta_1,\ldots,\beta_m}} A_{\alpha_1\ldots\alpha_m},$$

$$B^{\beta_1\ldots\beta_m} := \frac{\partial}{\partial \xi^{\beta_1,\ldots,\beta_m}} B,$$

and in the second term of the left hand side of (2.15) the summation over $\beta_1,\ldots,\beta_m$ is carried out before the computation of the norm $|\cdot|_Q$.

For further details see (Krylov and Rozovskij 1979).

§6. Stochastic Partial Differential Equations II.

The Cauchy Problem for Second-Order Linear Equations

6.1. We consider problem (2.9) under the assumption that $m = 1$, $G = E^\alpha$, $p = 2$, and A, B are linear functions of ξ which, generally speaking, are not equal to zero for $\xi = 0$. All the assumptions (2.12)–(2.13) and measurability properties are assumed to hold. Problem (2.9)–(2.11) is then converted to the following:

$$du(t,x) = \mathcal{D}^\alpha(a_{\alpha\beta}(t,x)\mathcal{D}^\beta u(t,x) + f_\alpha(t,x))dt$$
$$+ (b_\alpha(t,x)\mathcal{D}^\alpha u(t,x) + g(t,x))dW(t) + dz(t,x), \qquad (2.16)$$
$$u(t,\cdot) \in L_2(E^d), \quad u(0,x) = u_0(x), \ x \in E^d, \qquad (2.17)$$

where $a_{\alpha\beta}, f_\alpha$ are real functions and b_α, g are functions with values in E. Conditions (2.12), (2.13) are equivalent to the requirement that $a_{\alpha\beta}, |b_\alpha|_E$ be bounded and the inequality

$$\sum_\alpha \mathbf{E} \int_0^T \|f_\alpha\|_2^2 dt + \mathbf{E} \int_0^T \||g|_E\|_2^2 dt < \infty$$

hold.

The solution of problem (2.16), (2.17) is understood in the sense of the integral identity (2.14); for a V-solution it holds for almost all (t,ω); for an H-solution, for each t a.s.

Lemma 2.2. *Suppose that the inequality*

$$2\sum_{i,j=1}^d a_{ij}(t,x)\eta^i\eta^j - \left|\sum_{i=1}^d b_i(t,x)\eta^i\right|_Q^2 \geq \epsilon|\eta|^2 \qquad (2.18)$$

holds for all x, $\eta \in E^d$, $t \in [0,T]$, where $\epsilon > 0$ is a constant, $a_{ij} = a_{\alpha\beta}, b_i = b_\alpha$ if α is the ith coordinate vector and β is the jth. Then the strong-parabolicity algebraic condition (2.15) is satisfied.

The next result is a consequence of this lemma and the results of the previous sections.

Theorem 2.11. *Let condition (2.18) be satisfied. Then there exists a function $u(t,\omega)$ defined on $[0,T] \times \Omega$ with values in $L_2(E^d)$ that is strongly continuous with respect to t in $L_2(E^d)$, is $(\mathcal{F}_t)$-adapted and is such that*
 a) $u \in W_2^1(E^d)$ *a.e.* (t,ω),
 b) *for each $\eta \in W_2^1(E^d)$ we have a.s.*

Chapter 2.II. Stochastic Evolution Equations

$$(u(t), \eta)_0 = (u_0, \eta)_0 + \int_0^t (-1)^\alpha (\mathcal{D}^\beta u(s), a_{\alpha\beta} \mathcal{D}_\eta^\alpha)_0 ds$$
$$+ \int_0^t (-1)^\alpha (f_\alpha(s), \mathcal{D}_\eta^\alpha)_0 ds + \int_0^t (b_\alpha(s) \mathcal{D}^\alpha u(s)$$
$$+ g(s), \eta)_0 dW(s) + (z(t), \eta)_0 \quad (2.19)$$

for all $t \in [0, T]$.

Theorems 2.8 and 2.9 allow one to prove theorems on the uniqueness and the Markov property of the function u in Theorem 2.11. A result on the stability of the solution with respect to the initial data also holds.

In filtration theory (see Krylov and Rozovskij 1977) there is also the important question regarding the increase in smoothness of a solution of a problem of type (2.16), (2.17). (The scalar product $(\,\cdot\,,\,\cdot\,)_0$ in the filtration equations is replaced by $(\,\cdot\,,\,\cdot\,)_m$, the scalar product in $W_2^m(E^d)$.)

It is further assumed that $m \geq 0$ is an integer, $z(t)$ is a square-integrable martingale with values in $W_2^m(E^d)$ that is continuous with respect to t in $W_2^m(E^d)$ for all t, ω, the functions $a_{\alpha\beta}(b_\alpha)$ have m derivatives (or weak derivatives) with respect to x that are continuous (or weakly continuous) with respect to x and bounded (or bounded in the norm of E) uniformly in t, x, ω. Let $g \equiv 0$ (without loss of generality) and suppose that for all (t, ω) the functions $f_\alpha \in W_2^m(E^d)$, $u_0 \in W_2^m(E^d)$ and

$$\mathbf{E}\|u_0\|_{m,2}^2 + \mathbf{E} \int_0^T \|f_\alpha(t)\|_{m,2}^2 dt < \infty.$$

Theorem 2.12. *Suppose that (2.18) and all the assumptions from the preceding paragraph are satisfied. Then there exists a set $\Omega' \subseteq \Omega$ such that $\mathbf{P}(\Omega') = 1$ and for $\omega \in \Omega'$ the function $u(t)$ in Theorem 2.11 belongs to $W_2^m(E^d)$ and is continuous with respect to t in the norm of $W_2^m(E^d)$. Moreover, $u \in W_2^{m+1}(E^d)$ a.e. (t, ω), and*

$$\mathbf{E} \sup_{t \leq T} \|u\|_{m,2}^2 + \mathbf{E} \int_0^T \|u(s)\|_{m+1,2}^2 ds < \infty.$$

For the proof see (Krylov and Rozovskij 1979); (Rozovskij 1983) and also, regarding the more general problem, (Krylov and Rozovskij 1977).

References*

Alyushina, L.A., Krylov, N.V. (1988): On passage to the limit in Itô stochastic equations. Teor. Veroyatn. Primen. *33*, No. 1, 3–13. [English transl.: Theory Probab. Appl. *33*, No. 1, 1–10 (1988)] Zbl. 671.60052

Anulova, S.V. (1978): Processes with a Lévy generating operator in a half-space. Izv. Akad. Nauk SSSR, Ser. Mat. *42*, No. 4, 708–750. [English transl.: Mat. USSR, Izv. *13*, 9–51 (1979)] Zbl. 386.60043

Baklan, V.V. (1963): On the existence of solutions of stochastic equations in Hilbert space. Dopov. Akad. Nauk Ukr. SSR, Ser. A, No. 10, 1299–1303. (Ukrainian, Russian summary) Zbl. 129,106

Baklan, V.V. (1964): Variational differential equations and Markov processes in Hilbert space. Dokl. Akad. Nauk SSSR *159*, No. 4, 707–710. [English transl.: Sov. Math., Dokl. *5*, 1553–1556 (1965)] Zbl. 295.60040

Daletskij, Yu.L. (1967): Infinite-dimensional elliptic operators and the parabolic equations connected with them. Usp. Mat. Nauk *22*, No. 4, 3–54. [English transl.: Russ. Math. Surv. *22*, No. 4, 1–54 (1967)] Zbl. 164,413

Dunford, N., Schwartz, J.T. (1958): Linear Operators. I. General Theory. Interscience, New York. Zbl. 84,104

Gyöngy, I, Krylov, N.V. (1980): On stochastic equations with respect to semimartingales. I. Stochastics *4*, 1–21. Zbl. 439.60061

Gyöngy, I, Krylov, N.V. (1982): On stochastic equations with respect to semimartingales. II. Stochastics *6*, 153–173. Zbl. 481.60060

Krylov, N.V., Rozovskij, B.L. (1977): The Cauchy problem for linear stochastic partial differential equations. Izv. Akad. Nauk SSSR, Ser. Mat. *41*, No. 6, 1329–1347. [English transl.: Mat. USSR, Izv. *11*, 1267–1284 (1977)] Zbl. 371.60076

Krylov, N.V., Rozovskij, B.L. (1979): Stochastic evolution equations. Itogi Nauki Tekh. Ser. Sovrem. Probl. Mat., Fundam. Napravleniya *14*, 71–146. [English transl.: J. Sov. Math. *16*, 1233–1277 (1981)] Zbl. 436.60047

Métivier, M., Viot, M. (1988): On weak solutions of stochastic partial differential equations. Stochastic Anal., Proc. Jap.-Fr. Semin. Paris France 1987, Lect. Notes Math. *1322*, 139–150. Zbl. 647.60069

Nikol'skij, S.M. (1969): Approximation of Functions of Several Variables and Embedding Theorems. Nauka, Moscow. [English transl.: Springer, Berlin Heidelberg New York (1975)] Zbl. 185,379

Pardoux, E. (1975): Équations aux dérivées partielles stochastiques non linéaires monotones. Étude de solutions fortes de type Itô. Thèse Univ. de Paris Sud

Rozovskij, B.L (1983): Stochastic Evolution Systems. Nauka, Moscow. [English transl.: Kluwer, Dordrecht (1990)] Zbl. 525.60063

Viot, M. (1976): Solutions faibles d'équations aux dérivées partielles non linéaires. Thèse Univ. de Pierre et Marie Curie, Paris

Vishik, M.I. (1963): Quasilinear strongly-elliptic systems of differential equations in divergence form. Tr. Mosk. Mat. O.-va *12*, 125–184. [English transl.: Trans. Mosc. Math. Soc. *1963*, 140–208 (1965)] Zbl. 144.362

---

* For the convenience of the reader, references to reviews in Zentralblatt für Mathematik (Zbl.), compiled using the MATH database, have, as far as possible, been included in this bibliography.

III. Stochastic Calculus (Malliavin Calculus). Applications to Stochastic Differential Equations

A.Yu. Veretennikov

§1. Introduction

The stochastic calculus of variations or the Malliavin calculus is nowadays a powerful weapon in various problems of asymptotic analysis, in theoretical physics, ergodic theory *et al.* It initially arose as a probabilistic approach to hypoellipticity problems in the theory of partial differential equations. Hypoellipticity of a differential operator L means the C_{loc}^∞-smoothness of any function u if $Lu \in C_{\text{loc}}^\infty$. The hypoellipticity of second-order differential operators was analyzed in papers (Hörmander 1967) and (Olejnik and Radkevich 1971). It turned out that the hypoellipticity of such operators can also occur under diffusion degeneracy if Hörmander's conditions are satisfied by the Lie brackets of the vector fields forming the diffusion matrix. Malliavin (Malliavin 1976), (Malliavin 1982) proposed an approach for studying the smoothness of the density of the solution of a stochastic differential equation without relying upon partial differential equations, and also leading to Hörmander-type conditions. This approach allows one to study considerably more general problems on the smoothness of the distribution of functionals of a Wiener process, processes with jumps, processes with aftereffect, and so on. Different treatments of Malliavin's idea have been proposed. The interpretations of Stroock (Stroock 1983), Bismut (Bismut 1981) and Zakai (Zakai 1985) are the main of them. (We do not claim to offer a classification of these and other interpretations: see also, for example, (Shigekawa 1980); (Watanabe 1984); (Watanabe 1987); (Bell 1987); Zakai himself terms his exposition a variant of Stroock's, etc.) In this chapter we set forth an introduction to the Malliavin calculus based largely on the paper (Zakai 1985) which is devoted to a comparison of the versions of Stroock and Bismut. The exposition is oriented towards applications to the solutions of stochastic differential equations (SDEs), and does not contain many important aspects of the theory. Thus we do not touch upon the theory of extended stochastic integrals (see (Nualart and Zakai 1986); (Skorokhod 1975)), processes with jumps (see, for example, (Bichteler and Jacod 1983); (Bismut 1984b)), the theory of generalized Brownian functionals (see (Watanabe 1984), (Watanabe 1987)), applications to the index theorem (Bismut 1984a), the Schrödinger operator (Watanabe), oscillatory integrals (Malliavin 1982) and many other results. Questions of the smoothness of conditional densities (see, for example, (Bismut and Michel 1981), (Bismut and Michel 1982)) are not discussed. However, the beautiful Ikeda-Watanabe-Kusuoka-Stroock-Norris theory on the integrability of an inverse Malliavin matrix under general Hörmander conditions is set forth. Also a method is presented of establishing the smoothness of the transition density with respect to the initial variables, proposed by Stroock, which is very useful in ergodic problems.

§2. Stochastic Derivatives

2.1. Let W_t, $0 \leq t \leq T$, be a Wiener process, and H_2 the space of square-integrable $\sigma(W_t : t \leq T)$-measurable functionals $F \in H_2$. Then (see, for example, Hida 1980) F is representable in the form of a series of multiple (or repeated, see (Itô 1951) and the proof of Krylov in (Veretennikov and Krylov 1976)) Wiener-Itô integrals

$$F = \sum_{m=0}^{\infty} \int_0^T \int_0^{t_m} \cdots \int_0^{t_k} f_m(t_1, \ldots, t_m) dW_{t_1} \ldots dW_{t_m}$$
$$\equiv \sum_{m=0}^{\infty} F^{(m)}(W), \tag{3.1}$$

where the f_m are Borel functions, $F^{(m)}(W)$ denotes the mth term of the series, and the series itself converges in mean-square. The *Malliavin derivative* $\mathcal{L}F$ is defined as follows:

$$\mathcal{L}F := \sum_{m=1}^{\infty} m F^{(m)}(W), \tag{3.2}$$

if the series in (3.2) converges in mean-square. The term "derivative" can be clarified here by defining the family of functionals $(F[\lambda], |\lambda| \leq 1)$:

$$F[\lambda] := \sum_{m=0}^{\infty} \lambda^m F^{(m)}(W), \tag{3.3}$$

see (Zakai 1985). Since $\mathbf{E} F^{(m)}(W) F^{(n)}(W) = 0$ for $m \neq n$, it follows that the series in (3.3) also converges in mean-square. It turns out that the Malliavin derivative $\mathcal{L}F$ can be interpreted as the "present" derivative $\partial F[\lambda]/\partial \lambda \big|_{\lambda=1}$. More precisely, we have the following.

Lemma 3.1 (Zakai 1985). *Let $F \in H_2$. Then $\mathcal{L}F$ is defined if and only if there the limit*

$$\underset{\lambda \uparrow 1}{\text{l.i.m.}} \frac{F - F[\lambda]}{1 - \lambda}$$

exists and this limit coincides with the Malliavin derivative $\mathcal{L}F$.

Warning. If the functional F is given in an explicit form as a function of W, for example, $F = f(W_T)$, then it is not true, in general, that $F[\lambda] = f(\lambda W_T)$. Indeed, for $f(x) = x^2$ we have $f(\lambda W_T) = \lambda^2 W_T^2$, whereas it can be shown that

Chapter 2.III. Malliavin Calculus and Stochastic Differential Equations

$$F[\lambda] = 2\int_0^T (\lambda W_s)d(\lambda W_s) + T = \lambda^2 W_T^2 + T(1-\lambda^2).$$

The properties of the Malliavin derivative are contained in the next lemma.

Lemma 3.2 (Zakai 1985). *Let $F, F_1, F_2 \in H_2$, and suppose that the expressions $\mathcal{L}F, \mathcal{L}F_1, \mathcal{L}F_2$ are defined. Then:*
(a) $\mathbf{E}(\mathcal{L}F) = 0$,
(b) $\mathbf{E}(F_1\mathcal{L}F_2) = \mathbf{E}(F_2\mathcal{L}F_1)$ — *self-adjointness*,
(c) $\mathbf{E}(F\mathcal{L}F) \geq 0$ — *non-negativity*,
(d) $\mathcal{L}$ *is invertible to within a constant* $\mathbf{E}F$,
(e) *there exists the self-adjoint operator* $(\mathcal{L})^{1/2}$:

$$(\mathcal{L})^{1/2} F := \sum_{m=1}^{\infty} \sqrt{m} F^{(m)}(W),$$

(f) $\mathcal{L}$ *is a closed operator.*

In some papers the Malliavin derivative $\mathcal{L}$ is defined to within a sign or a constant factor.

2.2. We are heading towards the very important concept of a directional derivative. Let W_t, $0 \leq t \leq T$, be a Wiener process on a probability space $(\Omega, \mathcal{F}, \mathbf{P})$, and let u_t, $0 \leq t \leq T$, be an $(\mathcal{F}_t^W)$-adapted square-integrable process on $(\Omega, \mathcal{F}, \mathbf{P})$. (Here square-integrability, or even $\mathcal{F}_t^W$-adaptedness may not be required in general, but this will be mentioned briefly below.) We define the family of functionals ($F^{\epsilon,u}$, $\epsilon \geq 0$):

$$F^{\epsilon,u} := \sum_{m=0}^{\infty} F^{(m)}\left(W + \epsilon \int_0^{\cdot} u_s ds\right). \tag{3.4}$$

This definition is understood in the following sense. The process $W_t + \epsilon \int_0^t u_s ds$ is, by Girsanov's theorem, Wiener with respect to the new measure $d\mathbf{P}^{\epsilon,u} = \rho^{\epsilon,u} d\mathbf{P}$, where $\rho^{\epsilon,u}$ is the martingale exponential:

$$\rho^{\epsilon,u} = \exp\left(-\epsilon \int_0^T u_s dW_s - \frac{\epsilon^2}{2}\int_0^T |u_s|^2 ds\right).$$

This means that the square-integrable functional $F^{\epsilon,u}$ is well defined on $(\Omega, \mathcal{F}, \mathbf{P}^{\epsilon,u})$. Hence we can consider $F^{\epsilon,u}$ to be, for example, an integrable functional on $(\Omega, \mathcal{F}, \mathbf{P})$ if $\mathbf{E}(\rho^{\epsilon,u})^2 < \infty$, while in the general case we can consider the sequence $F_N^{\epsilon,u} \equiv (F^{\epsilon,u} \wedge N) \vee (-N)$, $N = 1, 2, \ldots$, where $F_N^{\epsilon,u} \in H_2$ and $F_N^{\epsilon,u}$ converges pointwise to $F^{\epsilon,u}$ as $N \to \infty$. The *directional derivative* is defined as

$$\mathcal{D}_u F := \partial F^{\epsilon,u}/\partial \epsilon\big|_{\epsilon=0}, \tag{3.5}$$

if this derivative exists (for example) in mean-square.

Next, let $(h_s^i,\ 0 \leq s \leq t)_{i=1}^{\infty}$ be an orthonormal basis of non-random functions in $L_2[0,T]$. We define the quadratic form

$$(\mathcal{D}F, \mathcal{D}F) := \sum_{i=1}^{\infty}(\mathcal{D}_{h^i}F)^2, \quad F \in H_2, \tag{3.6}$$

under the assumption that the series in (3.6) converges in L_1. One can prove (Zakai 1985) that the sum of the series in (3.6) does not depend on the choice of the basis $(h^i)_{i=1}^{\infty}$ under the additional assumption

$$\mathbf{E}(\mathcal{D}_u F)^2 \leq K\mathbf{E}\int_0^T u_s^2 ds \ (K = K_F > 0) \tag{3.7}$$

for any $F \in H_2$ and any u for which the derivative $\mathcal{D}_u F$ is defined; here $K = K_F$ does not depend on u.

The *bilinear form* $(\mathcal{D}F_1, \mathcal{D}F_2)$ is defined from the quadratic form $(\mathcal{D}F, \mathcal{D}F)$ in the following manner:

$$(\mathcal{D}F_1, \mathcal{D}F_2) = \frac{1}{4}((\mathcal{D}(F_1+F_2), \mathcal{D}(F_1+F_2)) - (\mathcal{D}(F_1-F_2), \mathcal{D}(F_1-F_2)))$$

or

$$(\mathcal{D}F_1, \mathcal{D}F_2) = \sum_{i=1}^{\infty}(\mathcal{D}_{h^i}F_1)(\mathcal{D}_{h^i}F_2), \quad F_1, F_2 \in H_2.$$

Remark. Let $F \in H_2$ be such that the variable $(\mathcal{D}F, \mathcal{D}F)$ is defined and inequality (3.7) holds. Let $\lambda_i := \mathcal{D}_h^i F$. Then $\mathbf{E}\sum_{i=1}^{\infty}\lambda_i^2 = \mathbf{E}(\mathcal{D}F, \mathcal{D}F) < \infty$; hence

$$\mathbf{E}\left|\int_0^T \sum_{i=1}^{\infty}\lambda_i h_t^i dt\right|^2 = \mathbf{E}\sum_{i=1}^{\infty}\lambda_i^2 < \infty.$$

As a consequence, the stochastic function $\tilde{u}_t := \sum_{i=1}^{\infty}\lambda_i h^i$ is defined, and $\mathbf{E}\int_0^T |\tilde{u}_t|^2 dt < \infty$. It can be proved (Zakai 1985) that the distribution on $C[0,T]$ generated by the process $W_t + \epsilon\int_0^t \tilde{u}_s ds$, $0 \leq t \leq T$, is absolutely continuous with respect to Wiener measure, therefore the functionals $F^{\epsilon,\tilde{u}}$ are defined and, as it turns out, the derivative $\mathcal{D}_{\tilde{u}}F$ is defined, although the process $\tilde{u}_t$ is not adapted in general. Condition (3.7) is assumed to be satisfied for all processes u for which $\mathcal{D}_u F$ is defined, including non-adapted ones. With the help of condition (3.7), one can prove the validity of the equalities

$$(\mathcal{D}F, \mathcal{D}F) \equiv \sum_{i=1}^{\infty}\lambda_i \mathcal{D}_{h^i}F = \mathcal{D}_{\sum_{i=1}^{\infty}\lambda_i h^i}F \equiv \mathcal{D}_{\tilde{u}}F \tag{3.8}$$

(Zakai 1985). By virtue of (3.8) we have

$$\int_0^T |\tilde{u}_t|^2 dt = \sum_{i=1}^{\infty}\lambda_i^2 = (\mathcal{D}F, \mathcal{D}F) = \mathcal{D}_{\tilde{u}}F,$$

therefore (3.8) can be rewritten in the form

$$(\mathcal{D}F, \mathcal{D}F) = \frac{(\mathcal{D}_{\tilde{u}}F)^2}{\int_0^T \tilde{u}_t^2 dt},$$

or, setting $\hat{u}_t := \tilde{u}_t / \|\tilde{u}\|_{L_2[0,T]}$, we obtain

$$(\mathcal{D}F, \mathcal{D}F) = (\mathcal{D}_{\hat{u}}F)^2. \tag{3.9}$$

Thus the variable $(\mathcal{D}F, \mathcal{D}F)$ under condition (3.7) is representable in the form of the square of the derivative of F in some random direction $\hat{u}$. On the other hand (Zakai 1985), for any (not necessarily adapted) process u_t such that $\mathbf{E}\int_0^T u_s^2 ds < \infty$ and for all $\epsilon \in [0, \epsilon_0]$ ($\epsilon_0 > 0$), the distribution $W_t + \epsilon \int_0^T u_s ds$ in $C[0,T]$ is absolutely continuous with respect to Wiener measure, and the variable $\mathcal{D}_u F$ is also defined. We denote the class of such processes by $\mathfrak{M}(F)$. We have the relations $\left(\rho_i := \int_0^T u_s h_s^i ds\right)$

$$\mathcal{D}_u F = \mathcal{D}_{\sum_i \rho_i h^i} F = \sum_{i=1}^{\infty} \mathcal{D}_{\rho_i h^i} F = \sum_{i=1}^{\infty} \rho_i \mathcal{D}_{h^i} F. \tag{3.10}$$

These equations are established by the same arguments as in (3.8). By (3.10) and the Cauchy-Bunyakovskij inequality, we obtain

$$|\mathcal{D}_u F|^2 \leq \left(\sum_{j=1}^{\infty}(\mathcal{D}_{h^j}F)^2\right)\left(\sum_{i=1}^{\infty}\rho_i^2\right) = (\mathcal{D}F, \mathcal{D}F)\int_0^T u_s^2 ds.$$

Hence

$$(\mathcal{D}F, \mathcal{D}F) \geq \frac{(\mathcal{D}_u F)^2}{\int_0^T u_s^2 ds}.$$

Thus we have the following result.

Theorem 3.1 (Zakai 1985). *Let $F \in H_2$, and suppose that condition (3.7) holds. Then*

$$(\mathcal{D}F, \mathcal{D}F) = \sup_{\mathfrak{M}(F)} (\mathcal{D}_u F)^2.$$

In other words, the variable $(\mathcal{D}F, \mathcal{D}F)$ behaves as if it were the square of the gradient of F in random directions.

§3. Rules of the Malliavin Calculus

3.1. The basic *rules of the Malliavin calculus* are the following:

$$\mathcal{L}(F_1 F_2) = F_1 \mathcal{L} F_2 + F_2 \mathcal{L} F_1 - 2(\mathcal{D}F_1, \mathcal{D}F_2), \qquad (3.11)$$

$$\mathcal{L}\phi(F) = \phi'(F)\mathcal{L}F - \phi''(F)(\mathcal{D}F, \mathcal{D}F), \qquad (3.12)$$

where $F, F_1, F_2 \in H_2$, $\phi \in C_b^2$. It is necessary, however, to define the domain of applicability of these equalities more precisely. For this, following Zakai, we introduce the following definitions. Let $\psi : E^n \to E^1$, $\psi \in C^3(E^n)$, and suppose that the function ψ along with all the derivatives up to order three grows no faster than some power of the modulus of the argument. Functionals of the form

$$F = \psi(W_{t_1}, \ldots, W_{t_n}),\ 0 \le t_i \le T,\ 1 \le i \le n$$

are called simple functionals. We set

$$K_p = \{F : \forall F_n \in H_2,\ F_n \text{ is simple and } \|F - F_n\|_{L_p}$$
$$+ \|(\mathcal{D}F_n, \mathcal{D}F_n) - (\mathcal{D}F, \mathcal{D}F)\|_{L_p} + \|\mathcal{L}F_n - \mathcal{L}F\|_{L_p} \to 0,\ n \to \infty.\}$$

All simple functionals are members of $\bigcap_{p \ge 2} K_p$ (Zakai 1985), and multiple Wiener integrals of any order also belong to $\bigcap_{p \ge 2} K_p$ (Shigekawa 1980).

Theorem 3.2 (see Zakai 1985, Proposition 1.4.1). *Let $F_1, F_2 \in K_p$, $p \ge 4$. Then $F_1 F_2 \in K_{p/2}$ and equality* (3.11) *is valid.*

Proof. If F_1, F_2 are simple $F_1 = \psi(W_{t_1}, \ldots, W_{t_n})$, $F_2 = \phi(W_{t_1}, \ldots, W_{t_n})$, then

$$(\mathcal{D}F_1, \mathcal{D}F_2) = \sum_{k=1}^{\infty} \Big(\sum_{i=1}^{n} \int_0^{t_i} h_s^k ds \psi_i'(W_{t_1}, \ldots, W_{t_n})\Big)$$
$$\times \Big(\sum_{j=1}^{n} \int_0^{t_j} h_s^k ds \phi_j'(W_{t_1}, \ldots, W_{t_n})\Big)$$
$$= \sum_{i,j=1}^{n} \Big(\sum_{k=1}^{\infty} \Big(\int_0^{t_i} h_s^k ds\Big)\Big(\int_0^{t_j} h_s^k ds\Big)\Big)$$
$$\times \psi_i'(W_{t_1}, \ldots, W_{t_n})\phi_j'(W_{t_1}, \ldots, W_{t_n}).$$

Since

$$\min(t_i, t_j) = \int_0^T e_{t_i}(s) e_{t_j}(s) ds = \sum_{k=1}^{\infty} (e_{t_i}, h^k)(e_{t_j}, h^k),$$

where $e_t(s) = I(s \le t)$, we obtain

$$(\mathcal{D}F_1, \mathcal{D}F_2) = \sum_{i,j=1}^{n} \min(t_i, t_j) \psi_i'(W_{t_1}, \ldots, W_{t_n}) \phi_j'(W_{t_1}, \ldots, W_{t_n}). \qquad (3.13)$$

Next, we establish the equality

Chapter 2.III. Malliavin Calculus and Stochastic Differential Equations 97

$$\mathcal{L}\psi(W_{t_1},\ldots,W_{t_n}) = \sum_{i=1}^{n} W_{t_i}\psi'_i(W_{t_1},\ldots,W_{t_n})$$
$$-\sum_{i,j=1}^{n} \min(t_i,t_j)\psi''_{ij}(W_{t_1},\ldots,W_{t_n}). \quad (3.14)$$

To this end it is convenient to use another definition of the Malliavin derivative $\mathcal{L}F$, $F \in H_2$, equivalent to the original one. Let $\widetilde{W}_t$, $0 \le t \le T$, be an instance of a Wiener process not depending on W. We define the functionals

$$\widetilde{F}^\epsilon := \sum_{m=0}^{\infty} F^{(m)}(\sqrt{1-\epsilon}W + \sqrt{\epsilon}\widetilde{W}), \ 0 \le \epsilon \le 1.$$

This is well defined since the process $\nu_t := \sqrt{1-\epsilon}W_t + \sqrt{\epsilon}\widetilde{W}_t$ is also Wiener. Then, in view of the equality

$$\mathbf{E}\int_0^T\int_0^{t_m}\cdots\int_0^{t_2} f_m(t_1,\ldots,t_m)d\nu_{t_1}\ldots d\nu_{t_m}|\mathcal{F}_T^W)$$
$$= (1-\epsilon)^{m/2}\int_0^T\int_0^{t_m}\cdots\int_0^{t_2} f_m(t_1,\ldots,t_m)dW_{t_1}\ldots dW_{t_m},$$

we find that

$$\operatorname*{L.I.M.}_{\epsilon\downarrow 0} \frac{\widetilde{F}-E(\widetilde{F}^\epsilon|\mathcal{F}_T^W)}{\epsilon} = \frac{1}{2}\mathcal{L}F. \quad (3.15)$$

The limit on the left hand side of (3.15) is computed for simple functionals with the help of the Itô formula and reduces to formula (3.14). A similar equality also holds for F_2, while for the product F_1F_2 formula (3.14) gives

$$\mathcal{L}(\psi\phi)(W_{t_1},\ldots,W_{t_n}) = \sum_{i=1}^{n} W_{t_i}(\psi\phi)'_i(W_{t_1},\ldots,W_{t_n})$$
$$-\sum_{i,j=1}^{n} \min(t_i,t_j)(\psi\phi)''_{ij}(W_{t_1},\ldots,W_{t_n}). \quad (3.16)$$

The desired relation (3.11) for simple functionals follows from (3.13), (3.16) (3.14) and a similar equality for ϕ. The general case is obtained by passing to the limit with the help of the closure property of the operator $\mathcal{L}$. □

The next result is proved in similar fashion.

Theorem 3.3 (Zakai). *Let $F \in K_p$, $p \ge 4$, $\phi \in C_b^2(E^1)$. Then $\phi(F) \in K_p$ and equation (3.12) is valid.*

§4. Smoothness of the Density (Scheme of the Proof)

4.1.

Theorem 3.4. *Let $F \in H_2$, $A := (\mathcal{D}F, \mathcal{D}F)$, and suppose that the variables $\mathcal{L}F$, A, $\mathcal{L}A$, $(\mathcal{D}A, \mathcal{D}A)$ exist, where $(\mathcal{D}A, \mathcal{D}A) \in L_2$, and $A \neq 0$ a.s. Then the distribution of F is absolutely continuous with respect to Lebesgue measure.*

Scheme of Proof (Zakai 1985). It follows from the theory of the Fourier transform that the estimate

$$|\mathbf{E}\phi'(\xi)| \leq C\|\phi\|_C \quad \forall \phi \in C_b^\infty(E^1), \tag{3.17}$$

entails the existence of a distribution density p of the random variable ξ in the class $L_2(E^1)$ (see, for example, Zakai 1985). We have

$$(\mathcal{D}\phi(F), \mathcal{D}F) = \phi'(F)A,$$

whence

$$\phi'(F) = \frac{1}{2}\frac{-\mathcal{L}(F\phi(F)) + \phi(F)\mathcal{L}F + F\mathcal{L}\phi(F)}{A}.$$

Hence

$$\begin{aligned}2|\mathbf{E}\phi'(F)| &= |\mathbf{E}\phi(F)(-F\mathcal{L}A^{-1} + A^{-1}\mathcal{L}F + \mathcal{L}(FA^{-1}))| \\ &\leq 2\|\phi\|_C \mathbf{E}(|A^{-1}\mathcal{L}F| + |(\mathcal{D}F, \mathcal{D}A^{-1})|).\end{aligned} \tag{3.18}$$

Here, if the expectation on the right hand side is finite, then we obtain (3.17), which implies absolute continuity. But if the right hand side of (3.18) is infinite, then with the help of suitable approximations (see Zakai 1985), we can establish the similar estimate

$$|\mathbf{E}\phi'(F)A(A+\epsilon)^{-1}| \leq C_\epsilon \|\phi\|_C$$

for all $\epsilon > 0$, and although C_ϵ can increase as $\epsilon \downarrow 0$, nevertheless the absolute continuity of the measures

$$v^\epsilon(\cdot) := \mathbf{E} I(F \in \cdot) A(A+\epsilon)^{-1}$$

follows from this estimate, and as $\epsilon \downarrow 0$, by monotonic convergence, we obtain the desired absolute continuity of F. □

Example (Zakai 1985). Let $F := \int_0^T W_s W_{T-s} ds$. Then

$$\begin{aligned}\mathcal{D}_u F &= \int_0^T W_{T-s}\left(\int_0^s u_r dr\right) ds + \int_0^T W_s\left(\int_0^{T-s} u_r dr\right) ds \\ &= 2\int_0^T W_s\left(\int_0^{T-s} u_r dr\right) ds = 2\int_0^T u_s\left(\int_0^{T-s} W_r dr\right) ds.\end{aligned}$$

Hence, by the Cauchy-Bunyakovskij inequality,

$$(\mathcal{D}_u F)^2 \leq 4 \left(\int_0^T u_s^2 ds \right) \left(\int_0^T \left(\int_0^{T-s} W_r dr \right)^2 \right) ds$$

and, moreover,

$$\sup_u \frac{(\mathcal{D}_u F)^2}{\int_0^T u_s^2 ds} = 4 \int_0^T \left(\int_0^{T-s} W_r dr \right) ds$$

(the supremum is attained at $u_t := \int_0^{T-t} W_r dr$). Hence by Theorem 3.1,

$$(\mathcal{D}F, \mathcal{D}F) = 4 \int_0^T \left(\int_0^{T-s} W_r dr \right) ds,$$

and the desired absolute continuity follows from Theorem 3.4.

Remark. Although up to now we have been considering the one-dimensional case, the same methods and arguments enable us to study the distribution density of the vector $F = (F^1, \ldots, F^n)$, $F^i \in H_2$. Here, $A = (\mathcal{D}F, \mathcal{D}F)$ is an $n \times n$ matrix, called the Malliavin matrix. In order that the distribution density p of F should not merely exist, but actually be smooth, $p \in C_b^\infty$, it is sufficient (see (Bismut 1981); (Ikeda and Watanabe 1981) that the inequalities

$$|\mathbf{E}\phi^{(n)}(F)| \leq C_n \|\phi\|_C, \ n = 1, 2, \ldots \quad (3.19)$$

be satisfied for some constants C_n $\forall \phi \in C_b^\infty$. Here $\phi^{(n)}$ denotes the partial derivative of order n along an arbitrarily chosen direction. The same idea that was used in Theorem 3.4 enables us to establish the following important result (see Ikeda and Watanabe 1981).

Theorem 3.5 (Malliavin). *Let $F = (F^1, \ldots, F^n) \in H$, and suppose that the matrix $A = (\mathcal{D}F, \mathcal{D}F)$ is defined and $\mathbf{E}\|A^{-1}\|^p < \infty$ $\forall p \geq 1$. Then the distribution of the vector F has density $p \in C_b^\infty(E^n)$.*

§5. The Bismut Approach 1

5.1. Let $W = (W^1, \ldots, W^m)$, $F = (F^1, \ldots, F^d)^* \in H_2$. In the approach of Zakai and also that of Stroock and Malliavin himself, the most important role in obtaining estimates of type (3.17) or (3.19) is played by the equation $EF_1 \mathcal{L} F_2 = EF_2 \mathcal{L} F_1$, which allows one to transfer the differentiation operator from one functional to another. In other words, this is a kind of integration by parts formula. In the Bismut approach as well the most important role is played by the following integration by parts formula:

$$\mathbf{E}\mathcal{D}_u F = \mathbf{E} F \int_0^T u_s dW_s. \quad (3.20)$$

Here we assume that the $d \times m$-matrix process u_t is $\mathcal{F}_t^W$-adapted and $u \in H_2$. Equality (3.20) is useful and is verified straightforwardly by direct computations for a functional of the form

$$F = \int_0^T \int_0^{t_m} \cdots \int_0^{t_2} f_m(t_1, \ldots, t_m) dW_{t_1} \ldots dW_{t_m}$$

and for a non-stochastic function u. In the general case, the proof of (3.20) is based on the Girsanov formula. The idea is as follows. Let $u. = (u^1, \ldots, u^d)^*$, where each u_t^i is a row vector of dimension $1 \times m$, $F^{\epsilon, u^i} := \sum_{m=0}^{\infty} F^{(m)}(W + \epsilon \int_0^{\cdot} u_s^i ds)$, $\rho^i(\epsilon) := \exp\left(-\epsilon \int_0^T u_s^i dW_s - \frac{\epsilon^2}{2} \int_0^T |u_s^i|^2 ds\right)$. Then by Girsanov's theorem,

$$\mathbf{E} F = \mathbf{E} F^{\epsilon, u^i} \rho^i(\epsilon). \tag{3.21}$$

Formal differentiation of (3.21) at the point $\epsilon = 0$ leads to (3.20). For a rigorous discussion it is sufficient to require in addition that the condition

$$\mathbf{E} \exp\left(c \int_0^T \|u_s\|^2 ds\right) < \infty \tag{3.22}$$

should hold for some $c > 0$ (Zakai 1985).

5.2. Let $G = (G_{ij})_{i,j=1}^d \in H_2$ be some auxiliary matrix functional, $\phi \in C_b^{\infty}$. Then it follows from (3.20) that

$$\mathbf{E} G \phi(F) \int_0^T u_s dW_s = \mathbf{E} \mathcal{D}_u(G\phi(F)) = \mathbf{E} \phi(F) \mathcal{D}_u G + \mathbf{E} \phi'(F) G \mathcal{D}_u F,$$

whence, letting $G := (\mathcal{D}_u F)^{-1}$, we obtain

$$|\mathbf{E} \phi'(F)| \leq \|\phi\|_C \mathbf{E}\left(|\mathcal{D}_u(\mathcal{D}_u F)^{-1}| + \left|(\mathcal{D}_u F)^{-1} \int_0^T u_s dW_s\right|\right). \tag{3.23}$$

This estimate, which is similar to the estimate (3.18) in Zakai's approach, allows one to obtain estimates of the form (3.17) and establish the following results, comparable with Theorems 3.4 and 3.5.

Theorem 3.6 (see Zakai 1985). *Let $F = (F^1, \ldots, F^n)^* \in H_2$, $u_t \in H_2$ an $\mathcal{F}_t^W$-adapted process, and suppose that (3.22) holds for some $c > 0$. Let the variables $\mathcal{D}_u F, \mathcal{D}_u(\mathcal{D}_u F)$ be defined and $\mathcal{D}_u F > 0$ (the matrix $\mathcal{D}_u F$ is positive definite) a.s. Then the distribution F has a density p with respect to Lebesgue measure.*

Theorem 3.7 (Bismut). *Let the conditions of Theorem 3.6 be satisfied, and $\mathbf{E} \|(\mathcal{D}_u F)^{-1}\|^p < \infty \ \forall p \geq 1$. Then $p \in C_b^{\infty}(E^n)$.*

For a comparison of the approaches of Zakai (Stroock) and Bismut, see (Zakai 1985).

The idea itself of this kind of use of Girsanov's change of measure was proposed in (Haussmann 1978) in another problem — the problem of the representation of a martingale.

§6. The Bismut Approach 2. Stochastic Differential Equations

6.1. Let us consider the d-dimensional SDE

$$dx_t = X_0(x_t)dt + \sum_{i=1}^{m} X_i(x_t) \circ dW_t^i, \quad x_0 = x, \quad (3.24)$$

or

$$dx_t = X_0(x_t)dt + X(x_t) \circ dW_t, \quad x_0 = x, \quad (3.25)$$

where $X_0(\cdot), \ldots, X_m(\cdot)$ are vector fields of class $C_b^\infty(E^d)$, $X(\cdot) = X_1(\cdot), \ldots, X_m(\cdot))$, $W = (W^1, \ldots, W^m)^*$ is an m-dimensional Wiener process and the functions $X_0, \ldots, X_m$ and any of their derivatives with respect to x are bounded; here $X_i \circ dW^i$ denotes the Stratonovich stochastic differential. An equivalent way of writing (3.24) and (3.25) in the Itô form is:

$$dx_t = \tilde{X}_0(x_t)dt + \sum_{i=1}^{m} X_t(x_t)dW_t^i, \quad x_0 = x,$$

or

$$dx_t = \tilde{X}_0(x_t)dt + X(x_t)dW_t, \quad x_0 = x,$$

where

$$\tilde{X}_0(x) = X_0(x) + \frac{1}{2} \sum_{i=1}^{m} X_t' X_i(x),$$

that is,

$$\tilde{X}_0^k(x) = X_0^k(x) + \frac{1}{2} \sum_{i=1}^{m} \sum_{j=1}^{d} (\partial X_i^k / \partial x^j) X_i^j(x).$$

The aim of this section is to show the existence of a smooth distribution density x_t, $t > 0$, under conditions of Hörmander-type. This result was established by Kusuoka and Stroock (see (Stroock 1983), also (Norris 1986); (Bell 1987)); in the non-autonomous case, a similar result has been proved by Michel and Chaleyot-Maurel, using the techniques of partial differential equations. We present a scheme of the proof of the theorem on the summability of the inverse Malliavin matrix (the main technical feature) due to Norris (Norris 1986), which significantly simplifies the method of Kusuoka and Stroock. The account is given within the framework of Bismut's approach (one can read about it in Russian in the papers (Veretennikov 1983), (Veretennikov 1984)), although in the present instance Zakai's approach leads to an equivalent result.

We set
$$S_1(z) = \{X_1(z), \ldots, X_m(z)\},$$
$$S_2(z) = \{[X_i(\,\cdot\,), X_j(\,\cdot\,)](z),\ 1 \le i \le m,\ 0 \le j \le m\},$$
$$\cdots\cdots\cdots\cdots\cdots\cdots\cdots\cdots\cdots\cdots$$
$$S_{k+1}(z) = \{[L_1, L_2](z),\ L_1 \in S_k(\,\cdot\,),\ L_2 = X_j(\,\cdot\,),\ 0 \le j \le m\},$$
$$\cdots\cdots\cdots\cdots\cdots\cdots\cdots\cdots\cdots\cdots$$
$$\hat{S}_k(z) = \bigcup_{i=1}^{k} S_k(z),$$
$$\Phi_N(z) = \sum_{L_i \in \hat{S}_N(z)} (\det(L_1, \ldots, L_d))^2, \quad t_0 > 0.$$

The following Hörmander-type condition (H) is assumed to hold: there exists $N > 0$ such that
$$\Phi_N(x) > 0 \qquad (3.26)$$
(x is the initial condition in (3.24)).

Theorem 3.8. *Let condition* (H) *be satisfied. Then for any $T > 0$ the variable x_t has a density $p(0, x; t, \,\cdot\,) \in C_b^\infty$.*

In preparation for this we establish several auxiliary results. In order to use Theorem 3.7, we find an explicit expression for the matrix $\mathcal{D}_u F$, where $F = x_t$, $t > 0$, and $u_s = (u_s^{ki},\ 1 \le k \le d,\ 1 \le i \le m)$, $0 \le s \le T$, is a $(d \times m)$ matrix process. Repeating in part the arguments of Sect. 4, we consider the $(d \times m)$ matrix SDE
$$d(x_s^{\epsilon,u})^k = X_0((x_s^{\epsilon,u})^k)ds + X((x_s^{\epsilon,u})^k)(\circ dW_s + \epsilon u_s^k ds),$$
$$(x_0^{\epsilon,u})^k = x,\ 1 \le k \le d, \qquad (3.27)$$

with $\epsilon \in [0, 1]$. We assume that the process u_s is $\mathcal{F}_s^W$-adapted and is of class $\bigcap_{p \ge 1} L_p([0,T] \times \Omega)$, that is $\mathbf{E} \int_0^T \|u_s\|^p ds < \infty$ for all $p \ge 1$. Under this condition, the legitimacy of differentiating with respect to $\epsilon|_{\epsilon=0}$ the equation
$$\mathbf{E}g(x_t) = \mathbf{E}g((x_t^{\epsilon,u})^k)\rho^k(\epsilon)\ (1 \le k \le d)$$
can be justified, where $g \in C_b^\infty(E^d)$, $\rho^k(\epsilon) = \exp(-\epsilon \int_0^T u_s^k dw_s - (\epsilon^2/2) \times \int_0^T |u_s^k|^2 ds)$, $W_s = (W_s^1, \ldots, W_s^m)^*$ (cf. (3.21)). Moreover, we obtain
$$\mathbf{E}\nabla g(x_t)y_t^u = \mathbf{E}g(x_t) \int_0^t u_s dW_s, \qquad (3.28)$$

where $y_t^u = (y_t^{u^k}) := \partial x_t^{\epsilon,u}/\partial \epsilon\big|_{\epsilon=0} = (\partial x_t^{\epsilon,u^k}/\partial \epsilon\big|_{\epsilon=0})$ is a $d \times d$ matrix. The process y_t^u satisfies the SDE
$$dy_t^u = \nabla X_0(x_t)y_t^u dt + \nabla X(x_t)y_t^u \circ dW_t + X(x_t)u_t dt,$$
$$y_0^u = 0\ (= 0_{d \times d}) \qquad (3.29)$$

Chapter 2.III. Malliavin Calculus and Stochastic Differential Equations   103

($0_{d \times d}$ is the zero element in $E^d \times E^d$). We set $z_t = \partial x_t/\partial x$ (the derivative with respect to the initial data). The matrix process z_t satisfies the equation

$$dz_t = \nabla X_0(x_t) z_t dt + \nabla X(x_t) z_t \circ dW_t, \quad z_0 = I_{d \times d}. \qquad (3.30)$$

By "varying the constants", we therefore obtain the representation (which can be directly verified)

$$y_t^u = z_t \int_0^t z_s^{-1} \sum_{i=1}^m X_i^{\cdot}(x_s) u_s^{\cdot i} ds \qquad (3.31)$$

(cf. the formula for $(\mathcal{D}F, \mathcal{D}F)$). Here the choice $u_s^{\cdot i} = X_i^*(x_s)(z_s^{-1})^*$, which is optimal in a certain sense (see Bismut 1981; Zakai 1985), agrees with the expression for $(\mathcal{D}F, \mathcal{D}F)$. For this choice we obtain

$$y_t^u = z_t C_t,$$

where

$$C_t = \int_0^t z_s^{-1} \sum_{i=1}^m X_i(x_s) X_i^*(x_s) (z_s^{-1})^* ds,$$

and since $\mathbf{E}\sup_{0 \le t \le T} \|z_t\|^p < \infty \; \forall p \ge 1$, the verification of $\mathbf{E}\|(y_t^u)^{-1}\|^p < \infty$ reduces to the verification of the inequality $\mathbf{E}\|C_t^{-1}\|^p < \infty$.

A key role in the proof of this inequality is played by the following.

Lemma 3.3 (Kusuoka and Stroock). *Let $x_0, y_0 \in \mathbb{R}^1$, $a_s, b_s = (b_s^1, \ldots, b_s^m)$, $v_s = (v_s^1, \ldots, v_s^m)$ $\mathcal{F}_s^W$-adapted processes,*

$$x_t := x_0 + \int_0^t a_s ds + \int_0^t b_s dW_s,$$

$$y_t := y_0 + \int_0^t x_s ds + \int_0^t v_s dW_s,$$

let $\tau \le t_0$ be a Markov time, and

$$|a_s|, |b_s|, |x_s|, |v_s| \le C, \quad s \le \tau.$$

Then for any $q > 17$ there exist constants K, c, $\epsilon_0 > 0$ such that for all $\epsilon \in (0, \epsilon_0)$

$$\mathbf{P}\left(\int_0^\tau y_t^2 dt < \epsilon^q, \; \int_0^\tau (x_t^2 + |v_t|^2) dt \ge \epsilon \right) \le K \exp(-c/\epsilon). \qquad (3.32)$$

Scheme of Proof. We set out the scheme of the proof of this lemma given by Norris. We set

$$X_t := \int_0^t x_s ds, \quad M_t := \int_0^t v_s dW_s,$$
$$N_t := \int_0^t y_s v_s dW_s, \quad Q_t := \int_0^t X_s b_s dW_s,$$

and for $\kappa, \delta > 0$

$$B_1(\kappa, \delta) := \left(\omega : \int_0^T y_s^2 |v_s|^2 ds < \kappa; \sup_{t \leq \tau} |N_t| \geq \delta\right),$$
$$B_2(\kappa, \delta) := \left(\omega : \int_0^T |v_s|^2 ds < \kappa; \sup_{t \leq \tau} |M_t| \geq \delta\right),$$
$$B_3(\kappa, \delta) := \left(\omega : \int_0^T X_s^2 |b_s|^2 ds < \kappa; \sup_{t \leq \tau} |Q_t| \geq \delta\right).$$

For any $\kappa, \delta > 0$ we can use martingale methods to establish the inequalities

$$\mathbf{P}(B_i(\kappa, \delta)) \leq 2\exp(-\delta^2/2\kappa), \ 1 \leq i \leq 3. \tag{3.33}$$

We set $q_1 = (q-1)/2$, $q_2 = (q-5)/8$, $q_3 = (q-9)/8$ (all $q_j > 1$), $\delta_i(\epsilon) = \epsilon^{q_i}$ ($1 \leq i \leq 3$) (meanwhile the ϵ_i are still not chosen), and prove that numbers $\kappa_1(\epsilon), \kappa_2(\epsilon), \kappa_3(\epsilon)$ can be chosen such that for some $c, d > 0$

$$\mathbf{P}(B_i) \leq 2\exp(-c/\epsilon) \ (B_i \equiv B_i(\kappa_i(\epsilon), \delta_i(\epsilon))), \tag{3.34}$$

and for $\epsilon \in (0, d)$

$$\left(\int_0^T y_t^2 dt < \epsilon^q; \int_0^T (x_t^2 + |v_t|^2) dt \geq \epsilon\right) \subset \bigcup_{i=1}^3 B_i, \tag{3.35}$$

whence we obtain the desired assertion.

We set $\kappa_1(\epsilon) = C^2 \epsilon^q$ (C is from the condition of the lemma). In view of (3.33), $\kappa_1(\epsilon)$ satisfies (3.34) for $i = 1$. It follows from the equality

$$\int_0^T y_t^2 dt \leq \epsilon^q$$

that

$$\int_0^T y_t^2 |v_t|^2 dt \leq C^2 \epsilon^q \equiv \kappa_1(\epsilon).$$

Hence, assuming that $\omega \notin B_1$, we obtain

$$\sup_{t \leq \tau} |N_t| = \sup_{t \leq \tau} \left|\int_0^t y_s x_s ds\right| \leq \delta_1(\epsilon) = \epsilon^{q_1},$$

and also

$$\sup_{t \leq \tau} \left|\int_0^T y_s x_s ds\right| \leq \left(t_0 \int_0^T y_s^2 x_s^2 ds\right)^{1/2} \leq t_0^{1/2} C \epsilon^{q/2}.$$

Chapter 2.III. Malliavin Calculus and Stochastic Differential Equations

Since
$$y_s dy_s = y_s x_s ds + y_s v_s dw_s,$$
it follows that
$$\sup_{t \leq \tau} \left| \int_0^t y_s dy_s \right| \leq (1 + t_0^{1/2} C \epsilon^{1/2}) \epsilon^{q_1}.$$

We have
$$y_t^2 = y_0^2 + 2 \int_0^t y_s dy_s + \int_0^t |v_s|^2 ds,$$
therefore
$$\int_0^\tau \int_0^t |v_s|^2 ds dt = \int_0^\tau y_t^2 dt - \tau y_0^2 - 2 \int_0^\tau \left(\int_0^t y_s dy_s \right) dt$$
$$\leq \epsilon^q + 2t_0 (1 + t_0^{1/2} C \epsilon^{1/2}) \epsilon^{q_1}.$$

Hence, for small enough $\epsilon > 0$
$$\int_0^\tau \int_0^t |v_s|^2 ds dt \leq (2t_0 + 1) \epsilon^{q_1}. \tag{3.36}$$

It follows from (3.36) that for all $t \in (0, \tau]$
$$t \int_0^{\tau-t} |v_s|^2 ds = \int_0^t dr \int_0^{\tau-t} |v_s|^2 ds \leq \int_0^t \int_0^{\tau-r} |v_s|^2 ds dr$$
$$\leq \int_0^\tau \int_0^{\tau-r} |v_s|^2 ds dr = \int_0^\tau \int_0^r |v_s|^2 ds dr \leq (2t_0 + 1) \epsilon^{q_1}.$$

Therefore for any $t \in (0, \tau]$
$$\int_0^\tau |v_s|^2 ds \leq (2t_0 + 1) \epsilon^{q_1} / t + C^2 t,$$
which for $t = (2t_0 + 1)^{1/2} \epsilon^{q_1/2}$ gives
$$\int_0^\tau |v_s|^2 ds \leq \kappa_2(\epsilon) := (1 + C^2)(2t_0 + 1)^{1/2} \epsilon^{q_1/2}. \tag{3.37}$$

Since $q_1/2 > 1$, (3.37) implies that for sufficiently small $\epsilon > 0$
$$\mathbf{P}\left(\int_0^\tau y_t^2 dt < \epsilon^q; \int_0^\tau |v_s|^2 ds \geq \epsilon \right) \leq K \exp(-c/\epsilon),$$
therefore, this part is a simplified version of the basic part of the proof in (Ikeda and Watanabe 1981, proof of Theorem 5.8.2).

Now let $\omega \notin B_2$. Then $\sup_{t \leq \tau} |M_t| < \delta_2 \equiv \epsilon^{q_2}$. We find from the inequality $\int_0^\tau y_t^2 dt < \epsilon^q$ that
$$\text{mes}(t : t \in [0, \tau], |y_t| \geq \epsilon^{q/3}) \leq \epsilon^{q/3}$$

and hence
$$\text{mes}(t : t \in [0,T], |y_0 + X_t| \geq \epsilon^{q/3} + \epsilon^{q_2}) \leq \epsilon^{q/3}.$$

This means that for any $t \in [0, \tau]$ there exists $s \in [0, \tau]$, $|s - t| \leq \epsilon^{q/3}$ such that
$$|y_0 + X_s| \leq \epsilon^{q/3} + \epsilon^{q_2}$$

(here the continuity of X_t is also used and the inequality sign is not strict). Consequently
$$|y_0 + X_t| \leq |y_0 + X_s| + \int_s^t |x_r| dr \leq (1 + C)\epsilon^{q/3} + \epsilon^{q_2}.$$

Since for $t = 0$ the above inequality yields $|y_0| \leq (1 + C)\epsilon^{q/3} + \epsilon^{q_2}$, it follows that
$$|X_t| \leq 2(1 + C)\epsilon^{q/3} + 2\epsilon^{q_2} \leq 3\epsilon^{q_2}$$

for sufficiently small $\epsilon > 0$.

By the Itô formula
$$\int_0^\tau x_t^2 dt = \int_0^\tau x_t dX_t = x_\tau X_\tau - \int_0^\tau X_t a_t dt - \int_0^\tau b_t dW_t. \tag{3.38}$$

We have ($\omega \notin B_2$)
$$|x_\tau X_\tau| \leq 3C\epsilon^{q_2},$$
$$\left|\int_0^\tau X_t a_t dt\right| \leq 3Ct_0 \epsilon^{q_2},$$
$$\int_0^\tau |X_t|^2 |b_t|^2 dt \leq 9C^2 t_0 \epsilon^{2q_2} =: \kappa_3(\epsilon).$$

For such $\kappa_3(\epsilon)$ and for $\omega \notin B_2 \cup B_3$ we obtain
$$|Q_\tau| = \left|\int_0^\tau X_t b_t dW_t\right| \leq \delta_3 = \epsilon^{q_3},$$

therefore it follows from (3.38) that
$$\int_0^\tau x_t^2 dt \leq 3C(1 + t_0)\epsilon^{q_2} + \epsilon^{q_3} \leq 2\epsilon^{q_3}$$

if $\epsilon > 0$ is sufficiently small. This, in fact, proves the "second part" of the required estimate. Finally, we find that for $\omega \notin B_1 \cup B_2 \cup B_3$
$$\int_0^\tau (x_t^2 + v_t^2) dt \leq 2\epsilon^{q_3} + \epsilon^{q_1/2}(C^2 + 2t_0 + 1) \leq \epsilon$$

if $\epsilon > 0$ is sufficiently small. Lemma 3.3 is proved. $\square$

As has already been said, this lemma is the crucial feature of the proof of the theorem, and we omit the conclusion of the proof (which can be read in many places, see (Stroock 1983); (Norris 1986); (Bell 1987), and also formally under more stringent conditions of Hörmander and Ikeda-Watanabe type in (Ikeda and Watanabe 1981); these more stringent conditions can be weakened by taking into account Lemma 3.3) the exposition of the theorem under general conditions with additional asymptotic assertions see in [N. Ikeda, S. Watanabe, Stochastic Differential Equations and Diffusion Processes, 2-nd edition. North-Holland, Amsterdam, 1989].

§7. Stochastic Differential Equations (Smoothness of the Density with Respect to Inverse Variables)

7.1. Stroock (Stroock 1981) proposed a method allowing one to establish the smoothness of the density distribution of a solution of an SDE with respect to inverse variables. Here we present a result that is not the most general in this direction.

Theorem 3.9. *Let condition* (H) *of Sect. 6 be satisfied. Then the density $p(t,x,y)$ of the solution of the SDE (3.24) for $t > 0$ is of class C_b^∞ with respect to x and each derivative $\mathcal{D}_x^\alpha p(t,x,y)$ is bounded uniformly with respect to $y \in E^d$.*

Scheme of Proof. Let $U \subset E^d \times E^d$ be a bounded domain, $\operatorname{diam} U \leq R$, $h \in C_0^\infty(E^d \times E^d)$, $\operatorname{supp} h \subset U$. By the Sobolev imbedding theorem, to prove the desired assertion, it is sufficient to establish the bound

$$\left| \int (\mathcal{D}_x^\alpha p(t,x,y)) h(x,y) dx dy \right| \leq C_{t,q,\alpha,R} \|h\|_{L_q(U)} \qquad (3.39)$$

with any multi-index α, any $q > 1$, any $h \in C_0^\infty * (U)$ and constant $C_{t,q,\alpha,R}$. By the commutativity of the integration operator and the generalized differentiation operator $\mathcal{D}_x^\alpha$, inequality (3.39) is equivalent to:

$$\left| \int (\mathcal{D}_x^\alpha \mathbf{E} h(z, x_t^x))\Big|_{z=x} dx \right| \leq C_{t,q,\alpha,R} \|h\|_{L_q(U)}, \qquad (3.40)$$

where x_t^x is a solution of the SDE (3.24). We can represent the expression $\mathcal{D}_x^\alpha \mathbf{E} h(z, x_t)$ for $t > 0$ in the form

$$\mathcal{D}_x^\alpha \mathbf{E} h(z, x_t^x) = \mathbf{E} h(z, x_t^x) G_\alpha(\bar{x}_t^\alpha), \qquad (3.41)$$

where $\bar{x}_t^\alpha$ is a process in some extended state space (it includes, in particular, the derivatives of x_t^x in certain directions with respect to x up to order $|\alpha|$ inclusively and the inverse matrix y_t^{-1} of the first order derivatives, $y_t = \partial x_t^x / \partial x$ (see, for example (Veretennikov 1984)) which is summable to any power with respect to the measure $\mathbf{P}$; and G_α is some function of class C^∞, growing no faster than polynomially, along with any derivative of it. By Hölder's inequality we obtain

$$|\mathbf{E}h(x, x_t^x)G_\alpha(\bar{x}_t^\alpha)|$$
$$\leq \left(\mathbf{E}|h(x, x_t^x)|^{\frac{q+1}{2}}\right)^{\frac{2}{q+1}} \left(\mathbf{E}|G_\alpha(\bar{x}_t^\alpha)|^{\frac{q+1}{q-1}}\right)^{\frac{q-1}{q+1}}$$
$$\leq C(t, q, \alpha)\left(\int |h(x,y)|^{\frac{q+1}{2}} p(t,x,y)dy\right)^{\frac{2}{q+1}}.$$

Here, again by Hölder's inequality,

$$\left|\int (D_x^\alpha \mathbf{E}h(z, x_t^x))\big|_{z=x} dx\right|$$
$$\leq C(t, q, \alpha) \int \left(\int |h(x,y)|^{\frac{q+1}{2}} p(t,x,y) dy\right)^{\frac{2}{q+1}} dx$$
$$\leq C(t, q, \alpha)(\operatorname{diam} U)^{\frac{q-1}{q+1}} \left(\int_U |h(x,y)|^{\frac{q+1}{2}} p(t,x,y) dx dy\right)^{\frac{2}{q+1}}$$
$$\leq C(t, q, \alpha)(\operatorname{diam} U)^{\frac{q-1}{q+1}} \left(\int_U |h(x,y)|^q dx dy\right)^{1/q}$$
$$\times \left(\int_U p(t, x, y)^{\frac{q+1}{q-1}} dx dy\right)^{\frac{q-1}{2(q+1)^2}}$$
$$\leq C_{t,q,\alpha,R} \|h\|_{L_q(U)},$$

which, in fact, proves the required estimate (3.40) and with it, Theorem 3.9 as well. □

Remark. The result of Theorem 3.9 turns out to be useful in studying certain asymptotics, for example, large deviations for recurrent processes satisfying Hörmander-type conditions. In these and other limit theorems, it is useful to have in mind also the results of Part I, Sect. 5 on the support of the measure of a diffusion process in the space of trajectories.

References[*]

Bell, D.R. (1987): The Malliavin Calculus. Longman, Harlow. Zbl. 678.60042
Bichteler, K., Jacod, J. (1983): Calcul de Malliavin pour les diffusions avec sauts: existence d'une densité dans le cas unidimensionel. Lect. Notes Math. **986**, 132–157. Zbl. 525.60067

---

[*] For the convenience of the reader, references to reviews in Zentralblatt für Mathematik (Zbl.), compiled using the MATH database, have, as far as possible, been included in this bibliography.

Bismut, J.-M. (1981): Martingales, the Malliavin calculus and hypo-ellipticity under general Hörmander's conditions. Z. Wahrscheinlichkeitstheorie verw. Gebiete 56, pp. 469–505. Zbl. 445.60049

Bismut, J.-M. (1984a): The Atiyah-Singer theorems: A probabilistic approach. I. The index theorem. II. The Lefschetz fixed point formulas. J. Funct. Anal. 57, 56–99. Zbl. 538.58033; Ibid 57, 329-348. Zbl. 556.58027

Bismut, J.-M. (1984b): Jump Processes and Boundary Processes. Stochastic Analysis. Proc. Taniguchi Int. Symp. on Stochastic Analysis, Katata and Kyoto, 1982. North-Holland, Amsterdam Math. Libr. 32, pp. 53–104. Zbl. 556.60057

Bismut, J.-M., Michel, D. (1981): Diffusions conditionnelles. I. Hypoellipticité partielle. J. Funct. Anal. 44, 174–211. Zbl. 475.60061

Bismut, J.-M., Michel, D. (1982): Diffusions conditionnelles. II. Générateur conditionnel. Application au filtrage. J. Funct. Anal. 45, 274–292. Zbl. 475.60062

Gaveau, B., Trauber, P. (1982): L'intégrale stochastique comme opérateur de divergence dans l'espace fonctionnel. J. Funct. Anal. 46, 230–238. Zbl. 488.60068

Haussmann, U.G. (1978): Functionals of Itô processes as stochastic integrals. SIAM J. Control and Optimization 16, 252–269. Zbl. 375.60070

Hida, T. (1980): Brownian Motion. Springer, Berlin Heidelberg New York (transl. from the Japanese original, Tokyo 1975). Zbl. 327.60049; Zbl. 432.60002

Hörmander, L. (1967): Hypo-elliptic second order differential equations. Acta Math. 119, 147–171. Zbl. 156,107

Ikeda, N., Watanabe, S.(1981): Stochastic Differential Equations and Diffusion Processes. North-Holland, Amsterdam. Zbl. 495.60005

Itô, K. (1951): Multiple Wiener integrals. J. Math. Soc. Japan 3, 157–169. Zbl. 44,122

Malliavin, P. (1978a): C^k-hypoellipticity with degeneracy. Parts I, II. Stochastic Analysis. Proc. Int. Conf. Evanston/Il. 1978. Academic Press, New York, pp. 199–214; 327–340. Zbl. 449.58022; Zbl. 449.58023

Malliavin, P. (1978b): Stochastic calculus of variation and hypoelliptic operators. Proc. Int. Symp. on Stochast. Different. Equations Kyoto, 1976. Kinokuniya, Tokyo, pp. 195–214. Zbl. 411.60060

Malliavin, P. (1982): Sur certaines intégrales stochastiques oscillantes. C.R. Acad. Sci., Paris, Ser. I 295, No. 3, 295–300. Zbl. 526.60050

Michel, D. (1984): Conditional laws and Hörmander's conditions. Stochastic Analysis. Proc. Taniguchi Internat. Symp. on Stochastic Analysis, Katata and Kyoto, 1982. North-Holland, Amsterdam Math. Libr. 32, pp. 387–408. Zbl. 573.60035

Norris, N. (1986): Simplified Malliavin calculus. Lect. Notes Math. 1204, pp. 101–130. Zbl. 609.60066

Nualart, D., Zakai, M. (1986): Generalized stochastic integrals and the Malliavin calculus. Probab. Theory Relat. Fields. 73, No. 2, 255–280. Zbl. 601.60053

Oleinik, O.A. (= Olejnik, O.A.), Radkevich, E.V. (1971): Second-order equations with non-negative characteristic form. Itogi Nauki, Mat. Anal. VINITI, Moscow, Zbl. 217,415. [English transl.: Plenum, New York London (1973)]

Shigekawa, I. (1980): Derivatives of Wiener functionals and absolute continuity of induced measures. J. Math. Kyoto Univ. 20, 263–289. Zbl. 476.28008

Skorokhod, A.V. (1975): On a generalization of the stochastic integral. Teor. Veroyatn. Primen. 20, No. 2, 223–238. [English transl.: Theor. Probab. Appl. 20, 219–233 (1975)] Zbl. 333.60060

Stroock, D.W. (1981): The Malliavin calculus and its application to second order parabolic differential equations. Part 1. Math. Syst. Theory 14, 25–65. Zbl. 474.60061

Stroock, D.W. (1983): Some applications of stochastic calculus to partial differential equations. Lect. Notes Math. 976, 267–382. Zbl. 494.60060

Veretennikov, A.Yu. (1983): A probabilistic approach to hypoellipticity. Usp. Mat. Nauk SSSR, Ser. Mat. *38*, No. 3, 113–125. [English transl.: Russ. Math. Surv. *38*, No. 3, 127–140 (1983)] Zbl. 528.60050

Veretennikov, A.Yu. (1984): Probabilistic problems in the theory of hypoellipticity. Izv. Akad. Nauk SSSR, Ser. Mat. *48*, No. 6, 1151–1170. [English transl.: Math. USSR, Izv. *25*, 455–473 (1985)] Zbl. 578.60056

Veretennikov, A.Yu., Krylov, N.V. (1976): Explicit formulas for solutions of stochastic equations. Mat. Sb., Nov. Ser. *100(142)*, No. 2, 266–284. [English transl.: Math. USSR, Izv. *29*, 239–256 (1978)] Zbl. 353.60059

Watanabe, S. (1984): Lectures on stochastic differential equations and Malliavin calculus. Tata Institute of Fundamental Research. Springer, Berlin Heidelberg New York. Zbl. 546.60054

Watanabe, S (1987): Analysis of Wiener functionals (Malliavin calculus) and its applications to heat kernels. Ann. Probab. *15*, No. 1, 1–39. Zbl. 633.60077

Zakai, M. (1985): The Malliavin calculus. Acta Appl. Math. 3, 175–207. Zbl. 553.600593

Chapter 3
Stochastic Calculus on Filtered Probability Spaces

R. Sh. Liptser and A. N. Shiryaev

I. Elements of the General Theory of Stochastic Processes

§1. Kolmogorov's Axioms and the Stochastic Basis

1.1. Kolmogorov's axioms of probability theory give an approach, generally accepted at the present time, to the mathematical description of probabilistic-statistical phenomena. The problem of axiomatizing probability theory was formulated in the Sixth problem of D. Hilbert in his famous address of 8 August 1900 at the Second International Congress of Mathematicians in Paris. Hilbert, who included probability theory in physics (as was the general acceptance at the time), formulated the 6th problem as "The mathematical statement of the axioms of physics" (Hilbert 1901) : "Closely connected with investigations into the foundations of geometry is the problem of axiomatizing the construction within that same framework of those physical disciplines in which mathematics already plays a distinguished role: in the first place, this is the theory of probability and mechanics.

As regards the axioms of probability theory, I would think it desirable that, in parallel with the logical foundation of this theory, there be a rigorous and satisfactory development of the method of mean values in mathematical physics, in particular, in the kinetic theory of gases."

Various attempts at an axiomatic statement of probability theory were made by many authors: Bohlmann (Bohlmann 1908), Bernshtein (Bernshtein 1917), von Mises (von Mises 1919), (von Mises 1928), and Lomnicki (Lomnicki 1923). In (Kolmogorov 1929) and, in a final form, in (Kolmogorov 1933), Kolmogorov, under the influence of the general ideas of set theory, measure and integration theory, and also the metric theory of functions, formulated the concept of a probability model, or a system of axioms (which from the logical point of view is not, generally speaking, the only possible one). This proved to be simple and at the same time sufficiently general to embrace the classical branches of probability theory and also to pave the way for the development of new chapters in it, in particular, the theory of stochastic processes.

At the basis of Kolmogorov's system of axioms lies the concept of a *probability space*
$$(\Omega, \mathcal{F}, \mathbf{P})$$

consisting of three objects, where $\Omega = \{\omega\}$ is the space of elementary events (outcomes); $\mathcal{F}$ is a collection of subsets $A \subseteq \Omega$ called events, forming a σ-algebra; $\mathbf{P}$ is a countably additive non-negative normalized set function, $\mathbf{P} = \mathbf{P}(A)$ $(\mathbf{P}(\sum^{\infty} A_i) = \sum^{\infty} \mathbf{P}(A_i)$ if $A_i \in \mathcal{F}$ and $A_i \cap A_j = \emptyset, i \neq j, 0 \leq \mathbf{P}(\cdot) \leq 1$, $\mathbf{P}(\Omega) = 1$), called the probability of event A.

1.2. At the basis of the stochastic calculus considered in this chapter lies the concept of a *stochastic basis*, which is the specification of a probability space $(\Omega, \mathcal{F}, \mathbf{P})$ by giving on it a non-decreasing flow of σ-algebras (a *filtration*) $\mathbf{F} = (\mathcal{F}_t)_{t \geq 0}$, $\mathcal{F}_0 \subseteq \mathcal{F}_s \subseteq \mathcal{F}_t \subseteq \mathcal{F}$, $s \leq t$, where $\mathcal{F}_t$ is interpreted as the σ-algebra of events "observable" on the time interval $[0, t]$.

Definition. A *stochastic basis*

$$\mathcal{B} = (\Omega, \mathcal{F}, \mathbf{F} = (\mathcal{F}_t)_{t \geq 0}, \mathbf{P})$$

is a probability space $(\Omega, \mathcal{F}, \mathbf{P})$ endowed with a filtration $\mathbf{F} = (\mathcal{F}_t)_{t \geq 0}$ which, by definition, is a non-decreasing family of σ-algebras, $\mathcal{F}_0 \subseteq \mathcal{F}_s \subseteq \mathcal{F}_t \subset \mathcal{F}$ that is right-continuous, i.e., $\mathcal{F}_t = \mathcal{F}_{t+}$, where $\mathcal{F}_{t+} = \bigcap_{s > t} \mathcal{F}_s$. A stochastic basis is said to be complete or to satisfy the usual conditions if the σ-algebra $\mathcal{F}$ is completed by the sets of $\mathbf{P}$-measure zero and each $\mathcal{F}_t$ contains the sets of $\mathcal{F}$ with $\mathbf{P}$-measure zero.

The presence of a filtration (or a flow of σ-algebras) $\mathbf{F} = (\mathcal{F}_t)_{t \geq 0}$ makes it possible to introduce a number of new concepts and specific objects, which comprise the foundation of stochastic calculus. Among them are: Markov times, adapted processes, optional and predictable σ-algebras, martingales and local martingales, semimartingales *et al.*

§2. Stopping Times, Adapted Stochastic Processes, Optional and Predictable σ-Algebras. Classification of Stopping Times

2.1. Let $\mathcal{B} = (\Omega, \mathcal{F}, \mathbf{F}, \mathbf{P})$ be a stochastic basis.

Definition 1. By a *Markov time* or a *stopping time* is meant a mapping $\tau : \Omega \to R_+$ such that for each $t \in R_+$

$$\{\omega : \tau(\omega) \leq t\} \in \mathcal{F}_t.$$

We associate with each stopping time the two σ-algebras:

$$\mathcal{F}_\tau = \{A \in \mathcal{F}, \text{ such that } A \cap \{\tau \leq t\} \in \mathcal{F}_t, t \in \mathbb{R}_+\},$$
$$\mathcal{F}_{t-} = \sigma\{\mathcal{F}_0, A \cap \{t < \tau\}, \text{ where } t \in \mathbb{R}_+ \text{ and } A \in \mathcal{F}_t\}.$$

If $\tau \equiv t$, then $\mathcal{F}_\tau = \mathcal{F}_t$ and $\mathcal{F}_{\tau-} = \mathcal{F}_{t-}$, where

$$\mathcal{F}_{t-} = \begin{cases} \mathcal{F}_0, & \text{if } t = 0 \\ \bigvee_{s<t} \mathcal{F}_s, & \text{if } t \in (0, \infty]. \end{cases}$$

If τ is a stopping time, then
1) $\tau + t$ is a stopping time, $t \in R_+$;
2) $\mathcal{F}_{\tau-} \subseteq \mathcal{F}_\tau$ and τ is $\mathcal{F}_{\tau-}$-measurable;
3) if $A \in \mathcal{F}_\tau$, then

$$\tau_A(\omega) = \begin{cases} \tau(\omega), & \omega \in A \\ +\infty, & \omega \notin A \end{cases}$$

is a Markov time.

The assumption of right continuity of the family $\mathbf{F}$ leads to the following assertion: $\tau = \tau(\omega)$ is a Markov time if and only if $\{\tau < t\} \in \mathcal{F}_t$ for each $t \in R_+$.

If (τ_n) is a sequence of stopping times, then $\sigma = \inf \tau_n$ and $\tau = \sup \tau_n$ are also stopping times and $\mathcal{F}_\sigma = \bigcap \mathcal{F}_{\tau_n}$.

Associated with any two stopping times σ and τ are the stochastic intervals:

$$[\sigma, \tau] = \{(\omega, t) : t \in R_+, \sigma(\omega) \leq t \leq \tau(\omega)\},$$
$$[\sigma, \tau[= \{(\omega, t) : t \in R_+, \sigma(\omega) \leq t < \tau(\omega)\},$$
$$]\sigma, \tau] = \{(\omega, t) : t \in R_+, \sigma(\omega) < t \leq \tau(\omega)\},$$
$$]\sigma, \tau[= \{(\omega, t) : t \in R_+, \sigma(\omega) < t < \tau(\omega)\}.$$

The set $[\tau] = [\tau, \tau]$ is called the *graph* of the stopping time τ.

Definition 2. A stochastic process is a family $X = (X_t(\omega))_{t \in R_+}$ of mappings of Ω to the set R (if instead of R, some set E is taken, then it is said that X is an E-valued stochastic process). For fixed $\omega \in \Omega$ the mapping $t \to X_t(\omega)$ is called a trajectory or a sample function of the process X. A stochastic process X given on a stochastic basis $\mathcal{B} = (\Omega, \mathcal{F}, \mathbf{F}, \mathbf{P})$ is called adapted (or $\mathbf{F}$-adapted) if X_t is $\mathcal{F}_t$-measurable for each $t \geq 0$.

Examples of Markov times:
a) Let $X = (X_t(\omega))_{t \in R_+}$ be a right-continuous adapted stochastic process, and let B be an open subset of $\bar{R}$. Then

$$\tau = \inf(t : X_t \in B)$$

is a Markov time.

b) If X is an adapted right-continuous process with non-decreasing trajectories and $a \in \bar{R}$, then the time

$$\tau = \inf(t : X_t \geq a)$$

is a Markov time.

2.2. In the general theory of stochastic processes an important role is played by the optional and the predictable σ-algebras of subsets of the space $\Omega \times \mathbb{R}_+ = \{(\omega, t) : \omega \in \Omega, t \in \mathbb{R}_+\}$.

Definition 3. The *optional* σ-algebra $\mathcal{O}$ of subsets of $\Omega \times \mathbb{R}_+$ is the σ-algebra generated by all the adapted processes $Y = Y(t, \omega)$, $t \in \mathbb{R}_+$, $\omega \in \Omega$, considered as mappings $Y : (\omega, t) \to \mathbb{R}$, with trajectories in the space $\mathbf{D}$ (of right-continuous functions having left-hand limits).

It is shown that the optional σ-algebra $\mathcal{O}$ can also be defined as the σ-algebra generated by all the stochastic intervals $[0, \tau[$, where the τ are Markov times.

Definition 4. The *predictable* σ-algebra $\mathcal{P}$ of subsets of $\Omega \times \mathbb{R}_+$ is the σ-algebra generated by all the adapted processes $Y = Y(\omega, t)$, $t \in \mathbb{R}_+$, $\omega \in \Omega$, with left-continuous trajectories.

The predictable σ-algebra $\mathcal{P}$ is also generated by any of the collections of sets:

a) $A \times \{0\}$, $A \in \mathcal{F}_0$ and $[0, \tau]$, where the τ are stopping times;
b) $A \times \{0\}$, $A \in \mathcal{F}_0$ and $A \times]s, t]$, where $s < t$ and $A \in \mathcal{F}_s$.

Clearly, $\mathcal{P} \subseteq \mathcal{O}$. An $\mathcal{O}$-measurable or a $\mathcal{P}$-measurable stochastic process is called, respectively, *optional* or *predictable*.

The role of optional and predictable sets and processes appears especially in the construction (presented below) of stochastic integrals, which is an essential ingredient of the theory of stochastic calculus.

2.3.

Definition 5. A Markov time τ is called *predictable* if the stochastic interval $[0, \tau[$ is a predictable set.

This definition is equivalent to the following: there exists a non-decreasing sequence of Markov times $(\tau_n)_{n \geq 1}$ such that

$$\tau(\omega) = \lim \tau_n(\omega)$$

and $\tau_n(\omega) < \tau(\omega)$, $\omega \in \{\omega : \tau(\omega) > 0\}$. (Such a sequence (τ_n) is called an *announcing* sequence for τ.)

Let us note a number of properties of predictable (Markov) times:

a) if (τ_n) are predictable times, then $\sup \tau_n$ is also a predictable time;
b) if (τ_n) are predictable times and $\sigma = \inf \tau_n$, where $\bigcup \{\sigma = \tau_n\} = \Omega$, then σ is also a predictable time;
c) if τ is a predictable time and $A = \mathcal{F}_{\tau-}$, then the time

$$\tau_A(\omega) = \begin{cases} \tau(\omega), & \omega \in A, \\ +\infty, & \omega \notin A \end{cases}$$

is also predictable;

Chapter 3.I. Elements of the General Theory of Stochastic Processes   115

d) if τ is a stopping time and is the début of the predictable set A, that is, $\tau(\omega) = \inf\{t : (t,\omega) \in A\}$ and $[\tau] \subseteq A$, then τ is a predictable time.

An important and difficult result of the general theory of stochastic processes is the following:

Theorem 3.1 (on sections). *Let $\mathcal{B}$ be a complete stochastic basis and let $A \in \mathcal{P}$ (or $A \in \mathcal{O}$). Then for each $\epsilon > 0$ a predictable (or Markov) time τ exists such that its graph $[\tau]$ belongs to the set A and*

$$\mathbf{P}(\omega : \tau(\omega) = \infty \text{ and there exists } t \in \mathbb{R}_+ \text{ with } (\omega, t) \in A) \leq \epsilon.$$

A typical example of an application of this theorem, which is the basic tool for proving results on "uniqueness", is the following.

Theorem 3.2. *Let X and Y be predictable (or optional) processes. For each predictable (Markov) time τ let*

$$X_\tau = Y_\tau(\{\tau < \infty\}; \mathbf{P}\text{-}a.s.),$$

that is, $\mathbf{P}(X_\tau \neq Y_\tau, \tau < \infty) = 0$. Then the processes X and Y are indistinguishable, in other words, the random set $\{X \neq Y\} = \{(\omega, t) : X_t(\omega) \neq Y_t(\omega)\}$ is negligible, that is, $\mathbf{P}(\omega : \exists t \in \mathbb{R}_+ \text{ with } (\omega, t) \in \{X \neq Y\}) = 0$.

2.4. With a view to classifying Markov times we introduce the following two concepts.

Definition 6. A Markov time τ is called *accessible* if there is a sequence $(\tau_n)_{n \geq 1}$ of predictable times such that $[\tau] \subseteq \bigcup_n [\tau_n]$.

Definition 7. A Markov time σ is called *totally inaccessible* if $\mathbf{P}(\sigma = \tau < \infty) = 0$ for each predictable time τ.

Theorem 3.3. *For each Markov time τ there exists one and only one (to within $\mathbf{P}$-negligibility) pair of Markov times σ and γ such that*
  (1) *σ is an accessible time,*
  (2) *γ is totally inaccessible,*
  (3) *$[\tau] = [\sigma] \cup [\gamma]$ and $[\sigma] \cap [\gamma] = \emptyset$.*

In the sense indicated in this theorem, one can say that the classes of predictable and totally inaccessible times are "orthogonal".

If a stochastic process X in the space $\mathbf{D}$ is predictable, then there exists a sequence of predictable times exhausting the entire set of jump times. Moreover, $\Delta X_\tau = 0(\{\tau < \infty\} : \mathbf{P}\text{-a.s.})$ for all totally inaccessible times $(\Delta X_\tau = X_\tau - X_{\tau-})$.

Definition 8. If for a process X in the space $\mathbf{D}$ we have $\Delta X_\sigma = 0$ ($\{\sigma < \infty\}$; $\mathbf{P}$-a.s.) for each predictable time σ, then we say that X is *left quasicontinuous*.

Theorem 3.4. *Let the process X belong to the space $\mathbf{D}$. Then the following three conditions are equivalent:*

a) X is left quasicontinuous;

b) there exists a sequence of totally inaccessible times exhausting the entire set of jumps of the process;

c) for each increasing sequence $(\tau_n)_{n \geq 1}$ with limit τ ($\tau = \lim \tau_n$) we have

$$\lim X_{\tau_n} = X_\tau \quad (\{\tau < \infty\}; \ \mathbf{P}\text{-a.s.}).$$

2.5. Let us assume that the stochastic process $X = (X_t)_{t \in \mathbb{R}_+}$, is $\mathcal{F} \otimes \mathcal{B}(\mathbb{R}_+)$-measurable. Clearly $\mathcal{P} \subseteq \mathcal{O} \subseteq \mathcal{F} \otimes \mathcal{B}(\mathbb{R}_+)$. It turns out that under natural assumptions (of the type $X \geq 0$, $|X| \leq C$) ensuring the existence of conditional mathematical expectations, one and only one (to within indistinguishability) optional process ${}^oX$) (or predictable process ${}^pX$) exists such that

$$({}^oX)_\tau = \mathbf{E}(X_\tau | \mathcal{F}_\tau) \quad (\{\tau < \infty\}; \ \mathbf{P}\text{-a.s.})$$

for each Markov time τ (or

$$({}^pX)_\sigma = \mathbf{E}(X_\sigma | \mathcal{F}_{\sigma-}) \quad (\{\sigma < \infty\}; \ \mathbf{P}\text{-a.s.})$$

for each predictable time σ).

The processes ${}^oX$ and ${}^pX$ are called the *optional* and the *predictable projections* of the process X.

If X is a left quasicontinuous process, then starting from Definition 8 it follows that ${}^pX = X_-$.

§3. Martingales and Local Martingales

3.1.

Definition 1. An adapted process $X = (X_t)_{t \geq 0}$ defined on a stochastic basis $\mathcal{B} = (\Omega, \mathcal{F}, \mathbf{F}, \mathbf{P})$ and with trajectories from the space $\mathbf{D}$ is called a *martingale*, (or *submartingale*, or *supermartingale*) if

$$X_s = \mathbf{E}(X_t | \mathcal{F}_s) \text{ (or } X_s \leq \mathbf{E}(X_t | \mathcal{F}_s), \text{ or } X_s \geq \mathbf{E}(X_t | \mathcal{F}_s)), \ s \leq t, \ \mathbf{P}\text{-a.s.}$$

A number of fundamental properties of the above processes are given below. These are mainly due to Doob.

Theorem 3.5. Let X be a supermartingale such that an integrable random variable Y with $X_t \geq \mathbf{E}(Y | \mathcal{F}_t)$, $t \in \mathbb{R}_+$, exists. Then

a) $X_t \to X_\infty(\mathbf{P}\text{-a.s.})$, where X_∞ is some (finite) random variable (called terminal);

b) if σ and τ are two Markov times, then the random variables X_σ and X_τ are integrable and

$$X_\sigma \geq \mathbf{E}(X_\tau | \mathcal{F}_\sigma)$$

Chapter 3.I. Elements of the General Theory of Stochastic Processes   117

on the set $\{\sigma \leq \tau\}$. In particular, the "stopped" process $X^\tau = (X_{t\wedge\tau})_{t\geq 0}$ is again a supermartingale.

We denote by $\overline{\mathcal{M}}(\mathcal{B})$ or $\overline{\mathcal{M}}$ the class of all martingales given on the stochastic basis $\mathcal{B}$. We denote by $\mathcal{M}$ the subclass of $\overline{\mathcal{M}}$ consisting of uniformly integrable martingales X, that is, martingales for which the family of random variables $(X_t)_{t\in\mathbb{R}_+}$ is uniformly integrable:

$$\sup_{t\in\mathbb{R}_+} \mathbf{E}(|X_t|I(|X_t| > N)) \to 0, \quad N \to \infty.$$

Theorem 3.6. 1) *If $X \in \mathcal{M}$, then an integrable random variable X_∞ exists, such that*

$$X_t \to X_\infty \ (\mathbf{P}\text{-}a.s.), \quad t \to \infty,$$
$$X_t = \mathbf{E}(X_\infty|\mathcal{F}_t), \quad t \geq 0, \ (\mathbf{P}\text{-}a.s.)$$
$$\mathbf{E}|X_t - X_\infty| \to 0, \quad t \to \infty,$$
$$X_{\tau\wedge\sigma} = \mathbf{E}(X_\sigma|\mathcal{F}_\tau) \ (\mathbf{P}\text{-}a.s.)$$

for any Markov times τ and σ. The process $(X_t)_{0\leq t\leq \infty}$ is a martingale.

2) *If Y is an integrable random variable, one and only one martingale $X \in \mathcal{M}$ exists such that*

$$X_t = \mathbf{E}(Y|\mathcal{F}_t), \quad t \geq 0 \ (\mathbf{P}\text{-}a.s.).$$

3) *If $(\tau_n)_{n\geq 1}$ is an increasing sequence of Markov times, then*

$$\lim_n X_{\tau_n} = \mathbf{E}\{X_{\lim \tau_n} | \vee \mathcal{F}_{\tau_n}\} \ (\mathbf{P}\text{-}a.s.).$$

In particular, if τ is a predictable time, then

$$X_{\tau-} = \mathbf{E}(X_\tau|\mathcal{F}_{\tau-}) \ (\mathbf{P}\text{-}a.s.).$$

4) *Each uniformly integrable martingale X belongs to the class $\mathbf{D}$, that is, the family of random variables $\{X_\tau : \tau$ are finite-valued Markov times$\}$ is uniformly integrable.*

In the general theory of stochastic processes an important role is played by the class $\mathcal{H}^2$ of square-integrable martingales, that is martingales such that

$$\sup_{t\geq 0} \mathbf{E}X_t^2 < \infty.$$

It is obvious that $\mathcal{H}^2 \subseteq \mathcal{M}$. A martingale $X \in \mathcal{H}^2$ if and only if the "terminal" variable X_∞ in Theorem 3.6 is square-integrable. In this case $\mathbf{E}|X_t - X_\infty|^2 \to 0$ as $t \to \infty$.

If $X \in \mathcal{H}^2$, then we have the inequality (Kolmogorov, Doob)

$$\mathbf{E}(\sup_{t\in\mathbb{R}_+} X_t^2) \leq 4 \sup_{t\in\mathbb{R}_+} \mathbf{E}X_t^2 = 4\mathbf{E}X_\infty^2.$$

An interesting and useful characterization of the class of uniformly integrable martingales is given by the following theorem.

Theorem 3.7. *Let X be an adapted process of class $\mathbf{D}$ with a terminal variable X_∞ (that is, X_∞ is the limit $\lim_{t\to\infty} X_t$ ($\mathbf{P}$-a.s.)). Then X is a uniformly integrable martingale if and only if for each Markov time τ the variable X_τ is integrable and $\mathbf{E} X_\tau = \mathbf{E} X_0$.*

3.2. To introduce the concept of a "local martingale" we present the following

Definition 2. Let $\mathcal{G}$ be some class of stochastic processes. We will say that a process X is of class $\mathcal{G}_{\text{loc}}$ if and only if a non-decreasing sequence of Markov times $(\tau_n)_{n\geq 1}$ (depending on X in general) exists such that $\lim \tau_n = \infty$ and each stopped process $X^{\tau_n} \in \mathcal{G}$. The sequence $(\tau_n)_{n\geq 1}$, in this case, is called *localizing* for the process X.

Proceeding from this definition and the classes $\mathcal{M}$ and $\mathcal{H}^2$, the classes $\mathcal{M}_{\text{loc}}$ and $\mathcal{H}^2_{\text{loc}}$ — the classes of local martingales and locally square-integrable martingales – are introduced. Clearly we have

$$\mathcal{M} \subseteq \overline{\mathcal{M}} \subseteq \mathcal{M}_{\text{loc}}, \quad \mathcal{H}^2 \subseteq \mathcal{H}^2_{\text{loc}}.$$

In the case of discrete time and an obviously defined "discrete" stochastic basis $\mathcal{B} = (\Omega, \mathcal{F}, \mathbf{F} = (\mathcal{F}_n)_{n\geq 0}, \mathbf{P})$, one can give the following characterization of the class of local martingales.

Theorem 3.8. *Let $X = (X_n)_{n\geq 0}$ be an adapted process. Then X is a local martingale if and only if*

a) $\mathbf{E}|X_0| < \infty$,
b) $\mathbf{E}(|X_n| \mid \mathcal{F}_{n-1}) < \infty$ ($\mathbf{P}$-a.s.), $n \geq 1$,
c) $\mathbf{E}(X_n \mid \mathcal{F}_{n-1}) = X_{n-1}$ ($\mathbf{P}$-a.s.), $n \geq 1$.

(Here $\mathbf{E}(X_n \mid \mathcal{F}_{n-1})$ is an "extended" conditional mathematical expectation, defined not only under the assumption that $\mathbf{E}|X_n| < \infty$, but also under the assumption that $\mathbf{E}(|X_n| \mid \mathcal{F}_{n-1}) < \infty$ ($\mathbf{P}$-a.s.).)

§4. Increasing Processes. Doob-Meyer Decomposition. Compensators

4.1. Along with martingales and local martingales, an essential role in the general theory of stochastic processes is played by the concept of an "increasing process".

Chapter 3.I. Elements of the General Theory of Stochastic Processes

Definition 1. An adapted process $A = (A_t)_{t \geq 0}$ of class **D** with $A_0 = 0$ is called an *increasing* process if each trajectory $t \to A_t(\omega)$ is a non-decreasing function. The class of increasing processes is denoted by $\mathcal{V}^+$.

We denote by $\mathcal{V}$ the class $\mathcal{V}^+ \ominus \mathcal{V}^+$, that is, the class of all adapted processes with trajectories in **D** having finite variation (on each interval $[0, t]$, $t \in \mathbb{R}_+$).

If $A \in \mathcal{V}$ and H is an optional process (hence $t \to H_t(\omega)$ is a Borel function for each ω), then one can define the ("integral") process $H \circ A$ or $\int_0^{\cdot} H_s dA_s$ by setting

$$(H \circ A)_t(\omega) = \begin{cases} \int_0^t H_s(\omega) dA_s(\omega) & \text{if } \int_0^t |H_s(\omega)| d[\text{Var} A]_s(\omega) < \infty, \\ \infty & \text{otherwise,} \end{cases}$$

where all the integrals under consideration are understood as Lebesgue-Stieltjes integrals.

If $A \in \mathcal{V}$ and H is an optional process, then $B = H \circ A$ is an optional process. If, in addition, A and H are predictable, then so is the process B.

In the class $\mathcal{V}^+$ (or $\mathcal{V}$) we distinguish the subclass $\mathcal{A}^+$ (or $\mathcal{A}$) of processes for which $\mathbf{E} A_\infty < \infty$ (or $\mathbf{E}[\text{Var} A]_\infty < \infty$). Clearly $\mathcal{A} = \mathcal{A}^+ \ominus \mathcal{A}^+$.

Using the localization procedure introduced above, one can introduce the classes $\mathcal{V}^+_{\text{loc}}$, $\mathcal{V}_{\text{loc}}$, $\mathcal{A}^+_{\text{loc}}$ and $\mathcal{A}_{\text{loc}}$.

In the next theorem we gather a number of properties of the above processes.

Theorem 3.9. 1) *If A is a predictable process of class $\mathcal{V}$, then there exists a localizing sequence $(\sigma_n)_{n \geq 1}$ of Markov times such that $[\text{Var} A]_{\sigma_n} \leq n$ (**P**-a.s.) and, in particular, $A \in \mathcal{A}_{\text{loc}}$.*

2) *If a local martingale X is of class $\mathcal{V}$, then $X \in \mathcal{A}_{\text{loc}}$.*

3) *If $A \in \mathcal{A}$, M is a bounded martingale and τ is a stopping time, then*

$$\mathbf{E}(M \circ A)_\tau = \mathbf{E}(M_\tau A_\tau).$$

If, in addition, A is a predictable process, then

$$\mathbf{E}(M_- \circ A)_\tau = \mathbf{E}(M_\tau A_\tau).$$

4) *If $A \in \mathcal{A}_{\text{loc}}$ and M is a locally bounded local martingale then $MA - M \circ A \in \mathcal{M}_{\text{loc}}$. If, in addition, A is a predictable process, then the process $MA - M_- \circ A \in \mathcal{M}_{\text{loc}}$.*

4.2. The result given below, known as the "Doob-Meyer decomposition", plays a crucial role in all of stochastic calculus.

Theorem 3.10. *If X is a submartingale of class D (that is, the family (X_τ), where the τ are finite-valued Markov times, is uniformly integrable), then there exists a unique (to within indistinguishability) increasing integrable predictable* process A with $A_0 = 0$ such that $m = X - A$ is a uniformly integrable martingale.

An important corollary of the "Doob-Meyer decomposition" is the following.

Theorem 3.11. *Let $A \in \mathcal{A}^+_{\text{loc}}$. Then there exists a predictable process A^p in $\mathcal{A}^+_{\text{loc}}$ such that each of the following equivalent assertions holds:*
  a) $A - A^p \in \mathcal{M}_{\text{loc}}$;
  b) $\mathbf{E}A^p_\tau = \mathbf{E}A_\tau$ *for all stopping times τ;*
  c) $\mathbf{E}[(H \circ A^p)_\infty] = \mathbf{E}[(H \circ A)_\infty]$ *for all non-negative predictable processes H.*

The process A^p is called the *compensator* of the process A. (Sometimes it is called the *predictable compensator*, or the *dual predictable projection* of the process A.)

Theorem 3.11 admits an obvious generalization also to the case of processes $A \in \mathcal{A}_{\text{loc}}$ for which the corresponding predictable process $A^p \in \mathcal{A}_{\text{loc}}$ (such that $A - A^p \in \mathcal{M}_{\text{loc}}$) is also called the compensator of the process A.

Let us note a number of simple properties of compensators.

Theorem 3.12.
1) *If the process $A \in \mathcal{A}_{\text{loc}}$ is also predictable, then $A^p = A$.*
2) *If $A \in \mathcal{A}_{\text{loc}}$ and τ is a stopping time, then $(A^\tau)^p = (A^p)^\tau$.*
3) *If $A \in \mathcal{A}_{\text{loc}}$, then the predictable projection ${}^p(\Delta A) = \Delta(A^p)$.*
4) *If $A \in \mathcal{A}_{\text{loc}}$, then A is a local martingale if and only if $A^p = 0$.*
5) *If $A \in \mathcal{M}_{\text{loc}} \cap \mathcal{V}$ and H is a predictable process with $H \circ A \in \mathcal{A}_{\text{loc}}$, then $H \circ A \in \mathcal{M}_{\text{loc}}$ and $A^p = 0$.*

Example. If $A = (A_t)_{t \geq 0}$ is a Poisson process with parameter λ, then A is in $\mathcal{A}^+_{\text{loc}}$ and is locally bounded. Its compensator $A^p_t = \lambda t$, $t \geq 0$.

§5. Random Measures. Integral Random Measures

5.1. The concept of a "random measure" is one of the basic concepts in stochastic calculus making possible a detailed study of the jumps of stochastic processes the trajectories of which belong to the space **D** of right-continuous functions with left-hand limits.

When defining the random measure (in what follows, $\mu = \mu(\omega; dt, dx)$) the stochastic basis $\mathcal{B} = (\Omega, \mathcal{F}, \mathbf{F} = (\mathcal{F}_t)_{t \geq 0}, \mathbf{P})$ is assumed given and an auxiliary space $(E, \mathcal{E})$ assumed to be a Blackwell space (in what follows it is sufficient to take $E = \mathbb{R}$: an example of such a space is a Polish space with its Borel σ-algebra).

Definition 1. A *random measure* on $\mathbb{R}_+ \times E$ is a family $\mu = (\mu(\omega; dt, dx); \omega \in \Omega)$ of non-negative measures (on $\mathbb{R}_+ \times E$, $\mathcal{B}(\mathbb{R}_+) \otimes \mathcal{E}$) satisfying identically the condition

Chapter 3.I. Elements of the General Theory of Stochastic Processes

$$\mu(\omega; \{0\} \times E) = 0.$$

We introduce the notation:

$$\tilde{\Omega} = \Omega \times \mathbb{R}_+ \times E, \quad \tilde{\mathcal{O}} = \mathcal{O} \otimes \mathcal{E}, \quad \tilde{\mathcal{P}} = \mathcal{P} \otimes \mathcal{E}.$$

A function $W = W(\omega, t, x)$ defined on $\tilde{\Omega}$ that is $\tilde{\mathcal{O}}$- (or $\tilde{\mathcal{P}}$-)measurable is called *optional* (or *predictable*).

We associate with each random measure μ and optional function W an (integral) process $W * \mu$ by defining

$$W * \mu_t = \begin{cases} \int_{[0,t] \times E} W(\omega, s, x) \mu(\omega; \, ds, dx), & \text{if } \int_{[0,t] \times E} |W| \mu(\omega; \, ds, dx) < \infty \\ \infty, & \text{otherwise.} \end{cases}$$

Definition 2. A random measure μ is called *optional* (or *predictable*) if for each optional (or predictable) function W the process $W * \mu$ is optional (respectively, predictable). A random measure μ is called *integrable* if $I * \mu \in \mathcal{A}^+$. An optional measure μ is called $\tilde{\mathcal{P}}$-σ-*finite* if a $\tilde{\mathcal{P}}$-measurable decomposition (A_n) of the space $\tilde{\Omega}$ exists such that $I_{A_n} * \mu \in \mathcal{A}^+$.

Example. We associate with each counting process $A \in \mathcal{V}^+$ a random measure μ by setting

$$\mu(\omega; \, dt \times \{1\}) = dA_t(\omega).$$

In this case the measure is μ-optional; it is predictable if and only if A is a predictable process.

The next theorem (on the existence of a "compensator" for a random measure) is a generalization of Theorem 3 from the preceding section.

Theorem 3.13. *Let μ be an optional $\tilde{\mathcal{P}}$-σ-finite random measure. Then there exists a predictable random measure ν, called the compensator of μ, such that each of the following equivalent assertions is satisfied:*

a) *for each $\tilde{\mathcal{P}}$-measurable function W on $\tilde{\Omega}$ with $|W| * \mu \in \mathcal{A}_{\text{loc}}^+$ the process $|W| * \nu \in \mathcal{A}_{\text{loc}}^+$ and $W * \nu$ is the compensator of $W * \mu$, that is, the process $W * \mu - W * \nu \in \mathcal{M}_{\text{loc}}$.*

b) $E(W * \nu)_\infty = E(W * \mu)_\infty$ *for each non-negative $\tilde{\mathcal{P}}$-measurable function W on $\tilde{\Omega}$.*

*Such a measure ν is unique (**P**-a.s.).*

5.2. In the class of random measures an especially important role is played by integer-valued random measures.

Definition 3. A random measure μ is said to be *integer-valued* if
a) $\mu(\omega; \{t\} \times E) \leq 1$ identically;
b) $\mu(\,\cdot\,, A) \in \{0, 1, \ldots, \infty\}$ for each $A \in \mathcal{B}(\mathbb{R}_+) \otimes \mathcal{E}$;
c) μ is optional and $\tilde{\mathcal{P}} - \sigma$-finite.

As an example of such a measure one can take the measure μ^X of jumps of an adapted process X with trajectories in $\mathbf{D}$:

$$\mu^X(\omega;\, dt, dx) = \sum_s I_{\{\Delta X_s(\omega) \neq 0\}} \epsilon_{(s, \Delta X_s(\omega))}(dt, dx),$$

where $\epsilon_{(a)}$ is the Dirac measure at the point a.

A fundamental example of an integer-valued random measure is a Poisson measure.

Definition 4. An integer-valued random measure μ is called an *extended Poisson measure* if

a) the measure $m(A) = E\mu(A)$, $A \in \mathcal{B}(\mathbb{R}_+) \otimes \mathcal{E}$ is σ-finite;

b) for each $s \in \mathbb{R}_+$ and each $A \in \mathcal{B}(\mathbb{R}_+) \otimes \mathcal{E}$ such that $A \subseteq (s, \infty) \times E$ and $\mu(A) < \infty$, the random variable $\mu(\,\cdot\,, A)$ does not depend on the σ-algebra $\mathcal{F}_s$.

The measure m is called the intensity of the measure μ and if m is such that $m(\{t\} \times E) = 0$ for each $t \in \mathbb{R}_+$, then μ is called a Poisson random measure, and in addition, homogeneous if $m(dt, dx) = dt \times F(dx)$, where F is a positive σ-finite measure on $(E, \mathcal{E})$.

It is not hard to verify that for an extended Poisson measure with an intensity m its compensator $\nu(\omega;\,\cdot\,) = m(\,\cdot\,)$.

Another example of an integer-valued measure is the measure $\mu^{(T,X)}$ of the multivariate point process

$$(T, X) = (T_n, X_n)_{n \geq 1},$$

defined by the formula

$$\mu(\omega; dt, dx) = \sum_{n \geq 1} I(T_n < \infty) \epsilon_{(T_n, X_n)}(dt, dx),$$

where the $(T_n)_{n \geq 1}$ are Markov times such that

$$T_1 > 0, \quad T_n < T_{n+1} \text{ on } \{T_n < \infty\}, \quad T_{n+1} = T_n \text{ on } \{T_n = \infty\},$$

and the X_n are random elements with values in E such that

$$X_n \in E \text{ on } \{T_n < \infty\} \text{ and } X_n = \delta \text{ on } \{T_n = \infty\},$$

where δ is some "fictitious" point not belonging to E and $\{X_n \in C\} \in \mathcal{F}_{T_n}$ for each $C \in \mathcal{E}$.

§6. Locally Square-Integrable Martingales. The Quadratic Characteristic

6.1. Let $\mathcal{H}^2$ and $\mathcal{H}^2_{\mathrm{loc}}$ be the classes of square and locally square-integrable martingales. If M and N are of class $\mathcal{H}^2_{\mathrm{loc}}$, then one can define a predictable process, denoted by $\langle M, N \rangle$, of class $\mathcal{V}$ such that $MN - \langle M, N \rangle \in \mathcal{M}_{\mathrm{loc}}$.

This process $\langle M, N \rangle$ is called the *predictable quadratic covariation* or *quadratic characteristic* of the pair (M, N).

If $M \in \mathcal{H}^2$, then in accordance with the Doob-Meyer decomposition applied to the submartingale M^2 there is a predictable process $\langle M \rangle \in \mathcal{A}^+$ (which is unique to within stochastic indistinguishability) such that $M^2 - \langle M \rangle \in \mathcal{M}$. Hence, by the localization procedure the existence is established of a predictable process $\langle M \rangle$ or $\langle M, M \rangle$ of class $\mathcal{V}^+$, called the quadratic characteristic of M, such that $M^2 - \langle M \rangle \in \mathcal{M}_{\mathrm{loc}}$. It is immediately deduced from this result that the quadratic characteristic $\langle M, N \rangle$ can be defined by the formula

$$\langle M, N \rangle = \frac{1}{4}(\langle M + N, M + N \rangle - \langle M - N, M - N \rangle).$$

The fundamental example of a continuous square-integrable martingale is a Wiener process $W = (W_t)_{t \geq 0}$ defined on some stochastic basis $\mathcal{B} = (\Omega, \mathcal{F}, \mathbf{F}, \mathbf{P})$.

Definition 2. A *Wiener process* W on $\mathcal{B}$ is an adapted continuous process with $W_0 = 0$, $\sigma^2(t) = \mathbf{E}W_t^2$, $\mathbf{E}W_t = 0$, $t \geq 0$, and $W_t - W_s$ not depending on $\mathcal{F}_s$, $0 \leq s \leq t$.

If the variance $\sigma^2(t) = t$, then W is called the *standard Wiener process*.

A Wiener process W is a square-integrable martingale with quadratic characteristic $\langle W \rangle_t = \sigma^2(t)$.

§7. Decomposition of Local Martingales

7.1.

Definition 1. Two local martingales M and N are said to be *orthogonal* if their product $MN \in \mathcal{M}_{\mathrm{loc}}$. If $M, N \in \mathcal{H}^2$, then this definition is equivalent to $\langle M, N \rangle = 0$, which in large part explains the term "orthogonality" in the general situation.

Definition 2. Let $\mathcal{M}^c_{\mathrm{loc}}$ be the class of continuous (that is, with continuous trajectories) local martingales. A local martingale N is called a *purely discontinuous* local martingale ($N \in \mathcal{M}^d_{\mathrm{loc}}$) if N is orthogonal to any continuous martingale M.

In the general theory of stochastic processes the next two decompositions of local martingales are well known.

First Decomposition. Let $a > 0$. Each local martingale M admits a (generally speaking, non-unique) decomposition

$$M = M_0 + M' + M'',$$

where $M', M'' \in \mathcal{M}_{\text{loc}}$, $M'_0 = M''_0 = 0$, with M' of finite variation and $|\Delta M''| \leq a$ (consequently, $M'' \in \mathcal{H}^2_{\text{loc}}$).

Second Decomposition. Each local martingale M admits a decomposition

$$M = M_0 + M^c + M^d$$

(unique to within stochastic indistinguishability), where $M^c_0 = M^d_0 = 0$, M^c is a continuous local martingale ($M^c \in \mathcal{M}^c_{\text{loc}}$), and M^d is a purely discontinuous local martingale ($M^d \in \mathcal{M}^d_{\text{loc}}$).

The process M^c is called the *continuous part* of M, and M^d the *purely discontinuous part* of M.

II. Semimartingales. Stochastic Integrals

§1. Semimartingales. Quadratic Variation. Quasimartingales

1.1.

Definition 1. An adapted stochastic process $X = (X)_{t\geq 0}$ given on a stochastic basis $\mathcal{B} = (\Omega, \mathcal{F}, \mathbf{F} = (\mathcal{F}_t)_{t\geq 0}, \mathbf{P})$ with trajectories in $\mathbf{D}$ is called a *semimartingale* $(X \in S)$ if it admits a representation in the form

$$X = X_0 + M + A, \tag{3.1}$$

where X_0 is a finite-valued $\mathcal{F}_0$-measurable random variable, M is a local martingale $(M \in \mathcal{M}_{\text{loc}})$ with $M_0 = 0$ and A is a process of bounded variation $(A \in \mathcal{V})$ with $A_0 = 0$.

In the case when the representation (3.1) exists with a predictable process A, the semimartingale X is called *special* $(X \in S_p)$. Let us note that for special semimartingales, the representation in the form (3.1) with a predictable process A is unique. It is often called the *canonical decomposition* of the special martingale. Every semimartingale X with bounded jumps $|\Delta X| \leq c$ is special and in its canonical decomposition $X = X_0 + A + M$ we have $|\Delta M| \leq 2c$, $|\Delta A| \leq c$. In particular, if X is a continuous semimartingale, then in its canonical decomposition the processes M and A are also continuous.

Although it is not immediately clear from the definition given above, the class of semimartingales possesses many "pleasant" properties. For example, it is stable with respect to many transformations: a "stopped" semimartingale is again a semimartingale, "localization" preserves a semimartingale, a semimartingale remains a semimartingale under a time change, with respect to an absolutely continuous change of measure, and under reduction of filtrations.

An important property of semimartingales is that they form the maximal class of processes with respect to which one can integrate bounded predictable processes with properties naturally demanded of an integral of the type satisfied in the Lebesgue theorem on passing to the limit under the integral sign. (For further details see (Dellacherie 1980)).

1.2. Although the representation (3.1) of a semimartingale is not unique, the continuous martingale part X^c is unique. In other words, if $X = X_0 + M_1 + A_1$ and $X = X_0 + M_2 + A_2$ are two representations, then $M_1^c = M_2^c = X^c$.

1.3. An important characteristic of a semimartingale X is its *quadratic variation*

$$[X, X]_t = \langle X^c \rangle_t + \sum_{s\leq t} (\Delta X_s)^2.$$

If X and Y are two semimartingales, then $[X,Y]$ denotes their *quadratic covariation*, defined by the formula

$$[X,Y] = \frac{1}{4}([X+Y, X+Y] - [X-Y, X-Y]).$$

In the next theorem we present a number of properties of $[X.X]$ and $[X,Y]$.

Theorem 3.14. *If X and Y are semimartingales, then*
1) $[X,Y] \in \mathcal{V}$, $[X,X] \in \mathcal{V}^+$;
2) $\Delta[X,Y] = \Delta X \Delta Y$;
3) *if* $Y \in \mathcal{V}$, *then* $[X,Y]^c = 0$;
4) *if* $Y \in \mathcal{V}$ *and is continuous, then* $[X,Y] = 0$;
5) *if* $X \in \mathcal{M}_{\mathrm{loc}}^c$, *then* $[X,X] = [X,X]^c = \langle X \rangle$;
6) *if* $X \in \mathcal{M}_{\mathrm{loc}}$, $|X| \leq c$, $X_0 = 0$ *and Y is a predictable process of bounded variation, then* $[X,Y] \in \mathcal{M}_{\mathrm{loc}}$;
7) *if* $X, Y \in \mathcal{M}_{\mathrm{loc}}$, *then* $XY - X_0 Y_0 - [X,Y] \in \mathcal{M}_{\mathrm{loc}}$;
8) *if* $X \in \mathcal{M}_{\mathrm{loc}}$, *then* $[X,X]^{1/2} \in \mathcal{A}_{\mathrm{loc}}^+$;
9) *if* $X \in \mathcal{M}_{\mathrm{loc}}^c$, $Y \in \mathcal{M}_{\mathrm{loc}}^d$, *then* $[X,Y] = 0$;
10) $[X,Y] = \langle X,Y \rangle = 0$ *if* $X \in \mathcal{M}_{\mathrm{loc}}^c$, $Y \in \mathcal{M}_{\mathrm{loc}}^c$, *moreover, X and Y are orthogonal.*

1.4. Related to the concept of a semimartingale is that of a *quasimartingale*. Let $X = (X_t)_{t \geq 0}$ be an adapted process with trajectories in the space $\mathbf{D}$. For $n \geq 1$ and $0 \leq t_1 < t_2 \ldots < t_n$ we set

$$\mathrm{Var}(X; t_1, \ldots, t_n) = \sum_{1 \leq i \leq n-1} |\mathbf{E}(X_{t_{i+1}} - X_{t_i} | \mathcal{F}_{t_i})| + |X_{t_n}|$$

and

$$\mathrm{Var}(X) = \sup_{n, t_1, \ldots, t_n} \mathbf{E}\, \mathrm{Var}(X; t_1, \ldots, t_n).$$

If $\mathrm{Var}(X) < \infty$, then the process X is called a *quasimartingale* $(X \in Q)$. Note that if $X \in \mathcal{M}$, then

$$\mathrm{Var}(X) = \sup_t \mathbf{E}|X_t| < \infty$$

and hence $\mathcal{M} \subseteq Q$. If $X \in \mathcal{A}$, then

$$\mathrm{Var}(X) \leq 2\mathbf{E} \int_0^\infty |dX_s| < \infty$$

and hence $\mathcal{A} \in Q$. It can be deduced from this that every special semimartingale is a local quasimartingale. Indeed, these two classes of processes coincide.

§2. Construction of Stochastic Integrals with Respect to Semimartingales

2.1. In the case when $X \in \mathcal{V}$ and H is a bounded process, the integral process $(H \circ X)_t = \int_0^t H_s dX_s$ has been defined in Part I, Sect. 4. The aim of the present section is to define the integral $H \circ X$ for the case when X is a semimartingale.

Since a semimartingale does not necessarily have locally bounded variation, the definition of the integral $(H \circ X)_t$ as a Lebesgue-Stieltjes integral for each elementary outcome is inapplicable.

If the semimartingale $X = X_0 + M + A$, where $M \in \mathcal{M}_{\text{loc}}$ and $A \in \mathcal{V}$, then in the definition of the "integral" $H \circ X$ for locally bounded processes H, one can proceed as follows: we set by definition

$$H \circ X = H \circ M + H \circ A, \tag{3.2}$$

where $H \circ M$ is a "stochastic integral", subject to further definition, with respect to a local martingale, and $H \circ A$ is the integral with respect to A already defined (in Part I, Sect. 4). Of course, we must make sure that (3.2) is well defined in the sense that it does not depend on the form of the representation $X = X_0 + M + A$.

In defining the integral $H \circ M$ with respect to a local martingale M, one can adopt two approaches, based on either the first or the second decomposition of M (see Part I, Sect. 7).

If it is based on the first decomposition, the fundamental difficulty will be in defining the stochastic integral $H \circ M$ with respect to a local martingale $M \in \mathcal{H}^2$.

Let us describe the corresponding construction of the integral $H \circ M$.

Let $\mathcal{E}$ be the set of all processes $H = (H_t(\omega))_{t \geq 0}$ of the form: either $H = Y I_{[0]}$, where Y is a bounded $\mathcal{F}_0$-measurable random variable or $H = Y I_{]r,s]}$, $r < s$, where Y is a bounded $\mathcal{F}_r$-measurable random variable.

For such functions H we set by definition

$$(H \circ M)_t = \begin{cases} 0, & \text{if } H = Y I_{[0]} \\ Y(M_{s \wedge t} - M_{r \wedge t}), & \text{if } H = Y I_{]r,s]}. \end{cases} \tag{3.3}$$

Along with the notation $(H \circ M)_t$ we shall also use the notation $\int_0^t H_s dM_s$ or $\int_{(0,t]} H_s dM_s$.

Theorem 3.15 *Let $M \in \mathcal{H}^2_{\text{loc}}$. Then the mapping $H \to H \circ M$ defined by formula (3.3) for functions of class $\mathcal{E}$ can be extended to the class of functions*

$$L^2_{\text{loc}}(M) = \{H : H \text{ predictable and } H^2 \circ \langle M \rangle \in \mathcal{A}^+_{\text{loc}}\}$$

so that this extension, also denoted $H \circ M$, possesses the following properties:
1) *$H \circ M$ is an adapted process with trajectories in $\mathbf{D}$;*
2) *the map $H \to H \circ M$ is linear, that is $(aH + K) \circ M = aH \circ M + K \circ M$;*

3) *if the sequence* (H^n) *of predictable processes converges pointwise to the limit* H *and* $|H^n| \leq K$, *where* $K \in L^2_{\text{loc}}(M)$, *then*

$$\sup_{s \leq t} |H^n \circ M - H \circ M| \xrightarrow{\text{P}} 0, \ t \in \mathbb{R}_+.$$

The extension $H \circ M$ *possesses the further properties:*
4) $H \circ M \in \mathcal{H}^2_{\text{loc}}$;
5) $H \circ M \in \mathcal{H}^2$ *iff* $H \in L^2(M) = \{H : H \text{ predictable}, H^2 \circ \langle M \rangle \in \mathcal{A}^+$;
6) *the map* $M \rightsquigarrow H \circ M$ *is linear*;
7) $(H \circ M)_0 = 0, \quad H \circ M = H \circ (M - M_0)$;
8) $\Delta(H \circ M) = H \Delta M$;
9) *if* $H \in L^2_{\text{loc}}(M)$ *and* $K \in L^2_{\text{loc}}(H \circ M)$, *then* $K \circ (H \circ M) = (KH) \circ M$;
10) *if* $M, N \in \mathcal{H}^2_{\text{loc}}$ *and* $H \in L^2_{\text{loc}}(M), K \in L^2_{\text{loc}}(N)$, *then* $\langle H \circ M, K \circ N \rangle = (HK) \circ \langle M, N \rangle$.

2.2. In the case of the definition of the stochastic integrals $H \circ X$ with respect to a semimartingale on the class of "integrands" H, it is necessary to impose restrictions making it possible to define simultaneously the integrals $H \circ M$ and $H \circ A$ while preserving the requirements naturally demanded of them (adaptatedness, linearity, ...). This can be done (using formula (3.2) as the definition of $H \circ X$) by requiring H to be a predictable locally bounded process.

Theorem 3.15'. *Let* X *be a semimartingale. Then the mapping defined by the formula*

$$(H \cdot X)_t = \begin{cases} 0, & \text{if } H = YI_{[0]}, \\ Y(X_{s \wedge t} - X_{r \wedge t}), & \text{if } H = YI_{]r,s]} \end{cases} \qquad (3.4)$$

for $H \in \mathcal{E}$ *can be extended to the class of all locally bounded predictable processes* H. *This extension, also denoted by* $H \circ X (\int_0^{\cdot} H_s dX_s, \int_{(0,\cdot]} H_s dX_s)$ *and called the stochastic integral of* H *with respect to the semimartingale* X, *possesses the following properties:*

1) $H \circ X$ *is an adapted process with trajectories in* **D**;
2) *the map* $H \rightsquigarrow H \circ X$ *is linear*;
3) *if a sequence* (H^n) *of predictable processes converges pointwise to the limit* H *and* $|H^n| \leq K$, *where* K *is a locally bounded predictable process, then* $H^n \circ X_t \xrightarrow{\text{P}} H \circ X_t$ *for all* $t \in \mathbb{R}_+$;

The integral $H \circ X$ *possesses the further properties:*
4) $H \circ X$ *is a semimartingale*;
5) *if* $x \in M_{\text{loc}}$, *then* $H \circ X \in M_{\text{loc}}$; *if* $X \in \mathcal{V}$, *then* $H \circ X \in \mathcal{V}$;
6) *the map* $X \rightsquigarrow H \circ X$ *is linear*;
7) $(H \circ X)_0 = 0, \quad H \circ X = H \cdot (X - X_0)$;
8) $\Delta(H \circ X) = H \Delta X$.

Chapter 3.II. Semimartingales. Stochastic Integrals

2.3. It is of interest to note that when the process H to be integrated is left-continuous the integral can be defined as a "limit of Riemann sums".

Namely, let $\tau = (\tau_n)$ be a sequence of Markov times, such that $\tau_0 = 0$, $\sup_n \tau_n < \infty$ and $\tau_n < \tau_{n+1}$ if $\tau_n < \infty$. We call the variable

$$\tau(H \circ X)_t = \sum_n H_{\tau_n}(X_{\tau_{n+1} \wedge t} - X_{\tau_n \wedge t})$$

the τ-Riemann approximation of the integral $(H \circ X)_t$.

We say that the "double" sequence $(\tau_n)_{n \geq 1} = ((\tau(n,m))_{m \geq 1})_{n \geq 1}$ is a Riemann sequence if

$$\sup_{m \geq 1} [\tau(n, m+1) \wedge t - \tau(n,m) \wedge t] \to 0, \quad n \to \infty.$$

Theorem 3.16. *Let X be a semimartingale, H a left-continuous adapted process, and $(\tau_n)_{n \geq 1}$ a Riemann sequence. Then the τ_n-Riemann approximation $\tau_n(H \circ X)$ converges to $H \circ X$ in probability uniformly on each compact interval:*

$$\sup_{s \leq t} |\tau_n(H \circ X)_s - \tau(H \circ X)_s| \xrightarrow{\mathsf{P}} 0.$$

2.4. In Part II, Sect. 1 the quadratic variation $[X, X]$ and covariation $[X, Y]$ of the semimartingales X and Y were defined and a number of their properties indicated. Now that we have the definition of a stochastic integral with respect to a semimartingale, we can establish the validity of the following formula:

$$[X, Y] = XY - X_0 Y_0 - X_- \circ Y - Y_- \circ X.$$

The explanation of the terminology for $[X, X]$ and $[X, Y]$ as the quadratic variation and covariation comes out of the following properties.

Let $(\tau_n)_{n \geq 1} = ((\tau(n,m))_{m \geq 1})_{n \geq 1}$ be a Riemann sequence, and

$$S_{\tau_n}(X, Y)_t = \sum_{m \geq 1} (X_{\tau(n,m+1) \wedge t} - X_{\tau(n,m) \wedge t})(Y_{\tau(n,m+1) \wedge t} - Y_{\tau(n,m) \wedge t}).$$

Then the process $S_{\tau_n}(X, Y)$ converges to the process $[X, Y]$ in measure, uniformly on each compact interval.

§3. The Itô Formula

3.1. Let $D_i f$ and $D_{ij} f$ be the partial derivatives $\partial f / \partial x^i$, $\partial^2 f / \partial x^i \partial x^j$ of the function $f = f(x^1, \ldots, x^d)$.

Theorem 3.17. *The Itô formula (change of variables). Let $X = (X^1, \ldots, X^d)$ be a d-dimensional semimartingale (in other words, each of the processes X^i, $i = 1, \ldots, d$ is a semimartingale), and $f = f(x^1, \ldots, x^d)$ a function of class C^2. Then the process $f(X)$ is a semimartingale and*

$$f(X_t) = f(X_0) + \sum_{i \leq d} D_i f(X_-) \circ X_t^i + \frac{1}{2} \sum_{i,j \leq d} D_{ij} f(X_-) \circ \langle X^{ic}, X^{jc} \rangle_t$$
$$+ \sum_{s \leq t} \left[f(X_s) - f(X_{s-}) - \sum_{i \leq d} D_i f(X_{s-}) \Delta X_s^i \right]. \quad (3.5)$$

Remark. In the case of discrete time, formula (3.5) is transformed into the trivial identity:

$$f(X_n) = f(X_0) + \sum_{1 \leq m \leq n} \sum_{i \leq d} D_i f(X_{m-1})(X_m^i - X_{m-1}^i)$$
$$+ \sum_{1 \leq m \leq n} \left[f(X_m) - f(X_{m-1}) \right.$$
$$\left. - \sum_{i \leq d} D_i f(X_{m-1})(X_m^i - X_{m-1}^i) \right]. \quad (3.6)$$

We give a number of examples on the application of the Itô formula.

Example 1. If X and Y are semimartingales and $f(x,y) = xy$, then it follows from (3.5) that

$$X_t Y_t = X_0 Y_0 + (X_- \circ Y)_t + (Y_- \circ X)_t + [X, Y]_t, \quad (3.7)$$

since

$$[X, Y] = \langle X^c, Y^c \rangle + \sum_{s \leq \cdot} \Delta X_s \Delta Y_s.$$

In particular,

$$X_t^2 = X_0^2 + 2(X_- \circ X)_t + [X, X]_t. \quad (3.8)$$

Example 2 (Doleans-Dade equation). Let X be a semimartingale. We consider the Doleans-Dade equation

$$Y_t = 1 + \int_0^t Y_{s-} dX_s \quad (3.9)$$

or in "differential" form

$$dY = Y_- dX, \quad Y_0 = 1.$$

Furthermore, this equation has a unique adapted solution $\mathcal{E}(X)$ in **D** which is a semimartingale and is given by the formula

$$\mathcal{E}(X)_t = e^{X_t - X_0 - \frac{1}{2} \langle X^c \rangle_t} \Pi_{s \leq t} (1 + \Delta X_s) e^{-\Delta X_s}. \quad (3.10)$$

(The function $\mathcal{E}(X)$ is called the *stochastic exponential*.) It can be verified directly that $\mathcal{E}(X)$ is indeed a solution of equation (3.9) by applying the Itô formula to the product $V_t U_t$ of the two semimartingales

$$V_t = e^{X_t - X_0 - \frac{1}{2}\langle X^c \rangle_t},$$
$$U_t = \Pi_{s \leq t}(1 + \Delta X_s)e^{-\Delta X_s}.$$

§4. Construction of Stochastic Integrals with Respect to Random Measures

4.1. In studying the jump components of local martingales and semimartingales we need to deal with integrals of types $W * \mu$ and $W * \nu$ (with respect to random measures μ and their compensators ν; see Part I, Sect. 5), and also integrals of type $W * (\mu - \nu)$ with respect to the "compensated" measures $\mu - \nu$.

As a "naïve definition" of the integral $W * (\mu - \nu)$ we could take the difference $W * \mu - W * \nu$. However, such a "definition" has the drawback that the class of functions W that can be integrated is too narrow.

In order to give a full account of the circle of questions involved here, we introduce a number of definitions and notions.

We assume throughout that we are given a stochastic basis $\mathcal{B} = (\Omega, \mathcal{F}, \mathbf{F} = (\mathcal{F}_t)_{t \geq 0}, \mathbf{P})$ and an integer-valued random measure $\mu = \mu(\omega; dt, dx)$ on $\mathbb{R}_+ \times E$, where $(E, \mathcal{E})$ is a Blackwell space and $\nu = \nu(\omega; dt, dx)$ is the compensator of the measure μ. (There always exists a version of ν with $\nu(\omega; \{t\} \times E) \leq 1$ identically, and this is the version that we consider in what follows.)

We set

$$a_t(\omega) = \nu(\omega; \{t\} \times E), \quad J = \{\omega, t : a_t(\omega) > 0\},$$
$$\nu^c(\omega; dt, dx) = \nu(\omega; dt, dx)I_{\bar{J}}(\omega, t).$$

We associate with each $\widetilde{\mathcal{P}}$-measurable function $W = W(\omega, t, x)$ the following two increasing predictable processes:

$$C(W)_t = (W - \widehat{W})^2 * \nu_t + \sum_{s \leq t}(1 - a_s)(\widehat{W}_s)^2,$$
$$\bar{C}(W)_t = |W - \widehat{W}| * \nu_t + \sum_{s \leq t}(1 - a_s)|\widehat{W}_s|,$$

where

$$\widehat{W}_t(\omega) = \begin{cases} \int_E W(\omega, t, x)\nu(\omega; \{t\} \times dx), & \text{if } \int_E |W(\omega, t, x)|\nu(\omega; \{t\} \times dx) < \infty, \\ \infty, & \text{otherwise.} \end{cases}$$

In addition we set

$$W' = (W - \widehat{W})I_{\{|W - \widehat{W}| \leq 1\}} + \widehat{W}I_{\{|\widehat{W}| \leq 1\}},$$
$$W'' = (W - \widehat{W})I_{\{|W - \widehat{W}| > 1\}} + \widehat{W}I_{\{|\widehat{W}| > 1\}}.$$

Definition 1. We say that a $\widetilde{\mathcal{P}}$-measurable function $W = W(\omega, t, x)$ is of class $G_{\text{loc}}(\mu)$ if
$$C(W') + \overline{C}(W'') \in \mathcal{A}_{\text{loc}}^+.$$

Definition 2. By the stochastic integral of a $\widetilde{\mathcal{P}}$-measurable function W in $G_{\text{loc}}(\mu)$ with respect to $\mu - \nu$, denoted by $W * (\mu - \nu)$, is meant a purely discontinuous local martingale X such that
$$\Delta X_t = \int_E W(\omega, t, x)\mu(\omega; \{t\} \times dx) - \widehat{W}_t(\omega).$$

It can be shown that this definition is correct in the sense that for functions $W \in G_{\text{loc}}(\mu)$, such a process $X \in \mathcal{M}_{\text{loc}}^{\text{d}}$ actually exists and is unique to within stochastic indistinguishability.

The meaning of the definitions and concepts introduced above is revealed by the next theorem.

Theorem 3.18. *Let W be a $\widetilde{\mathcal{P}}$-predictable function.*
1) The function $W \in G_{\text{loc}}(\mu)$ and $W * (\mu - \nu) \in H^2$ (or $\mathcal{H}_{\text{loc}}^2$) if and only if $C(W) \in \mathcal{A}^+$ (or $\mathcal{A}_{\text{loc}}^+$). In this case
$$\langle W * (\mu - \nu) \rangle = C(W).$$

2) The function $W \in G_{\text{loc}}(\mu)$ and $W * (\mu - \nu) \in \mathcal{A}$ (or $\mathcal{A}_{\text{loc}}$) if and only if $\overline{C}(W) \in \mathcal{A}^+$ (or $\mathcal{A}_{\text{loc}}^+$).

3) The function $W \in G_{\text{loc}}(\mu)$ if and only if the process
$$\widetilde{W}_t(\omega) = \int_E W(\omega, t, x)\mu(\omega; \{t\} \times dx) - \widehat{W}_t(\omega)$$

is such that
$$\sqrt{\sum_{s \leq \cdot} (\widetilde{W}_s)^2} \in \mathcal{A}_{\text{loc}}^+.$$

In Theorem 3.19 presented below, a case is described when the integral $W * (\mu - \nu)$ equals $W * \mu - W * \nu$.

Theorem 3.19. *Let W be $\widetilde{\mathcal{P}}$-measurable, and $|W| * \mu \in \mathcal{A}_{\text{loc}}^+$ (equivalently: $|W| * \nu \in \mathcal{A}_{\text{loc}}^+$). Then $W \in G_{\text{loc}}(\mu)$ and*
$$W * (\mu - \nu) = W * \mu - W * \nu.$$

4.2. Let us consider the problem of representing purely discontinuous local martingales as stochastic integrals with respect to the measure $\mu - \nu$.

Let $M \in \mathcal{M}_{\text{loc}}$, $E = R \setminus \{0\}$, let
$$\mu(\omega; dt, dx) = \sum_{s > 0} I(\Delta M_s \neq 0) \epsilon_{(s, \Delta M_s)}(dt, dx)$$

be the measure of jumps of the process M, and ν its compensator. Then for $W(\omega, t, x) = x$

$$\widehat{W}_t(\omega) = \int_E x\nu(\omega; \{t\} \times dx) = 0,$$

$C(W) = x^2 * \nu$, $\overline{C}(W) = |x| * \nu$, $(x^2 \wedge |x|) * \nu \in \mathcal{A}_{\text{loc}}^+$. Hence $W \in G_{\text{loc}}(\mu)$ and the integral $x * (\mu - \nu)$ is defined. Moreover, $x * (\mu - \nu)$ coincides with the purely discontinuous component M^{d} of the process M:

$$M_t^{\text{d}} = \int_0^t \int_E x \, d(\mu - \nu).$$

If, moreover, $M^{\text{d}} \in \mathcal{M}_{\text{loc}}^{2,\text{d}}$, then $\langle M^{\text{d}} \rangle = x^2 * \nu$.

§5. Characteristics of Semimartingales. The Triple of Predictable Characteristics $T = (B, C, \nu)$. Martingale and Semimartingale Problems. Examples

5.1. Let $X = (X^1, \ldots, X^d)$ be a d-dimensional semimartingale defined on the stochastic basis $\mathcal{B} = (\Omega, \mathcal{F}, \mathbf{F} = (\mathcal{F}_t)_{t \geq 0}, \mathbf{P})$. In this section we give the definition of the important concept of a triple of predictable characteristics of a semimartingale X, in terms of which various properties of them are described.

Let $\mathcal{G}_{\text{tr}}^d$ be the class of all truncation functions $h : E^d \to E^d$ that are bounded, have compact support and satisfy the property $h(x) = x$ in a neighborhood of zero.

If $h \in \mathcal{G}_{\text{tr}}^d$, then $\Delta X_s - h(\Delta X_s) \neq 0$ only if $|\Delta X_s| > b$ for some $b > 0$. Let $\mu = \mu(\omega; dt, dx)$ be the measure of the jumps of X,

$$\check{X}(h)_t = \sum_{s \leq t} [\Delta X_s - h(\Delta X_s)] \left(= \int_0^t \int (x - h(x)) d\mu \right),$$

$$X(h) = X - \check{X}(h).$$

The process $\check{X}(h) \in \mathcal{V}^d$ (that is, its components are of class $\mathcal{V}$) and the process $X(h)$ is a semimartingale with bounded jumps and is therefore a special semimartingale. According to Sect. 1, each special semimartingale admits the canonical decomposition

$$X(h) = X_0 + M(h) + B(h), \qquad (3.11)$$

where the $M(h) \in \mathcal{M}_{\text{loc}}$, $M_0(h) = 0$ and $B(h)$ is a predictable process of class $\mathcal{V}^d$. Taking into account the formula

$$M^{\text{d}}(h)_t = \int_0^t \int h(x) d(\mu - \nu),$$

we obtain the following canonical representation for a semimartingale X:

$$X_t = X_0 + B_t(h) + X_t^c + \int_0^t \int h(x) d(\mu - \nu) + \int_0^t \int (x - h(x)) d\mu. \quad (3.12)$$

Definition 1. Let h be a fixed truncation function $h \in \mathcal{G}_{tr}^d$. By a *triple of predictable characteristics* (with respect to h), $T = (B, C, \nu)$, is meant a collection consisting of:

1) $B = (B^i)_{i \leq d}$, a predictable process $B = B(h)$;
2) $C = (c^{ij})_{i,j \leq d}$, a continuous process in $\mathcal{V}^d \times \mathcal{V}^d$ with $c^{ij} = \langle X^{ic}, X^{jc} \rangle$;
3) ν, the predictable random measure on $\mathbb{R}_+ \times E^d$ that is the compensator of the measure μ of the jumps of the process X.

It is important to note that the second and third characteristics C and ν are "internal" characteristics of the semimartingale (in the sense that they do not depend on the choice of the truncation function h. As regards the first characteristic, the truncation functions h and h' satisfy the relation

$$B(h) - B(h') = (h - h') * \nu.$$

5.2. It is convenient for many purposes to introduce the concept of the *second modified* characteristic $\widetilde{C} = (\tilde{c}^{ij})_{i,j \leq d}$, defined by the formula

$$\tilde{c}^{ij} = \langle M(h)^i, M(h)^j \rangle.$$

The connection of this characteristic with (B, C, ν) is described by the formula:

$$\tilde{c}^{ij} = c^{ij} + (h^i h_j) * \nu - \sum_{s \leq \cdot} \left(\int h^i(x) \nu(\{s\} \times dx) \right) \left(\int h^i(x) \nu(\{s\} \times dx) \right)$$
$$= c^{ij} + (h^i h^j) * \nu - \sum_{s \leq \cdot} \Delta B_s^i(h) \Delta B_s^j(h). \quad (3.13)$$

5.3. The next result is useful from the point of view of alternative (and often more convenient) criteria for X to be a semimartingale with triple $T = (B, c, \nu)$.

Theorem 3.20. *The following conditions are equivalent:*
1) X *is a semimartingale with triple* $T(B, C, \nu)$;
2) *for each* $\lambda \in E^d$ *the process* $e^{i\lambda \circ X} - e^{i\lambda X_-} \circ G(\lambda)$ *is a (complex-valued) local martingale, where*

$$G(\lambda)_t = (i\lambda \circ B)_t - \frac{1}{2}(\lambda \circ C_t \circ \lambda)$$
$$+ \int (e^{i\lambda x} - 1 - i\lambda h(x)) \nu([0,t] * dx); \quad (3.14)$$

3) *for each bounded function* $f = f(x^1, \ldots, x^d)$ *of class* C^2 *the process*

$$f(X) - f(X_0) - \sum_{j \le d} D_j f(X_-) \circ B^j - \frac{1}{2} \sum_{j,k \le d} D_{jk} f(X_-) \circ C^{jk}$$
$$- \left[f(X_- + x) - f(X_-) - \sum_{j \le d} D_j f(X_-) h^j(x) \right] \times \nu \quad (3.15)$$

is a local martingale;
4) *Each of the three processes*

$$M(h) = X(h) - B - X_0,$$
$$M(h)^i M(h)^j - \tilde{c}^{ij}, \quad i, j \le d \quad (3.16)$$
$$g * \mu - g * \nu, \quad g \in \mathcal{G}^+$$

is a local martingale (where $\mathcal{G}^+$ is the family of bounded Borel functions on E^d vanishing in a neighborhood of 0);
5) *Under the additional assumption that $\Delta G(\lambda) \ne -1$ (where $G(\lambda)$ is the function in 2)) the process*

$$e^{i\lambda \cdot X} / \mathcal{E}(G(\lambda)), \quad (3.17)$$

where

$$\mathcal{E}(G)_s = e^{G_s} \prod_{u \le s} (1 + \Delta G_u) e^{-\Delta G_u}, \quad (3.18)$$

is (for each $\lambda \in E^d$) a local martingale.

5.4. In connection with the characterization presented in 4), we recall the so-called semimartingale and martingale problems in the following formulation.

Let $(\Omega, \mathcal{F}, \mathbf{F} = (\mathcal{F}_t)_{t \ge 0})$ be a measurable space with a filtration $\mathbf{F}$, and $\mathbf{P}_{\mathcal{F}_0}$ a probability measure on $(\Omega, \mathcal{F}_0)$. Also let $X = (X^i)_{i \le d}$ be a d-dimensional process given on $(\Omega, \mathcal{F}, \mathbf{F})$ with trajectories in $\mathbf{D}$ (the candidate being a semimartingale), and $T = (B, C, \nu)$ a set (the candidate being the triple of the semimartingale X), where $B = (B^i)_{i \le d}$ is an $\mathbf{F}$-predictable process of finite variation on each finite interval, $B_0 = 0$; $C = (c^{ij})$, $i, j \le d$, is an $\mathbf{F}$-predictable continuous matrix such that the matrix $C_t - C_s$, $s \le t$, is positive semidefinite and symmetric; ν is an $\mathbf{F}$-predictable random measure on $\mathbb{R}_+ \times E^d$, such that

$$\nu(\omega; \mathbb{R}_+ \times \{0\}) = \nu(\omega; \{0\} \times E^d) = 0, \quad |x|^2 \wedge 1 * \nu_t(\omega) < \infty,$$
$$\int \nu(\omega; \{t\} \times dx) h(x) = \Delta B_t(\omega), \quad \nu(\omega; \{t\} \times E^d) \le 1$$

identically.

Definition 1. By a *solution of the semimartingale problem* associated with $(\mathcal{F}_0, X)$ and $(\mathbf{P}_{\mathcal{F}_0}; B, C, \nu)$ is meant a probability measure $\mathbf{P}$ on $(\Omega, \mathcal{F})$ such that

1) $\mathbf{P}|\mathcal{F}_0 = \mathbf{P}_{\mathcal{F}_0}$,

2) X is a semimartingale on $(\Omega, \mathcal{F}, \mathbf{F}, \mathbf{P})$ with a triple of (given) characteristics $T = (B, C, \nu)$ (with respect to a truncation function h).

We denote by $S(\mathcal{F}_0, X|\mathbf{P}_{\mathcal{F}_0}; B, C, \nu)$ the set of all solutions $\mathbf{P}$.

Definition 2. Let $\mathcal{X}$ be the family of optional $\bar{\mathbb{R}}$-valued processes on $(\Omega, \mathcal{F}, \mathbf{F} = (\mathcal{F}_t)_{t \geq 0})$, $\mathcal{H}$ a σ-subalgebra of $\mathcal{F}_0$, and $\mathbf{P}_{\mathcal{H}}$ a probability measure on $\mathcal{H}$. By a solution of the martingale problem associated with $\mathcal{X}$ and $\mathbf{P}_{\mathcal{H}}$ is meant a probability measure $\mathbf{P}$ on $(\Omega, \mathcal{F})$ such that $\mathbf{P}|\mathcal{H} = \mathbf{P}_{\mathcal{H}}$ and each of the processes $X \in \mathcal{X}$ is a local martingale on the stochastic basis $(\Omega, \mathcal{F}, \mathbf{F}, \mathbf{P})$.

Assertion 4) of Theorem 3.20 states that the semimartingale problem associated with the triple $T = (B, C, \nu)$ is equivalent to the corresponding martingale problem (with set $\mathcal{X}$ consisting of the three processes (3.16)).

It can be shown that the set of solutions $S(\mathcal{F}_0, X|\mathbf{P}_{\mathcal{F}_0}; B, C, \nu)$ is a convex set. (Regarding the existence and uniqueness of martingale and semimartingale problems, see (Jacod and Shiryaev 1987, Chap. III).)

5.5. We now give some examples of semimartingales and their triples.

Example 1. Let $\widetilde{B} = (\Omega, \mathcal{F}, \widetilde{\mathbf{F}} = (\mathcal{F}_n)_{n \geq 0}, \mathbf{P})$ be a "discrete" stochastic basis, and $\xi = (\xi_n)_{n \geq 0}$ a sequence of random variables such that ξ_n is $\mathcal{F}_n$-measurable, $n \geq 0$. On the "discrete" stochastic basis $\mathcal{B} = (\Omega, \mathcal{F}, \mathbf{F} = (\mathcal{F}_t)_{t \geq 0}, \mathbf{P})$ with $\mathcal{F}_t = \mathcal{F}_{[t]}$ we consider the process

$$X_t = \sum_{n \leq t} \xi_n, \quad t \geq 0.$$

This process is a semimartingale and its triple (with respect to the truncation function $h = h(x)$) $T = (B, C, \nu)$ is defined by the formula

$$B_t = \sum_{k \leq t} \mathbf{E}(h(\xi_k)|\mathcal{F}_{k-1}),$$

$$C_t \equiv 0,$$

$$\nu([0, t] \times g) = \sum_{k \leq t} \mathbf{E}[g(\xi_k) I(\xi_k \neq 0)|\mathcal{F}_{k-1}],$$

where

$$\nu([0, t] \times g) = \int_0^t \int g(x) \nu(\omega; ds, dx).$$

Example 2. The Wiener process $W = (W_t)_{t \geq 0}$ with $\mathbf{E}W_t = 0$, $\mathbf{E}W_t^2 = \sigma^2(t)$ is a continuous local martingale with $B = 0$, $C_t \equiv \sigma^2(t)$, $\nu = 0$.

Example 3. Let $X = (X_t)_{t \geq 0}$ be a process with independent increments with characteristic function

$$g(\lambda)_t = \mathbf{E} \exp(i \lambda \circ X_t), \quad \lambda \in E^d. \tag{3.19}$$

A process X with independent increments is a semimartingale if and only if for each $\lambda \in E^d$ the function $t \to g(\lambda)_t$ has finite variation on each finite interval. On the other hand, let X be a d-dimensional semimartingale with $X_0 = 0$. Then X is a process with independent increments if and only if there exists a deterministic version of the triple of characteristics. If $T = (B, C, \nu)$ is such a triple, then

$$g(\lambda)_t = \mathcal{E}(G(\lambda))_t, \qquad (3.20)$$

where $G(\lambda)$ is the function defined in terms of (B, C, ν) by formula (3.14) and $\mathcal{E}(G)$ is defined by formula (3.18). In particular, if X is a process with independent increments without fixed times of jumps (or, equivalently, $\nu(\{t\} \times E^d) = 0$, $t \geq 0$), then the functions B_t, C_t, $\nu([0, t] \times A)$ are continuous (deterministic) functions, $\Delta G(\lambda)_t = 0$ and formula (3.20) is transformed into the Lévy-Khinchin formula:

$$g(\lambda)_t = \exp(G(\lambda)_t),$$

where $G(\lambda)_t$ is given by formula (3.14).

§6. Integral Representation of Local Martingales

6.1. Let $X = (X^i)_{i \leq d}$ be a d-dimensional semimartingale with characteristics $T = (B, C, \nu)$, given on the stochastic basis $(\Omega, \mathcal{F}, \mathbf{F}, \mathbf{P})$. Let X^c be its continuous martingale part, and μ the measure of jumps of X.

Definition 1. A local martingale M given on $(\Omega, \mathcal{F}, \mathbf{F}, \mathbf{P})$ is said to have an integral representation (with respect to a semimartingale X) if M can be represented in the form

$$M = M_0 + H \circ X^c + W * (\mu - \nu), \qquad (3.21)$$

where $H = (H^i)_{i \leq d}$ belongs to $L^2_{\text{loc}}(X^c)$ and $W \in G_{\text{loc}}(\mu)$.

The next two examples are classical examples when the representation proves to be possible with $\mathcal{F}_t = \bigcap_{s > t} \mathcal{F}^0_s$, where $\mathcal{F}^0_s = \sigma\{X_r, r \leq s\}$.

Example 1. Let X be a Wiener process on $(\Omega, \mathcal{F}, \mathbf{F} = (\mathcal{F}_t)_{t \geq 0}, \mathbf{P})$. Then each local martingale M admits the representation

$$M = M_0 + H \circ X \qquad (3.22)$$

with $H \in L^2_{\text{loc}}(X)$ and is therefore continuous.

Example 2. Let $X = (X_t)_{t \geq 0}$ be a Poisson process on $(\Omega, \mathcal{F}, \mathbf{F} = (\mathcal{F}_t)_{t \geq 0}, \mathbf{P})$. Then each local martingale M admits a representation in the form (3.21).

6.2. The question of the possibility of an integral representation of local martingales with respect to a semimartingale X, given on a stochastic basis $(\Omega, \mathcal{F}, \mathbf{F}, \mathbf{P})$, is closely related to the semimartingale problem. The basic result in this direction can be stated in broad outline as follows: the representation (3.21) holds if and only if the measure **P** is an extremal point in the convex set $S = (\mathcal{F}_0, X|\mathbf{P}_H; B, C, \nu)$. For more detail, see (Jacod and Shiryaev 1987, Chap. III, §4) and ((Liptser and Shiryaev 1986, Chap. 4, §8).

§7. Stability of the Class of Semimartingales with Respect to a Series of Transformations

7.1. It was noted in Sect. 1 that the class of semimartingales is stable with respect to a series of transformations, in particular, with respect to an absolutely continuous change of measure, reduction of filtration and a random change of time. We consider this circle of questions in more detail.

Let $X = (X^i)_{i \leq d}$ be a d-dimensional semimartingale given on a stochastic basis $\mathcal{B} = (\Omega, \mathcal{F}, \mathbf{F} = (\mathcal{F}_t)_{t \geq 0}, \mathbf{P})$ with triple $T = (B, C, \nu)$. We consider a new stochastic basis $\widetilde{\mathcal{B}} = (\Omega, \mathcal{F}, \mathbf{F} = (\mathcal{F}_t)_{t \geq 0}, \widetilde{\mathbf{P}})$, where $\widetilde{\mathbf{P}}$ is some (new) probability measure such that $\widetilde{\mathbf{P}} \ll \mathbf{P}$, that is, $\widetilde{\mathbf{P}}$ is absolutely continuous with respect to **P**.

Theorem 3.21. *The process $X = (X^i)_{i \leq d}$ considered on the stochastic basis $\widetilde{\mathcal{B}} = (\Omega, \mathcal{F}, \mathbf{F}, \widetilde{\mathbf{P}})$ is also a semimartingale with triple $\widetilde{T} = (\widetilde{B}, \widetilde{C}, \widetilde{\nu})$, where*

$$\widetilde{B}^i = B^i + \left(\sum_{j \leq d} c^{ij} \beta^j\right) \circ A + h^i(x)(Y-1) * \nu,$$

$$\widetilde{C} = C,$$

$$\widetilde{\nu} = Y\nu,$$

Y is a $\widetilde{\mathcal{P}}$-measurable non-negative function, $\beta = (\beta^i)_{i \leq d}$ is a predictable process, and c^{ij} and the predictable increasing process A are such that $C^{ij} = c^{ij} \circ A$.

7.2. Let X be a semimartingale on a stochastic basis $\mathcal{B} = (\Omega, \mathcal{F}, \mathbf{F} = (\mathcal{F}_t)_{t \geq 0}, \mathbf{P})$, $\mathcal{F}_{t+}^X = \bigcap_{\epsilon > 0} \sigma\{X_s, 0 \leq s \leq t + \epsilon\} \vee \mathcal{N}$, where $\mathcal{N}$ is a system of sets in $\mathcal{F}$ of **P**-measure zero, and $\mathcal{G} = (\mathcal{G}_t)_{t \geq 0}$ is a non-decreasing flow of σ-algebras satisfying the usual conditions and such that

$$\mathcal{F}_{t+}^X \subseteq \mathcal{G}_t \subseteq \mathcal{F}_t, \ t \geq 0.$$

Theorem 3.22. *The process X considered on the (reduced) stochastic basis $\widetilde{\mathcal{B}} = (\Omega, \mathcal{F}, \mathcal{G} = (\mathcal{G}_t)_{t \geq 0}, \mathbf{P})$ is also a semimartingale.*

7.3. Again let X be a semimartingale on a stochastic basis

$$\mathcal{B} = (\Omega, \mathcal{F}, \mathbf{F} = (\mathcal{F}_t)_{t \geq 0}, \mathbf{P}).$$

Definition. A stochastic process $\hat{\tau} = (\hat{\tau}_t)_{t \geq 0}$ of class $\mathcal{V}^+$ such that $\hat{\tau}_t$ is a stopping time for each $t \geq 0$ is called a *random change of time*. We form the new process
$$\widehat{X}_t(\omega) = X_{\hat{\tau}_t(\omega)}(\omega), \ t \geq 0$$
and the new flow $\widehat{\mathbf{F}} = (\widehat{\mathcal{F}}_t)_{t \geq 0}$ with $\widehat{\mathcal{F}}_t = \mathcal{F}_{\hat{\tau}_t}$.

Theorem 3.23. *The process $\widehat{X}$ considered on the stochastic basis $\hat{\mathcal{B}} = (\Omega, \mathcal{F}, \widehat{\mathbf{F}}, \mathbf{P})$ is also a semimartingale.*

III. Absolute Continuity and Singularity of Probability Distributions

§1. Local Density. Lebesgue Decomposition

1.1. Let $(\Omega, \mathcal{F}, \mathbf{F}, \mathbf{Q})$ be a stochastic basis,

$$\mathbf{Q} = \frac{1}{2}(\mathbf{P}' + \mathbf{P}),$$

where $\mathbf{P}'$ and $\mathbf{P}$ are probability measures on $(\Omega, \mathcal{F})$. In this definition of $\mathbf{Q}$ the measures $\mathbf{P}'$ and $\mathbf{P}$ are absolutely continuous with respect to $\mathbf{Q}$ ($\mathbf{P}' \ll \mathbf{Q}$, $\mathbf{P} \ll \mathbf{Q}$). Let $\mathfrak{z}' = (\mathfrak{z}'_t)_{t \geq 0}$ be the optional projection of $d\mathbf{P}'/d\mathbf{Q}$ with respect to $(\mathbf{F}, \mathbf{Q})$, which one can choose in such a manner that the trajectories of $\mathfrak{z}'$ are right-continuous with left limits. The process $\mathfrak{z}'$ chosen in this manner is a non-negative uniformly integrable martingale with respect to $(\mathbf{F}, \mathbf{Q})$ and is called the local density of the measure $\mathbf{P}'$ with respect to $\mathbf{Q}$. Moreover, $\mathfrak{z}'_\infty = \lim_{t\to\infty} \mathfrak{z}'_t$ coincides $\mathbf{Q}$-a.s. with $d\mathbf{P}'/d\mathbf{Q}$. The local density process $\mathfrak{z} = (\mathfrak{z}_t)_{t\geq 0}$ of the measure $\mathbf{P}$ with respect to $\mathbf{Q}$ with $\mathfrak{z}_\infty = d\mathbf{P}/d\mathbf{Q}$ ($\mathbf{Q}$-a.s.), is defined in similar fashion. For any $t \in \mathbb{R}^+$ we have

$$\mathbf{Q}(\mathfrak{z}'_t = 0, \ \mathfrak{z}_t = 0) = 0, \tag{3.23}$$

that is, the processes $\mathfrak{z}'$ and $\mathfrak{z}$ do not simultaneously vanish.

If τ is a Markov time (or predictable Markov time), then $\mathbf{P}'_\tau$ and $\mathbf{P}_\tau$ (or $\mathbf{P}'_{\tau-}$ and $\mathbf{P}_{\tau-}$) are the restrictions of $\mathbf{P}'$ and $\mathbf{P}$ to $\mathcal{F}_\tau$ (or $\mathcal{F}_{\tau-}$).

1.2. The following decomposition of the measure $\mathbf{P}'$ with respect to $\mathbf{P}$ is called the *Lebesgue decomposition*: for any set $A \in \mathcal{F}_\tau$ (τ is a Markov time)

$$\mathbf{P}'(A) = \int_A (\mathfrak{z}'_\tau/\mathfrak{z}_\tau)d\mathbf{P} + \mathbf{P}'(A, \mathfrak{z}_\tau = 0). \tag{3.24}$$

If τ is a predictable Markov time and $A \in \mathcal{F}_{\tau-}$, then

$$\mathbf{P}'(A) = \int_A (\mathfrak{z}'_{\tau-}/\mathfrak{z}_{\tau-})d\mathbf{P} + \mathbf{P}'(A, \mathfrak{z}_{\tau-} = 0). \tag{3.25}$$

The variable $\mathfrak{z}'_\tau/\mathfrak{z}_\tau$ (or $\mathfrak{z}'_{\tau-}/\mathfrak{z}_{\tau-}$ in the predictable case) is called the derivative of the absolutely continuous part of $\mathbf{P}'_\tau$ (or $\mathbf{P}'_{\tau-}$) with respect to $\mathbf{P}_\tau$ (or $\mathbf{P}_{\tau-}$).

Definition 1. We will say that a measure $\mathbf{P}'$ is *locally absolutely continuous* with respect to $\mathbf{P}$ (and write $\mathbf{P}' \overset{\text{loc}}{\ll} \mathbf{P}$) if $\mathbf{P}'_t \ll \mathbf{P}_t$ for each $t \in \mathbb{R}_+$.

In the case $\mathbf{P}' \overset{\text{loc}}{\ll} \mathbf{P}$ property (3.23) of the processes $\mathfrak{z}'$ and $\mathfrak{z}$ allows one to define a process $Z = (Z_t)_{t\geq 0}$ with

Chapter 3.III. Absolute Continuity and Singularity of Distributions   141

$$Z_t = \mathfrak{z}'_t/\mathfrak{z}_t,$$

where $Z_t = d\mathbf{P}'_t/d\mathbf{P}_t$. The process so defined, called the *local density process* of the measure $\mathbf{P}'$ with respect to $\mathbf{P}$, is a non-negative local martingale with respect to $(\mathbf{F},\mathbf{P})$ and possesses the following property: for any Markov time τ (or predictable Markov time τ)

$$I(\tau < \infty)Z_\tau = I(\tau < \infty)d\mathbf{P}'_\tau/d\mathbf{P}_\tau,$$
$$(\text{or } I(\tau < \infty)Z_{\tau-} = I(\tau < \infty)d\mathbf{P}'_{\tau-}/d\mathbf{P}_{\tau-}),$$

where $d\mathbf{P}'_\tau/d\mathbf{P}_\tau$ (or $d\mathbf{P}'_{\tau-}/d\mathbf{P}_{\tau-}$) is the derivative of the absolutely continuous part of $\mathbf{P}'_\tau$ (or $\mathbf{P}'_{\tau-}$) with respect to $\mathbf{P}_\tau$ (or $\mathbf{P}_{\tau-}$).

In the case of absolute continuity of $\mathbf{P}'$ with respect to $\mathbf{P}$ ($\mathbf{P}' \ll \mathbf{P}$) the local density process Z is a uniformly integrable martingale with respect to $(\mathbf{F},\mathbf{P})$.

§2. Girsanov's Theorem and its Generalization. Transformation of Predictable Characteristics

2.1. Let the measure $\mathbf{P}'$ be absolutely continuous with respect to $\mathbf{P}$. Let us assume that the local density process Z, which in this case is a uniformly integrable martingale, has continuous trajectories. This means that Z is a locally square-integrable martingale with respect to $(\mathbf{F},\mathbf{P})$.

Let $W = (W_t)_{t \geq 0}$ be a Wiener process on a stochastic basis $(\Omega, \mathcal{F}, \mathbf{F}, \mathbf{P})$. Since W is a locally square-integrable martingale with respect to $(\mathbf{F},\mathbf{P})$, it follows that the mutual quadratic characteristic $\langle W, Z \rangle = (\langle W, Z \rangle_t)_{t \geq 0}$ is defined.

The following classical result is due to Girsanov.

Theorem 3.24. *Let* $\langle W, Z \rangle_t = \int_0^t a(\omega, s)ds$, *where* $a = (a(\omega, t))_{t \geq 0}$ *is an $\mathbf{F}$-adapted process such that* $\int_0^t a^2(\omega, s)ds < \infty$ $\mathbf{P}$-*a.s.*, $t > 0$.
Then the stochastic process $W' = (W'_t)_{t \geq 0}$ *with*

$$W'_t = W_t - \int_0^t a(\omega, s)ds$$

is Wiener on the stochastic basis $(\Omega, \mathcal{F}, \mathbf{F}, \mathbf{P}')$.

2.2. We present a generalization of this classical result. Let $\mathbf{P}$ and $\mathbf{P}'$ be probability measures on $(\Omega, \mathcal{F})$, and $\mathbf{Q} = \frac{1}{2}(\mathbf{P}'+\mathbf{P})$. We denote the completion of the σ-algebra $\mathcal{F}$ and the filtration $\mathbf{F}$ with respect to the measure $\mathbf{Q}$ by $\mathcal{F}^{\mathbf{Q}}$ and $\mathbf{F}^{\mathbf{Q}}$, respectively.

Theorem 3.25. *Let* $M = (M_t)_{t \geq 0}$ *be a local martingale with respect to* $(\mathbf{F}^{\mathbf{Q}}, \mathbf{P})$. *Let* $\mathbf{P}'$ *be a locally absolutely continuous measure with respect to* $\mathbf{P}$ ($\mathbf{P}' \overset{\text{loc}}{\ll} \mathbf{P}$), *and* $Z = (Z_t)_{t \geq 0}$ *the local density process.*

Suppose that the mutual quadratic variation $[M,Z] = ([M,Z]_t)_{t\geq 0}$ has locally integrable variation with respect to the measure $\mathbf{P}$ (in this case $[M,Z]$ has a compensator $\widetilde{[M,Z]} = (\widetilde{[M,Z]}_t)_{t\geq 0}$ with respect to $(\mathbf{F}^\mathbf{Q}, \mathbf{P})$).
Then the process $M' = (M'_t)_{t\geq 0}$ with

$$M'_t = M_t - I(Z_- > 0) Z_-^{-1} \circ \widetilde{[M,Z]}_t$$

is a local martingale with respect to $[\mathbf{F}^\mathbf{Q}, \mathbf{P}']$. Moreover, the processes $\langle M^c \rangle$ and $\langle M'^c \rangle$ are $\mathbf{P}'$-indistinguishable, and the processes $I(Z_- > 0) \circ \langle M^c \rangle$ and $I(Z_- > 0) \circ \langle M^c \rangle$ are $\mathbf{P}$-indistinguishable.

We now present a result showing how the compensators of an integer-valued random measure are changed under a locally absolutely continuous change of measure.

Let $\mu = \mu(dt, dx)$ be an integer-valued random measure on $(\mathbb{R}_+ \times E, \mathcal{B}(\mathbb{R}_+) \otimes \mathcal{E})$ (E is a Blackwell space). We denote the compensators of μ with respect to $(\mathbf{F}^\mathbf{Q}, \mathbf{P}')$ and $(\mathbf{F}^\mathbf{Q}, \mathbf{P})$ by ν' and ν, respectively.

Let $\mathbf{P}' \overset{\text{loc}}{\ll} \mathbf{P}$, and let $Z = (Z_t)_{t\geq 0}$ be the local density process of the measure $\mathbf{P}'$ with respect to $\mathbf{P}$.

We set

$$Y(t,x) = Z_{t-}^{-1} I(Z_{t-} > 0) M_\mu^\mathbf{P}(Z|\tilde{\mathcal{P}})(t,x),$$

where $M_\mu^\mathbf{P}(\cdot|\tilde{\mathcal{P}})$ is the conditional expectation of the Doleans-Dade measure $M_\mu^\mathbf{P}(d\omega, dt, dx) = \mathbf{P}(d\omega)\mu(\omega; dt, dx)$ given the σ-algebra $\tilde{\mathcal{P}} = \mathcal{P} \otimes \mathcal{E}$.

Theorem 3.26. Let $\mathbf{P}' \overset{\text{loc}}{\ll} \mathbf{P}$. Then
1) $\nu'(\omega;\ dt, dx) = Y(\omega, t, x)\nu(\omega;\ dt, dx)$ ($\mathbf{P}'$-a.s.);
2) $I(Z_-(\omega) > 0)\nu'(\omega;\ dt, dx) = I(Z_-(\omega) > 0)Y(\omega,t,x)\nu(\omega;\ dt, dx)$ ($\mathbf{P}$-a.s).

Theorems 3.25 and 3.26 enable one to establish that under a locally absolute change of measure a semimartingale remains a semimartingale, and to derive a rule for transforming a triple of predictable characteristics.

Theorem 3.27. Let $X = (X_t)_{t\geq 0}$ be a semimartingale with respect to $(\mathbf{F}^\mathbf{Q}, \mathbf{P})$ with triple $T = (B, C, \nu)$ of predictable characteristics (with truncation function $h(x) = I(|x| \leq 1)$).

If $\mathbf{P}' \overset{\text{loc}}{\ll} \mathbf{P}$, then the process X is a semimartingale with respect to $(\mathbf{F}^\mathbf{Q}, \mathbf{P}')$ and its triple $T' = (B', C', \nu')$ is defined by the formulas ($\mathbf{P}'$-a.s.):

$$B' = B + \beta \circ C + I(|x| \leq 1)x(Y-1) * \nu,$$
$$C' = C,$$
$$\nu' = Y\nu,$$

where

$$Y = I(Z_- > 0)Z_-^{-1} M_\mu^P(Z|\tilde{\mathcal{P}}),$$
$$\beta = I(Z_- > 0)Z_-^{-1} \frac{d\langle X^c, Z^c \rangle}{dC},$$

Z is the local density process, Z^c its continuous martingale part, and X^c the continuous martingale part of X.

2.3. We now consider another situation in which the semimartingale property is preserved under a change of measure.

Let $(\Omega, \mathcal{F}, \mathbf{F}, \mathbf{P})$ be a stochastic basis, where the measure $\mathbf{P}$ is a convex combination of the probability measures $\mathbf{P}'$ and $\mathbf{P}''$: $\mathbf{P} = \alpha' \mathbf{P}' + \alpha'' \mathbf{P}''$ for some $\alpha', \alpha'' > 0$, $\alpha' + \alpha'' = 1$. We assume that $X = (X_t)_{t \geq 0}$ is a semimartingale with respect to $(\mathbf{F}, \mathbf{P}')$ and $(\mathbf{F}, \mathbf{P}'')$ with triples of predictable characteristics $T' = (B', C', \nu')$ and $T'' = (B'', C'', \nu'')$, respectively. Since $\alpha' > 0$ and $\alpha'' > 0$, it follows that $\mathbf{P}' \ll \mathbf{P}$ and $\mathbf{P}'' \ll \mathbf{P}$. We denote by Z' and Z'', respectively, the local density processes of $\mathbf{P}'$ and $\mathbf{P}''$ with respect to $\mathbf{P}$.

Theorem 3.28. *The process $X = (X_t)_{t \geq 0}$ is a semimartingale with respect to $(\mathbf{F}, \mathbf{P})$ with triple $T = (B, C, \nu)$ of predictable characteristics defined by the formulas:*
$$B = \alpha' Z'_- \circ B' + \alpha'' Z''_- \circ B'',$$
$$C = \alpha' Z'_- \circ C' + \alpha'' Z''_- \circ C'',$$
$$\nu = \alpha' Z'_- \nu' + \alpha'' Z''_- \nu''.$$

§3. The Hellinger Integral and the Hellinger Process

3.1. Let $\mathbf{P}, \mathbf{P}'$ and $\mathbf{Q}$ be probability measures on $(\Omega, \mathcal{F})$, where $\mathbf{Q}$ dominates $\mathbf{P}$ and $\mathbf{P}'$ ($\mathbf{P} \ll \mathbf{Q}$, $\mathbf{P}' \ll \mathbf{Q}$). We denote by Z and Z' the Radon-Nikodým derivatives of the measures $\mathbf{P}$ and $\mathbf{P}'$ with respect to $\mathbf{Q}$:
$$Z = d\mathbf{P}/d\mathbf{Q}, \quad Z' = d\mathbf{P}'/d\mathbf{Q}.$$

Definition 1. By the *Hellinger integral* for the measures $\mathbf{P}$ and $\mathbf{P}'$ is meant the variable
$$H(\mathbf{P}, \mathbf{P}') = \mathbf{E}_\mathbf{Q} \sqrt{ZZ'},$$
where $\mathbf{E}_\mathbf{Q}$ is the expectation with respect to the measure $\mathbf{Q}$.

Clearly, the Hellinger integral does not depend on the dominating measure $\mathbf{Q}$. In this connection the following notation is often used:
$$H(\mathbf{P}, \mathbf{P}') = \int_\Omega \sqrt{d\mathbf{P}\, d\mathbf{P}'}.$$

Clearly, $H(\mathbf{P}, \mathbf{P}') \leq 1$ and $H(\mathbf{P}, \mathbf{P}) = 1$. Furthermore, the quantity

$$\rho(\mathbf{P},\mathbf{P}') = 1 - H(\mathbf{P},\mathbf{P}') = \frac{1}{2}\mathbf{E}_\mathbf{Q}(\sqrt{Z} - \sqrt{Z'})^2$$

$$= \frac{1}{2}\int_\Omega \left(\sqrt{\frac{d\mathbf{P}}{d\mathbf{Q}}} - \sqrt{\frac{d\mathbf{P}'}{d\mathbf{Q}}}\right)^2 d\mathbf{Q}$$

does not depend on the dominating measure $\mathbf{Q}$ and is the Kakutani-Hellinger distance between $\mathbf{P}$ and $\mathbf{P}'$.

In studying questions of the absolute continuity and singularity of probability measures, an essential role is played by the Hellinger integral of order α, defined as follows:

$$H(\alpha;\mathbf{P},\mathbf{P}') = \mathbf{E}_\mathbf{Q}(Z^\alpha Z'^{1-\alpha}), \ \alpha \in (0,1),$$

that is, $H(\mathbf{P},\mathbf{P}') = H(\frac{1}{2};\mathbf{P},\mathbf{P}')$. Also as in the case $\alpha = \frac{1}{2}$, the Hellinger integral $H(\alpha;\mathbf{P},\mathbf{P}')$ does not depend on the dominating measure $\mathbf{Q}$. In this connection, it is convenient to take as $\mathbf{Q}$ the measure $\frac{1}{2}(\mathbf{P} + \mathbf{P}')$. In this case we have the convenient equation:

$$Z + Z' = 2.$$

3.2. Let $\mathbf{P}$ and $\mathbf{P}'$ be probability measures given on a space with a filtration $(\Omega,\mathcal{F},\mathbf{F})$, and $\mathbf{Q}$ a probability measure on $(\Omega,\mathcal{F})$ such that

$$\mathbf{P} \stackrel{\text{loc}}{\ll} \mathbf{Q}, \quad \mathbf{P}' \stackrel{\text{loc}}{\ll} \mathbf{Q}.$$

We denote by $Z = (Z_t)_{t\geq 0}$ and $Z' = (Z'_t)_{t\geq 0}$ local density processes of the measures $\mathbf{P}$ and $\mathbf{P}'$ with respect to $\mathbf{Q}$. The processes Z and Z' are martingales with respect to $(\mathbf{F},\mathbf{Q})$.

We set

$$R_n = \inf(t: Z_t < 1/n), \quad R = \lim_n R_n, \quad \Gamma = \bigcup_n [0,R_n],$$

$$R'_n = \inf(t: Z'_t < 1/n), \quad R' = \lim_n R'_n, \quad \Gamma' = \bigcup_n [0,R'_n],$$

$$S_n = R_n \wedge R'_n, \quad S = R \wedge R', \quad \Gamma'' = \Gamma \cap \Gamma' = \bigcup_n [0,S_n].$$

Let $\alpha \in (0,1)$ and $Y(\alpha) = Z^\alpha Z'^{1-\alpha}$. The process $Y(\alpha)$ is a supermartingale with respect to $(\mathbf{F},\mathbf{Q})$. We denote by $h(\alpha) = (h(\alpha)_t)_{t\geq 0}$ the compensator of $Y(\alpha)$ with respect to $(\mathbf{F},\mathbf{Q})$. The process $h(\alpha)$ is an increasing predictable process and possesses the following properties:

$$h(\alpha) = I_{\Gamma''} \circ h(\alpha),$$
$$M(\alpha) = Y(\alpha) + Y(\alpha)_- \circ h(\alpha)$$

is a martingale with respect to $(\mathbf{F},\mathbf{Q})$. If instead of local absolute continuity $\mathbf{P} \stackrel{\text{loc}}{\ll} \mathbf{Q}, \mathbf{P}' \stackrel{\text{loc}}{\ll} \mathbf{Q}$, we have absolute continuity $\mathbf{P} \ll \mathbf{Q}, \mathbf{P}' \ll \mathbf{Q}$, then $M(\alpha)$

Chapter 3.III. Absolute Continuity and Singularity of Distributions

is a uniformly integrable martingale. The process $h(\alpha)$ is independent of $\mathbf{Q}$ in the following sense: if $\overline{\mathbf{Q}}$ is another measure and $\overline{\mathbf{Q}} \overset{\text{loc}}{\ll} \mathbf{Q}$ and $\bar{h}(\alpha)$ is a process defined similarly to $h(\alpha)$ with respect to $\overline{\mathbf{Q}}$, then the processes $h(\alpha)$ and $\bar{h}(\alpha)$ are $\mathbf{Q}$-indistinguishable.

Definition 2. The increasing predictable $\mathbf{P}$- and $\mathbf{P}'$-unique process $h(\alpha)$ is called the *Hellinger process* (in the strict sense) *of order* α for the measures $\mathbf{P}$ and $\mathbf{P}'$. By a Hellinger process of order α for the measures $\mathbf{P}$ and $\mathbf{P}'$ is meant any increasing process $h'(\alpha)$ such that the processes $I_{\Gamma''} \circ h(\alpha)$ and $I_{\Gamma''} \circ h'(\alpha)$ are $\mathbf{P}$ and $\mathbf{P}'$-indistinguishable.

In what follows, in order to emphasize the role of the measures $\mathbf{P}$ and $\mathbf{P}'$, the notation $h(\alpha; \mathbf{P}, \mathbf{P}')$ will be used for $h(\alpha)$.

Since the Hellinger process is associated with the density processes Z and Z' of the measures $\mathbf{P}$ and $\mathbf{P}'$ with respect to $\mathbf{Q}$, one of its versions can be expressed in predictable terms, connected with the processes Z and Z'. We assume for the sake of simplicity that $\mathbf{Q} = \frac{1}{2}(\mathbf{P}' + \mathbf{P})$. In this case

$$Z + Z' = 2$$

and hence, to represent the Hellinger process, one can restrict oneself to the predictable characteristics of the density process Z.

To this end, we define the following objects:

$$\phi_\alpha(u,v) = \alpha u + (1-\alpha)v - u^\alpha v^{1-\alpha},$$

Z^c, the continuous martingale part of the density process Z, and $\nu^z(dt, dx)$, the compensator of the jump measure of Z.

Theorem 3.29. *Let* $\mathbf{Q} = \frac{1}{2}(\mathbf{P}' + \mathbf{P})$. *Then the Hellinger process of order* α *is given in the strict sense by the formula:*

$$h(\alpha : \mathbf{P}, \mathbf{P}') = \frac{\alpha(1-\alpha)}{2}\left(\frac{1}{Z_-} + \frac{1}{2-Z_-}\right) \circ \langle Z^c \rangle$$
$$+ \phi_\alpha(1 + x/Z_-, 1 + x/(2 - Z_-)) * \nu^Z.$$

For $\mathbf{Q} = \frac{1}{2}(\mathbf{P}' + \mathbf{P})$ one can consider the Hellinger process $h(0; \mathbf{P}, \mathbf{P}')$.

Definition 3. By the *Hellinger process of order zero* for the measures $\mathbf{P}$ and $\mathbf{P}'$ is meant the process

$$h(0; \mathbf{P}, \mathbf{P}') = \left(1 - \frac{x}{2-Z_-}\right)\psi\left(\frac{1+x/Z_-}{1-x/(2-Z_-)}\right) * \nu^Z,$$

where

$$\psi(x) = \begin{cases} 1, & x = 0, \\ 0, & x > 0, \end{cases}$$

Z is the density process of $\mathbf{P}$ with respect to $\mathbf{Q}$ and ν^z is the compensator.

By the Hellinger process of order zero (in the strict sense) is meant a version $h(0; \mathbf{P}, \mathbf{P}')$ for which $h(0; \mathbf{P}, \mathbf{P}') = I_{\Gamma''} \circ h(0; \mathbf{P}, \mathbf{P}')$ (to within $\mathbf{Q}$-indistinguishability).

In the case of discrete time, the Hellinger process is expressed simply enough. Let $(\Omega, \mathcal{F}, \mathbf{F} = (\mathcal{F}_n)_{n \geq 0})$ be a measurable space with a discrete filtration, and $Z = (Z_n)_{n \geq 0}$, $Z' = (Z'_n)_{n \geq 0}$ local density processes of the measures $\mathbf{P}$ and $\mathbf{P}'$ with respect to $\mathbf{Q}$.

We set
$$\beta_n = Z_n/Z_{n-1}, \quad \beta'_n = Z'_n/Z'_{n-1}$$
(setting $0/0 = 0$ and bearing in mind that $Z_n = 0$ if $Z_{n-1} = 0$). Then the Hellinger process $h(\alpha; \mathbf{P}, \mathbf{P}') = (h_n(\alpha; \mathbf{P}, \mathbf{P}')_n)_{n \geq 1}$ is given by the formulas:
$$h(\alpha; \mathbf{P}, \mathbf{P}')_n = \sum_{k=1}^n \mathbf{E}_\mathbf{Q}(1 - \beta_k^\alpha \beta_k'^{1-\alpha}|\mathcal{F}_{k-1})$$
or
$$h(\alpha; \mathbf{P}, \mathbf{P}')_n = \sum_{k=1}^n \mathbf{E}_\mathbf{Q}(\phi_\alpha(\beta_k, \beta'_k)|\mathcal{F}_{k-1}).$$

We end this section by noting that in the case of local absolute continuity $\mathbf{P}' \overset{\text{loc}}{\ll} \mathbf{P}$ the Hellinger process is defined in the following manner:
$$h(\alpha; \mathbf{P}, \mathbf{P}') = \frac{\alpha(1-\alpha)}{2} \frac{1}{Z_-^2} \circ \langle Z^c \rangle$$
$$+ \{\alpha + (1-\alpha)(1 + x/Z_-) - (1 + x/Z_-)^{1-\alpha}\} * \nu^Z, \quad \alpha \in (0,1),$$
where Z is the local density process of $\mathbf{P}'$ with respect to $\mathbf{P}$, $\langle Z^c \rangle$ is the quadratic characteristic of the continuous martingale part of Z and ν^Z is the compensator of the jump measure of Z.

In particular,
$$h(\tfrac{1}{2}; \mathbf{P}, \mathbf{P}') = \frac{1}{8Z_-^2} \circ \langle Z^c \rangle + \tfrac{1}{2}\{1 - \sqrt{1 + x/Z_-}\}^2 * \nu^Z.$$

For $\alpha = 0$ we have
$$h(0; \mathbf{P}, \mathbf{P}') = (1 + x/Z_-)\psi\left(\frac{1}{1 + x/Z_-}\right) * \nu^Z,$$
where
$$\psi(x) = I(x = 0).$$

§4. General and Predictable Criteria of Absolute Continuity and Singularity of Probability Measures

4.1. Let $\mathbf{P}$ and $\mathbf{P}'$ be probability measures on $(\Omega, \mathcal{F})$ and $\mathbf{Q}$ be a probability measure on the same measurable space such that $\mathbf{P} \ll \mathbf{Q}$, $\mathbf{P}' \ll \mathbf{Q}$. We set $\mathfrak{z} = d\mathbf{P}/d\mathbf{Q}$ and $\mathfrak{z}' = d\mathbf{P}'/d\mathbf{Q}$. By the Lebesgue decomposition (see (3.24)), for any set A we have, in fact:

$$\mathbf{P}'(A) = \int_A \mathfrak{z}'/\mathfrak{z}\, d\mathbf{P} + \mathbf{P}'(A, \mathfrak{z} = 0).$$

Hence it is easy to derive criteria for the absolute continuity and singularity of the measures $\mathbf{P}$ and $\mathbf{P}'$. Namely,

$$\mathbf{P}' \ll \mathbf{P} \Leftrightarrow \mathbf{P}'(\mathfrak{z} > 0) = 1,$$
$$\mathbf{P}' \perp \mathbf{P} \Leftrightarrow \mathbf{P}'(\mathfrak{z} > 0) = 0.$$

One can give criteria for the absolute continuity and singularity of the measures $\mathbf{P}$ and $\mathbf{P}'$ in terms of the Hellinger integral $H(\alpha; \mathbf{P}, \mathbf{P}')$ of order α.

Theorem 3.30. a) *The following conditions are equivalent:*
(i) $\mathbf{P}' \ll \mathbf{P}$,
(ii) $\mathbf{P}'(\mathfrak{z} > 0) = 1$
(iii) $H(\alpha; \mathbf{P}, \mathbf{P}') \to 1$, $\alpha \downarrow 0$.
b) *The following conditions are equivalent:*
(i) $\mathbf{P}' \perp \mathbf{P}$,
(ii) $\mathbf{P}'(\mathfrak{z} > 0) = 0$
(iii) $H(\alpha; \mathbf{P}', \mathbf{P}) \to 0$, $\alpha \downarrow 0$.
(iv) $H(\alpha; \mathbf{P}', \mathbf{P}) = 0$ *for all* $\alpha \in (0, 1)$,
(v) $H(\alpha; \mathbf{P}', \mathbf{P}) = 0$ *for some* $\alpha \in (0, 1)$.

4.2. In analyzing the conditions of absolute continuity and singularity of measures corresponding to semimartingales, an essential role is played by the so-called predictable criteria. With the help of these criteria, the conditions of absolute continuity and singularity are formulated in terms of triples of predictable characteristics.

Theorem 3.31. *Let $(\Omega, \mathcal{F}, \mathbf{F} = (\mathcal{F}_t)_{t \geq 0}, \mathbf{Q})$ be a given stochastic basis, $\mathcal{F} = \mathcal{F}_{\infty-}$, $\mathbf{Q} = \frac{1}{2}(\mathbf{P}' + \mathbf{P})$; let $h(\alpha; \mathbf{P}, \mathbf{P}')$ be a Hellinger process of order $\alpha \in (\alpha \in [0, 1))$, and T a Markov time.*

The following conditions are equivalent:
(i) $\mathbf{P}'_T \ll \mathbf{P}_T$,
(ii) $\mathbf{P}'_0 \ll \mathbf{P}_0$ *and* $\mathbf{P}'(h(\frac{1}{2}; \mathbf{P}, \mathbf{P}')_T < \infty) = 1$, $\mathbf{P}'(h(0; \mathbf{P}, \mathbf{P}')_T = 0) = 1$,
(iii) $\mathbf{P}'_0 \ll \mathbf{P}_0$ *and* $h(\alpha; \mathbf{P}', \mathbf{P}) \xrightarrow{\mathbf{P}'} 0$, $\alpha \downarrow 0$ *(where $\xrightarrow{\mathbf{P}'}$ denotes convergence in $\mathbf{P}'$-probability)*.

Corollary. *If* $\mathbf{P}' \overset{\text{loc}}{\ll} \mathbf{P}$, *then for the absolute continuity of* $\mathbf{P}' \ll \mathbf{P}$ *it is necessary and sufficient that*

$$\mathbf{P}'(h(\tfrac{1}{2}; \mathbf{P}, \mathbf{P}')_\infty < \infty) = 1.$$

Remark. In the statement of the theorem and corollary $h(\tfrac{1}{2}; \mathbf{P}, \mathbf{P}')$ can be replaced by $h(\beta; \mathbf{P}, \mathbf{P}')$ for any $\beta \in (0,1)$.

For the formulation of the results connected with the singularity of the measures $\mathbf{P}$ and $\mathbf{P}'$, we define the following objects.

Let Z and Z' be density processes of the measures $\mathbf{P}$ and $\mathbf{P}'$ with respect to $\mathbf{Q}$. We set

$$G_0 = \{Z_0 > 0,\ Z'_0 > 0\},$$
$$C_T = G_0 \cap \{h(\tfrac{1}{2}; \mathbf{P}, \mathbf{P}')_T < \infty\} \cap \{h(0; \mathbf{P}, \mathbf{P}')_T = 0\},$$
$$\widetilde{G}_T = G_0 \cap \{\limsup_{\alpha \downarrow 0} h(\alpha; \mathbf{P}, \mathbf{P}^\alpha)_T = 0\},$$

where T is a Markov time and $G_T = \widetilde{G}_T = G_0$ on the set $\{T = 0\}$.

Theorem 3.32. *In the notation of Theorem 3.31 we have the following:*
a) *There is the implication:*

$$\mathbf{P}'_T \perp \mathbf{P}_T \Rightarrow \mathbf{P}'(G_T) = 0 \text{ and } \mathbf{P}'(\widetilde{G}_T).$$

b) *If at least one of the conditions* $\mathbf{P}'_0 \perp \mathbf{P}_0$ *or* $\mathbf{P}'(h(\tfrac{1}{2}; \mathbf{P}, \mathbf{P}')_T < \infty) = 0$ *holds, then*

$$\mathbf{P}'_T \perp \mathbf{P}_T.$$

Corollary. *Let* $\mathbf{P}' \overset{\text{loc}}{\ll} \mathbf{P}$. *Then for the singularity of the measures* $\mathbf{P}'$ *and* $\mathbf{P}$ ($\mathbf{P}' \perp \mathbf{P}$) *it is necessary and sufficient that*

$$\mathbf{P}'(h(\tfrac{1}{2}; \mathbf{P}, \mathbf{P}')_\infty < \infty) = 0.$$

§5. Particular Cases

In this section we look at the realizations of the results stated in Theorems 3.31 and 3.32 and the corollary to them, for a number of particular cases corresponding to distributions of certain stochastic processes.

Throughout this section we suppose that the stochastic process X is defined on a stochastic basis $(\Omega, \mathcal{F}, \mathbf{F}, \tfrac{1}{2}(\mathbf{P}' + \mathbf{P}))$, where $\mathbf{P}'$ and $\mathbf{P}$ are probability measures, $\mathbf{F}$ is $\mathbf{F}_+^{(\mathbf{P}'+\mathbf{P})/2}(X)$ and $\mathcal{F} = \mathcal{F}_{\infty-}^{(\mathbf{P}'+\mathbf{P})/2}$.

5.1. Discrete Time. Let $(\Omega, \mathcal{F}, \mathbf{F} = (\mathcal{F}_n)_{n \geq 0})$ be a measurable space with a discrete filtration $\mathbf{F} = (\mathcal{F}_n)_{n \geq 0}$. The restrictions $\mathbf{P}_n$ and $\mathbf{P}'_n$ of the measures $\mathbf{P}$

Chapter 3.III. Absolute Continuity and Singularity of Distributions   149

and $\mathbf{P}'$ to the σ-algebra $\mathcal{F}_n$ can be treated as the distribution of the random sequences $(\xi_0, \xi_1, \ldots, \xi_n)$ and $(\xi'_0, \xi'_1, \ldots, \xi'_n)$. We assume that for each $n \geq 0$, $\mathbf{P}'_n \ll \mathbf{P}_n$, that is, $\mathbf{P}' \overset{\text{loc}}{\ll} \mathbf{P}$. In this case,

$$h(\tfrac{1}{2}; \mathbf{P}, \mathbf{P}')_n = \sum_{k=0}^{n} \tfrac{1}{2}\mathbf{E}[(1 - \sqrt{\beta_k})^2 | \mathcal{F}_{k-1}],$$

where $\beta_k = Z_k/Z_{k-1}$, $Z_k = d\mathbf{P}'_k/d\mathbf{P}_k$. Hence, by the corollaries to Theorems 3.31 and 3.32,

$$\mathbf{P}' \ll \mathbf{P} \Leftrightarrow \mathbf{P}'\Big(\sum_{k \geq 0} \mathbf{E}[1 - \sqrt{\beta_k}]^2 | \mathcal{F}_{k-1}] < \infty\Big) = 1,$$

$$\mathbf{P}' \perp \mathbf{P} \Leftrightarrow \mathbf{P}'\Big(\sum_{k \geq 0} \mathbf{E}[1 - \sqrt{\beta_k}]^2 | \mathcal{F}_{k-1}] < \infty\Big) = 0.$$

5.2. Point Process. Let the semimartingale X be a point process with compensators A and A' with respect to the measures $\mathbf{P}$ and $\mathbf{P}'$, respectively.

We obtain from Theorem 3.31 and 3.32 the following results.

Theorem 3.33. *For the absolute continuity of $\mathbf{P}' \ll \mathbf{P}$ it is necessary and sufficient that the following conditions hold:*
1) *there exists a non-negative predictable process such that $A' = \lambda \circ A$,*
2) $\Delta A_t = 1 \Rightarrow \Delta A'_t = \lambda_t \Delta A_t = 1$,
3) $\mathbf{P}'\Big((1 - \sqrt{\lambda}) \circ A_\infty + \sum_{t>0}(\sqrt{1 - \Delta A_t} - \sqrt{1 - \Delta A'_t})^2 < \infty\Big) = 1$.

Theorem 3.34. *Under the condition 1) of Theorem 3.33 and the condition*

$$\mathbf{P}'\Big((1 - \sqrt{\lambda})^2 \circ A_\infty + \sum_{t>0}(\sqrt{1 - \Delta A_t} - \sqrt{1 - \Delta A'_t})^2 < \infty\Big) = 0$$

the measures $\mathbf{P}'$ and $\mathbf{P}$ are singular: $\mathbf{P}' \perp \mathbf{P}$.

5.3. A Semimartingale with a Gaussian Martingale Part. We suppose that the semimartingale X with $X_0 = 0$ admits the representation:

$$X = A + M' \quad \mathbf{P}'\text{-a.s.}$$
$$X = M \quad \mathbf{P}\text{-a.s.},$$

where A is a predictable process of finite variation on each finite interval and M' and M are Gaussian martingales with respect to $\mathbf{P}'$ and $\mathbf{P}$, respectively.

Let us note that, in view of the Gaussian nature of the processes M' and M, we have

$$\langle M \rangle_t = \mathbf{E}M_t^2, \quad \langle M^c \rangle_t = \langle M \rangle_t^c, \quad \langle M^d \rangle_t = \langle M \rangle_t^d,$$

where $\mathbf{E}$ is the expectation with respect to the measure $\mathbf{P}$ (the quadratic characteristics of M' are defined similarly by replacing $\mathbf{E}$ by the expectation $\mathbf{E}'$ with respect to $\mathbf{P}'$).

Theorem 3.35. *For the absolute continuity of $\mathbf{P}' \ll \mathbf{P}$ it is necessary and sufficient that the following conditions hold:*

1) *there exist predictable processes $\gamma = (\gamma_t)_{t\geq 0}$ and $\rho = (\rho_t)_{t\geq 0}$ ($\rho \geq 0$) such that*

$$A = \gamma \circ \langle M \rangle, \quad \langle M' \rangle^d = \rho \circ \langle M \rangle^d, \quad \langle M \rangle^d = \rho^{-1} \circ \langle M' \rangle^d,$$

2) $\langle M \rangle^c = \langle M' \rangle^c$,

3) $\mathbf{P}'\left(\gamma^2 \circ \langle M' \rangle_\infty + \sum_{t>0} I(\Delta\langle M \rangle_t > 0)(1 - \rho_t)^2 < \infty\right) = 1.$

If instead of condition 3) we have

(3') $\mathbf{P}'\left(\gamma^2 \circ \langle M' \rangle_\infty + \sum_{t>0} I(\Delta\langle M \rangle_t > 0)(1 - \rho_t)^2 < \infty\right) = 0,$ *then* $\mathbf{P}' \perp \mathbf{P}.$

Example. Let X be a solution with respect to $\mathbf{P}'$ of the Itô stochastic equation

$$dX_t = a(t, X_t)dt + dW_t, \; X_0 = 0$$

with respect to a Wiener process $W = (W_t)_{t \geq 0}$, and let $\mathbf{P}$ be Wiener measure. Then $\mathbf{P}' \ll \mathbf{P}$ if and only if

$$\int_0^\infty a^2(t, X_t)dt < \infty \quad \mathbf{P}'\text{-a.s.}$$

In the case when $\int_0^T a^2(t, X_t)dt < \infty$ $\mathbf{P}'$-a.s., where T is a Markov time, we have $\mathbf{P}'_T \ll \mathbf{P}_T$.

5.4. Processes with Independent Increments. The semimartingale X is a process with independent increments and triples of predictable characteristics $T = (B, C, \nu)$ and $T' = (B', C', \nu')$ with respect to $\mathbf{P}$ and $\mathbf{P}'$, respectively. Since X is a process with independent increments, the triples T and T' are deterministic. We set

$$a_t = \nu(\{t\} \times \mathbb{R}\setminus\{0\}), \; a'_t = \nu'(\{t\} \times \mathbb{R}\setminus\{0\}).$$

The next result follows from Theorem 3.31.

Theorem 3.36. *The following conditions are necessary and sufficient for the absolute continuity of $\mathbf{P}' \ll \mathbf{P}$:*

1) $\mathbf{P}'_0 \ll \mathbf{P}_0$, *where $\mathbf{P}'_0$ and $\mathbf{P}_0$ are the restrictions of $\mathbf{P}'$ and $\mathbf{P}$ to the σ-algebra $\mathcal{F}_0$),*

2) *there exists a non-negative measurable deterministic function $Y = Y(t, x)$ such that*

$$\nu'(dt, dx) = Y(t, x)\nu(dt, dx),$$

3) $a_t = 1 \Rightarrow a'_t = 1$,

4) $I(|x| \leq 1)|x(Y-1)| * \nu_t < \infty, \; \forall t > 0,$

5) *there exists a measurable deterministic function $\beta = (\beta_t)_{t \geq 0}$ such that $\beta^2 \circ C_t < \infty, \; \forall t > 0$ and*

Chapter 3.III. Absolute Continuity and Singularity of Distributions 151

$$B' = B + \beta \circ C + I(|x| \leq 1)x(Y-1) * \nu,$$

6) $C = C'$,
7) $\beta^2 \circ C_\infty + (1 - \sqrt{Y})^2 * \nu_\infty + \sum_{t>0}(\sqrt{1-a_t} - \sqrt{1-a'_t})^2 < \infty$.

Remark. If conditions 1)–6) and the condition

$$L_t = \beta_0^2 C_t + (1 - \sqrt{Y})^2 * \nu_t + \sum_{s \leq t}(\sqrt{1-a_s} - \sqrt{1-a'_s})^2 = \infty, \ \forall t > 0$$

are satisfied, then

$$\mathbf{P'} \perp \mathbf{P} \Leftrightarrow L_\infty = \infty.$$

5.5. Markov Processes with a Countable Set of States. Suppose that $X = (X_t)_{t \geq 0}$ takes values in the set $J = (\alpha, \beta, \gamma, \ldots)$ and is a Markov process with respect to $\mathbf{P}$ and $\mathbf{P'}$ with intensity transition matrices $\|\lambda_{\alpha\beta}(t)\|$ and $\|\lambda'_{\alpha\beta}(t)\|$, respectively, where $\lambda_{\alpha\beta}(t)$ and $\lambda'_{\alpha\beta}(t)$ are measurable functions possessing the following properties:

$$\sum_{\beta \in J} \lambda_{\alpha\beta}(t) = 0, \ \sum_{\beta \in J} \lambda'_{\alpha\beta}(t) = 0, \ \alpha \in J,$$

$$\int_0^t \sup_\alpha |\lambda_{\alpha\alpha}(s)| ds < \infty, \ \int_0^t \sup_\alpha |\lambda'_{\alpha\alpha}(s)| ds < \infty, \ t > 0.$$

Let us define the stochastic processes $X_\beta = (X_\beta(t))_{t \geq 0}$, $\beta \in J$, with

$$X_\beta(t) = I(X_t = \beta).$$

The process X_β is a semimartingale with respect to $\mathbf{P}$ and $\mathbf{P'}$ with decompositions with respect to $\mathbf{P}$ and $\mathbf{P'}$:

$$X_\beta(t) = X_\beta(0) + \int_0^t \lambda_{X_s,\beta}(s)ds + M_\beta(t),$$

$$X_\beta(t) = X_\beta(0) + \int_0^t \lambda'_{X_s,\beta}(s)ds + M'_\beta(t),$$

where $M_\beta = (M_\beta(t))_{t \geq 0}$ and $M'_\beta = (M'_\beta(t))_{t \geq 0}$ are square-integrable martingales (with respect to $\mathbf{P}$ and $\mathbf{P'}$).

Using these decompositions, Theorem 3.31 and the corollary to it, we obtain the next result.

Theorem 3.37. *For the absolute continuity of $\mathbf{P'} \ll \mathbf{P}$ it is necessary and sufficient that the following conditions hold:*
a) $\mathbf{P}(X_0 = \alpha) = 0 \Rightarrow \mathbf{P'}(X_0 = \alpha) = 0, \ \forall \alpha \in J$,
b) $\int_0^t I(X_s = \alpha)\lambda'_{\alpha\beta}(s)ds = \int_0^t I(X_s = \alpha)\lambda'_{\alpha\beta}(s)I(\lambda_{\alpha\beta}(s) > 0)ds$,
$\mathbf{P}$-a.s., $t > 0$, $\alpha, \beta \in J$,

c) $\mathbf{P}'\left(\int_0^\infty \sum_{\alpha\neq\beta}\left(1-\sqrt{\frac{\lambda'_{\alpha\beta}(s)}{\lambda_{\alpha\beta}(s)}}I(\lambda_{\alpha\beta}(s)>0)\right)^2 I(X_s=\alpha)\lambda_{\alpha\beta}(s)ds<\infty\right)=1.$

5.6. Semimartingales with a Local Uniqueness Condition. We suppose that $X=(X_t)_{t\geq 0}$ is a semimartingale with respect to $\mathbf{P}$ and $\mathbf{P}'$ with triples of predictable characteristics $T=(B,C,\nu)$ and $T'=(B',C',\nu')$.

Theorem 3.38 (cf. Theorem 3.35). *In order that* $\mathbf{P}'\ll\mathbf{P}$ *be absolutely continuous, it suffices that the following conditions hold:*
1) $\mathbf{P}'_0\ll\mathbf{P}_0$,
2) there exists a non-negative $\widetilde{\mathcal{P}}$-measurable function $Y=Y(\omega,t,x)$ such that
$$\nu'(\omega;dt,dx)=Y(\omega,t,x)\nu(\omega;dt,dx)\ \mathbf{P}'\text{-a.s.}$$
3) $\nu(\{t\}\times\mathbb{R}\setminus\{0\})=1 \Rightarrow \nu'(\{t\}\times\mathbb{R}\setminus\{0\})=1\ \mathbf{P}'\text{-a.s.}$,
4) $I(|x|\leq 1)|x(Y-1)|*\nu_t<\infty\ \mathbf{P}'\text{-a.s.},\ t>0$,
5) there exists a $\mathcal{P}$-measurable function $\beta=\beta(\omega,t)$ such that
$$\beta^2\circ C_t<\infty\ \mathbf{P}'\text{-a.s.},\ t>0$$
and
$$B'=B+\beta\circ C+I(|x|\leq 1)x(Y-1)*\nu\ \mathbf{P}'\text{-a.s.}$$
6) $C=C'$,
7) $\mathbf{P}'(\beta^2\circ C_\infty+(1-\sqrt{Y})^2*\nu_\infty+\sum_{t>0}(\sqrt{1-a_t}-\sqrt{1-a'_t})^2<\infty)=1$,
8) the martingale problem $\mathcal{S}(\mathcal{F}_0,X|\mathbf{P}'_0;B',C',\nu')$ possesses the local uniqueness property for the measure $\mathbf{P}'$.

The absolute continuity of $\mathbf{P}'\ll\mathbf{P}$ implies conditions 1)–7) (see (Kabanov, Liptser and Shiryaev 1978/1979); (Jacod and Shiryaev 1987)).

Commentary to Chapter 3

I.
§1. The axioms of Kolmogorov were presented in his books (Kolmogorov 1933, 1974). After the introduction of the Kolmogorov axioms, the next important step in the development of stochastic calculus was the introduction of a filtration (a flow of σ-algebras), which made it possible to consider with greater precision the structure of stochastic objects and obtain deeper results. Doob (Doob 1953) initiated this analysis.
§2. The material laid out here forms the basis of the general theory of stochastic processes and is based mainly on the results of the "Strasbourg school of probabilists" (Meyer, Dellacherie, Doleans-Dade ...). The material presented here was borrowed from the books: (Meyer 1966); (Dellacherie and Meyer 1975, 1980, 1983);

Chapter 3. Stochastic Calculus on Probability spaces with Filtrations

(Dellacherie 1972); (Jacod 1979); (Métivier 1982); (Elliott 1982); (Gikhman and Skorokhod 1982).

§3. The concept of a local martingale was introduced in the paper (Itô and Watanabe 1965).

§4. Increasing processes as an important and independent class were systematically considered in the monographs (Dellacherie 1972); (Meyer 1966). Regarding the "Doob-Meyer decomposition", see (Meyer 1972), (Dellacherie and Meyer 1975).

§5. Here the exposition of the theory of random measures and their compensators follows, in the main, the scheme proposed by Jacod (Jacod 1975); see also (Jacod 1979); (Jacod and Shiryaev 1987); (Liptser and Shiryaev 1986), Liptser and Shiryaev 1978); (Kabanov, Liptser and Shiryaev 1975).

§6. The role of a square-integrable martingale in martingale theory was largely clarified in the paper (Kunita and Watanabe 1967); see also (Dellacherie and Meyer 1975). The quadratic characteristic of a local martingale was introduced by Meyer (Meyer 1962), and the mutual quadratic characteristic by Kunita and Watanabe (Kunita and Watanabe 1967).

§7. The first decomposition of a local martingale in the case of discrete time is called the "Gundy decomposition". The second decomposition and its properties were established by Meyer (Meyer 1967); see also (Jacod 1979).

II.

§1. Semimartingales as an independent class were introduced in (Doleans-Dade and Meyer 1970), and special semimartingales in (Yoeurp 1976) and (Meyer 1967). A complete exposition of the theory of semimartingales can be found in the books (Dellacherie and Meyer 1975, 1980, 1983); (Jacod 1979), (Métivier 1982); (Elliott 1982); (Gikhman and Skorokhod 1982); (Jacod and Shiryaev 1987); (Liptser and Shiryaev 1986). The quadratic variation of a local martingale was introduced by Meyer (Meyer 1967). Information concerning the quadratic variation of a semimartingale is contained in (Dellacherie and Meyer 1975, 1980, 1983); (Elliott 1982); (Jacod, Shiryaev 1987); (Liptser and Shiryaev 1986). The concept of a quasimartingale was introduced in (Fisk 1965); (Orey 1965); see also (Métivier 1982); (Jacod 1979); (Rao 1969); (Liptser and Shiryaev 1986).

§2. The first construction of a stochastic integral with respect to a Wiener process of deterministic integrands was given in 1923 by Wiener. The general definition of a stochastic integral with respect to a Wiener process is due to Itô (Itô 1944). Integration with respect to square-integrable martingales was considered in (Doob 1953); (Meyer 1967); (Courrège 1963); (Kunita and Watanabe 1967). The construction of a stochastic integral with respect to a local martingale was given in (Jacod 1979); (Doleans-Dade and Meyer 1970); (Meyer 1967); see also (Métivier 1982); (Dellacherie and Meyer 1980); (Elliott 1982); (Jacod and Shiryaev 1987), (Liptser and Shiryaev 1986).

§3. The Itô formula was established in (Itô 1951) in the case of a semimartingale of diffusion type. Regarding the Itô formula for a semimartingale of general form, see (Dellacherie, Meyer 1980); (Gikhman and Skorokhod 1982); (Elliott 1982).

§4. The construction of a stochastic integral with respect to an integer-valued random martingale measure was given in (Kabanov, Liptser and Shiryaev 1978, 1979); (Jacod 1979); (Liptser and Shiryaev 1986).

§5. Triples of predictable characteristics for locally unboundedly divisible processes were introduced by Grigelionis (Grigelionis 1971a, 1975a, 1975b). Triples were systematically considered in (Jacod and Mémin 1976); (Kabanov, Liptser and Shiryaev 1978, 1979) in connection with questions of an absolutely continuous change of measure. Regarding the canonical representation, see (Dellacherie, Meyer 1983); (Jacod 1979); (Jacod and Shiryaev 1987); (Liptser and Shiryaev 1986). Regarding the martingale problem, see (Jacod and Shiryaev 1987).

§6. See (Venttsel' 1961); (Kunita, Watanabe 1967); (Dellacherie 1974); (Grigelionis 1974); (Liptser 1976); (Jacod 1977); (Liptser and Shiryaev 1986).

§7. The preservation of the semimartingale property under an absolutely continuous change of measure was first established by Girsanov (Girsanov 1960); see also (Grigelionis 1969, 1971b); (Kabanov, Liptser and Shiryaev 1978, 1979); (Jacod, Mémin 1976); (Jacod and Shiryaev 1987), (Liptser and Shiryaev 1986). The preservation of the semimartingale property under a random change of measure was presented in (Dellacherie and Meyer 1975); (Jacod 1979). In greatest generality, the preservation of the semimartingale property under the reduction of the filtration was set forth in (Stricker 1977); see also (Liptser and Shiryaev 1986).

III. §1. The concept of local absolute continuity and local density was introduced in (Kabanov, Liptser and Shiryaev 1978, 1979).

§2. Girsanov's theorem can be found in (Girsanov 1960). Its generalization is due to Van Schuppen and Wong (Van Schuppen and Wong 1974). The transformation of predictable characteristics under an absolutely continuous change of measure was given in (Grigelionis 1969, 1971b, 1973); (Kabanov, Liptser and Shiryaev 1978, 1979); (Jacod and Mémin 1976); (Jacod and Shiryaev 1987), (Liptser and Shiryaev 1986).

§3. The Hellinger integral is used in solving the problem of absolute continuity and singularity; see (Kakutani 1948); (Liese 1982, 1983). The Hellinger process was introduced by Liptser and Shiryaev (Liptser and Shiryaev 1983).

§4. See (Jacod and Shiryaev 1987); (Kabanov, Liptser and Shiryaev 1978, 1979); (Jacod and Shiryaev 1987).

References[*]

Bernshtein, S.N. (1917): An attempt to axiomatize the foundation of probability theory. Soobshch. Khar'k Mat. O.-va *15*, 209–274
Bohlmann, G. (1909): Die Grundbergriffe der Wahrscheinlichkeitsrechnung in ihrer Anwendung auf die Lebensversicherung. Atti IV. Congresso Int. Math. Roma 1908, III, Sec. 116, 244–278. Jbuch 40,291
Courrège, Ph. (1963): Intégrales stochastiques associées à une martingale de carré intégrable. C. R. Acad. Sci., Paris, Ser. I *256*, 867–870. Zbl. 113,333
Dellacherie, C. (1972): Capacités et Processus Stochastiques. Springer, Berlin Heidelberg New York. Zbl. 246.60032
Dellacherie, C. (1974): Intégrales stochastiques par rapport aux processus de Wiener ou de Poisson. Sémin Probab. VIII. Lect. Notes Math. *381*, pp. 25–26. (Correction; Lect. Notes Math. *465*, p. 494 (1975). Zbl. 302.60049, Zbl. 311.60036
Dellacherie, C. (1980): Un survol de la théorie de l'intégrale stochastique. Stochastic Processes Appl. *10*, 115–144. Zbl. 436.60043
Dellacherie, C., Meyer, P. (1975): Probabilités et Potentiel. I. Hermann, Paris. Zbl. 323.60039
Dellacherie, C., Meyer, P. (1980): Probabilités et Potentiel. II. Hermann, Paris. Zbl. 464.60001

---

[*] For the convenience of the reader, references to reviews in Zentralblatt für Mathematik (Zbl.), compiled using the MATH database, have, as far as possible, been included in this bibliography.

Dellacherie, C., Meyer, P. (1983): Probabilités et Potentiel. III. Hermann, Paris. Zbl. 526.60001

Doleans-Dade, C., Meyer, P.A. (1970): Intégrales stochastiques par rapport aux martingales locales. Sémin Probab. IV. Lect. Notes Math. *124*. pp. 77–107. Zbl. 211,219

Doob, J.L. (1953): Stochastic Processes. John Wiley, New York. Zbl. 53,268

Elliott, R.J. (1982): Stochastic Calculus and Applications. Springer, Berlin Heidelberg New York. Zbl. 503.60062

Fisk, D.L. (1965): Quasimartingales. Trans. Am. Math. Soc. *120*, 369–389. Zbl. 133,403

Gikhman, I.I., Skorokhod, A.V. (1982): Stochastic Differential Equations and Their Applications. Naukova Dumka, Kiev. Zbl. 557.60041

Girsanov, I.V. (1960): Transformation of a class of stochastic processes by means of an absolutely continuous change of measure. Teor. Veroyatn. Primen. *5*, 314–330. [English transl.: Theory Probab. Appl. *5*, 285–301 (1962)] Zbl. 100,340

Grigelionis, B.I. (1969): On absolutely continuous change of measure and the Markov property of stochastic processes. Litov. Mat. Sb. *9*, No. 1, 57–71. Zbl. 221.60052

Grigelionis, B.I. (1971a): On the representation of integer-valued random measures by means of stochastic integrals with respect to a Poisson measure. Litov. Mat. Sb. *11*, No. 3, 93–108. Zbl. 239.60050

Grigelionis, B.I. (1971b): On the absolute continuity of measures corresponding to stochastic processes. Litov. Mat. Sb. *11*, No. 3, 783–794. Zbl. 231.60024

Grigelionis, B.I. (1973): Structure of densities of measures corresponding to stochastic processes. Litov. Mat. Sb. *13*, No. 1, 71–78. [English transl.: Math. Trans. Acad. Sci. Lith. SSR *13*, 48–52 (1974)] Zbl. 277.60017

Grigelionis, B.I. (1974): Representation by stochastic integrals of square-integrable martingales. Litov. Mat. Sb. *14*, No. 3, 53–69. [English transl.: Lith. Math. Trans. *14*, 573–584 (1975)] Zbl. 301.60036

Grigelionis, B.I. (1975a): Random point processes and martingales. Litov. Mat. Sb. *15*, No. 3, 101–114. [English transl.: Lith. Math. J. *15*, 444-453 (1976)] Zbl. 332.60035

Grigelionis, B.I. (1975b): Characterization of stochastic processes with conditionally independent increments. Litov. Mat. Sb. *15*, No. 4, 53–58. [English transl.: Lith. Math. J. *15*, 562–567 (1976)] Zbl. 353.60079

Hilbert, D. (1901): Mathematische Probleme. Arch. Math. Phys., III. *1*, 44–63, 213–237. Jbuch 32,84

Itô, K. (1944): Stochastic integral. Proc. Imp. Acad. Tokyo *20*, 519–524. Zbl. 60,291

Itô, K. (1951): On a formula concerning stochastic differentials. Nagoya Math. J. *3*, 55–65. Zbl. 45,76

Itô, K., Watanabe, S. (1965): Transformation of Markov processes by multiplicative functionals. Colloq. Int. Cent. Nat. Rech. Sci. *146*, 13–30. Zbl. 141,151

Jacod, J. (1975): Multivariate point processes: predictable projection, Radon-Nikodym derivatives, representation of martingales. Z. Wahrscheinlichkeitstheorie verw. Gebiete *31*, 235–253. Zbl. 302.60032

Jacod, J. (1977): A general theorem of representation for martingales. Proc. Symp. Pure Math. Urbana 1976, 37–53. Zbl. 362.60068

Jacod, J. (1979): Calcul Stochastique et Problèmes de Martingales. Lect. Notes Math. *714*. Zbl. 414.60053

Jacod, J., Mémin, J. (1976): Caractéristiques locales et conditions de continuité absolue pour les semimartingales. Z. Wahrscheinlichkeitstheorie verw. Gebiete *35*, 1–37. Zbl. 315.60026

Jacod, J., Shiryaev, A.N. (1987): Limit Theorems for Stochastic Processes. Springer, Berlin Heidelberg New York. Zbl. 635.60021

Kabanov, Yu.M., Liptser, R.Sh., Shiryaev, A.N. (1975): Martingale methods in the theory of point processes. Proc. School Semin. Theory of Stochastic Processes, Druskininkai, 1974. Part II, Vil'nyus, pp. 269–354 (Russian)

Kabanov, Yu.M., Liptser, R.Sh., Shiryaev, A.N. (1978, 1979): Absolute continuity and singularity of locally absolutely continuous probability distributions. Mat. Sb., Nov. Ser. *107* (*149*), No. 3, 364–415; *108* (*150*), No. 1, 32–61. [English transl.: Math. USSR, Sb. *35*, 631–680 (1979); *36*, 31–58 (1980)] Zbl. 426.60039; Zbl. 427.60037

Kakutani, S. (1948): On equivalence of infinite product measures. Ann. Math. II. Ser. *49*, 214–224. Zbl. 30,23

Kolmogorov, A.N. (1929): General measure theory and probability calculus. Tr. Kommunist. Akad., Mat. Razdel *1*, 8–21 (Russian)

Kolmogorov, A.N. (1933): Grundbegriffe der Wahrscheinlichkeitsrechnung. Springer, Berlin Heidelberg New York. Zbl. 7,216

Kolmogorov, A.N. (1974): Fundamental Concepts of Probability Theory. Nauka, Moscow. Zbl. 278.60001

Kunita, H., Watanabe, S. (1967): On square-integrable martingales. Nagoya Math. J. *30*, 209–245. Zbl. 167,466

Liese, F. (1982): Hellinger integrals of Gaussian processes with independent increments. Stochastics *6*, 81–96. Zbl. 476.60041

Liese, F. (1983): Hellinger integrals of diffusion processes. Forschungsergeb., Friedrich-Schiller-Univ. Jena N/83/89; appeared also in Statistics *17*, 63–78 (1986). Zbl. 517.62006; Zbl. 598.60042

Liptser, R.Sh. (1976): The representation of local martingales. Teor. Veroyatn. Primen. *21*, No. 4, 718–726. [English transl. Theory Probab. Appl. *21*, 698–705 (1977)] Zbl. 385.60051

Liptser, R.Sh., Shiryaev, A.N. (1977/78): Statistics of Random Processes I. General Theory; II. Applications. Springer, Berlin Heidelberg New York. [transl. from the Russian original. Nauka, Moscow, 1974] Zbl. 364.60004; Zbl. 369.60001; Zbl. 279.60021

Liptser, R.Sh., Shiryaev, A.N. (1983): On the problem of "predictable" criteria of contiguity. Proc. 5th Japan-USSR Symp. Lect. Notes Math *1021*, pp 384–418

Liptser, R.Sh., Shiryaev, A.N. (1986): Theory of Martingales. Nauka, Moscow. [English transl.: Kluwer, Dordrecht, 1989] Zbl. 654.60035; Zbl. 728.60048

Lomnicki, A. (1923): Nouveaux fondements du calcul des probabilités. Fundam. Math. *4*, 34–71. Jbuch 49,360

Métivier, M. (1982): Semimartingales. De Gruyter, Berlin New York. Zbl. 503.60054

Meyer, P.A. (1962): A decomposition theorem for supermartingales. Ill. J. Math. *6*, 193–205; *7*, 1–17. Zbl. 133,403; Zbl. 133,404

Meyer, P.A. (1966): Probabilités et Potentiel. Hermann, Paris. Zbl. 138,104

Meyer, P.A. (1967): Intégrales stochastiques I–IV, Sémin. Probab. I. Lect. Notes Math. *39*, pp. 72–162. Zbl. 157,250

von Mises, R. (1919): Grundlage der Wahrscheinlichkeitsrechnung. Math. Z. *5*, 52–99. Jbuch 47,483

Orey, S. (1965): F-processes. Proc. 5th Berkeley Symp. 2. Univ. California 1965/66, No. 2, Pt. 1, Berkeley, pp. 301–313. Zbl. 201,499

Rao, K.M. (1969): Quasimartingales. Math. Scand. *24*, 79–92. Zbl. 193,455

Stricker, C. (1977): Quasimartingales, martingales locales, semimartingales et filtration naturelle. Z. Wahrscheinlichkeitstheorie verw. Gebiete *39*, 55–63. Zbl. 362.60069

Van Schuppen, J.H., Wong, E. (1974): Transformation of local martingales under a change of law. Ann. Probab. *2*, 879–888. Zbl. 321.60040

Venttsel', A.D. (1961): Additive functionals of a multidimensional Wiener process. Dokl. Akad. Nauk SSSR *139*, No. 1, 13–16. [English transl.: Sov. Math. Dokl. *2*, 848–851 (1961)] Zbl. 107,127

Yoeurp, Ch. (1976): Décompositions des martingales locales et formules exponentielles. Sémin. Probab. X. Lect. Notes Math. *511*, 432–480. Zbl. 346.60033

Chapter 4
Martingales and Limit Theorems for Stochastic Processes

R. Sh. Liptser and A.N. Shiryaev

I. Theory: Weak Convergence of Probability Measures on Metric Spaces

§1. Introduction

1.1. In contemporary probability theory, martingales are a wide class of processes to which such fundamental processes as Brownian motion and the centered Poisson process are related. If one pursues an analogy with mathematical physics and potential theory, then the concept corresponding to the martingale is the concept of the harmonic function. Related to the martingale are submartingales and supermartingales, and their corresponding concepts in analysis are subharmonic and superharmonic functions.

The classical example of a sum $S_n = \xi_1 + \ldots + \xi_n$ of independent integrable random variables with zero means, considered in its evolution, $n \geq 1$, $S_0 = 0$, is a typical example of a martingale, since

$$\mathbf{E}(S_n|\xi_1,\ldots,\xi_{n-1}) = \mathbf{E}(S_{n-1} + \xi_n|\xi_1,\ldots,\xi_{n-1})$$
$$= S_{n-1} + \mathbf{E}(\xi_n|\xi_1,\ldots,\xi_{n-1}) = S_{n-1} + \mathbf{E}\xi_n = S_{n-1},$$

that is, if $\mathcal{F}_n = \sigma\{\xi_1,\ldots,\xi_n\}$ is the collection of events generated by the variables $\xi_1,\ldots,\xi_n$, then
1) S_n are $\mathcal{F}_n$-measurable
2) $\mathbf{E}(S_n|\mathcal{F}_{n-1}) = S_{n-1}$,

in accordance with the definition of a martingale.

From the formal point of view, it is convenient to give the definition of a martingale in the following manner.

Let $(\Omega, \mathcal{F}, \mathbf{P})$ be some probability space and $\mathbf{F} = (\mathcal{F}_n)_{n\geq 1}$ be a non-decreasing family of σ-algebras such that $\mathcal{F}_0 \subseteq \mathcal{F}_1 \subseteq \mathcal{F}_2 \subseteq \ldots \subseteq \mathcal{F}$. (The sets in $\mathcal{F}_n$ are treated as the events observable up to and including time n). A random sequence or a process with discrete time $X = (X_n)_{n\geq 0}$ is called a martingale if $\mathbf{E}|X_n| < \infty$, $n \geq 0$,
1) X_n are $\mathcal{F}_n$-measurable, (4.1)
2) $\mathbf{E}(X_n|\mathcal{F}_{n-1}) = X_{n-1}$, $n \geq 1$. (4.2)

It is property 2) that gave rise to the name "martingale" — a word denoting a casino gaming system in which the stake is doubled in the event of a loss. Namely, let $\eta_1, \eta_2, \ldots$ be independent random variables with $\mathbf{P}(\eta_i = 1) = p$,

$\mathbf{P}(\eta_i = -1) = q$. Here p is treated as the probability of winning by the player in the ith game, and q that of losing. If $V_i = V_i(\eta_1, \ldots, \eta_{n-1})$ is the stake in the ith game, based naturally on the data of the preceding games, then the total winnings after n games will be

$$X_n = \sum_{i=1}^n Y_i \eta_i \ (= X_{n-1} + V_n \eta_n).$$

The game is fair if $\mathbf{E}(X_n | \eta_1, \ldots, \eta_{n-1}) = X_{n-1}$, which will be the case when $p = q = \frac{1}{2}$, since then $\mathbf{E}(V_n \eta_n | \eta_1, \ldots, \eta_{n-1}) = V_n \mathbf{E} \eta_n = V_n(p - q) = 0$.

If $p > q$, then the game becomes profitable for the player, since

$$\mathbf{E}(X_n | \eta_1, \ldots, \eta_{n-1}) \geq X_{n-1}.$$

This property of the sequence $X = (X_n)$ is called the submartingale property. If, on the other hand,

$$\mathbf{E}(X_n | \eta_1, \ldots, \eta_{n-1}) \leq X_{n-1},$$

then we say that X is a supermartingale.

Thus, if the process X is a martingale, then $\mathbf{E}X_n = \mathbf{E}X_0 = $ const; if X is a submartingale, then $\mathbf{E}X_n \geq \mathbf{E}X_{n-1}$; if X is a supermartingale, then $\mathbf{E}X_n \leq \mathbf{E}X_{n-1}$.

Let us consider the special class of "strategies" $V = (V_n)_{n \geq 1}$ with $V_1 = 1$ and

$$V_n = \begin{cases} 2^{n-1}, & \text{if } \eta_1 = -1, \ldots, \eta_{n-1} = -1, \\ 0, & \text{otherwise.} \end{cases}$$

This means that the player, starting with a stake $V_1 = 1$, each time increases the stake by double after a loss and ceases to play after a win.

If $\eta_1 = -1, \ldots, \eta_n = -1$, then the total loss after n games is given by

$$\sum_{i=1}^n 2^{i-1} = 2^n - 1.$$

Therefore if, in addition, $\eta_{n+1} = 1$, then

$$X_{n+1} = X_n + V_{n+1} = -(2^n - 1) + 2^n = 1.$$

Hence if $\tau = \inf\{n \geq 1 : X_n = 1\}$, then $\mathbf{P}(\tau = n) = (\frac{1}{2})^n$, $\mathbf{P}(\tau < \infty) = 1$, $\mathbf{P}(X_\tau = 1) = 1$ and $\mathbf{E}X_\tau = 1$, although for each $n \geq 0$ $\mathbf{E}X_n = \mathbf{E}X_0 = 0$. Indeed, even in a fair game, one can increase one's capital if one adheres to the gaming system of doubling the stake, called a martingale. (Let us observe that this game requires unrestricted initial capital and the possiblity of unrestricted stakes, which is, of course, physically non-realizable.)

1.2. Of course, the above example does not explain why martingales play an important role in the limit theorems of probability theory. Our object is to show precisely how martingale methods "work" in limit theorems.

We note straight away that when we speak of martingale methods, we have in mind not so much the techniques of operating with martingales, but rather the stochastic calculus, which provides a powerful method of studying stochastic processes trajectorywise. Stochastic calculus, in particular, the stochastic differential calculus of Itô, which forms one of the basic methods of constructing complex processes from simple ones, plays an important role in the theory of optimal stochastic control and the non-linear analysis of stochastic processes. Stochastic calculus has also provided a powerful method for proving limit theorems, especially in the case of dependent observations, where traditional methods, such as the method of characteristic functions, do not work.

With a view to setting forth the results under the rubric "martingales and limit theorems for stochastic processes", we set ourselves the following goals: First, a simple presentation of the appropriate results and the technical side of the subject; second, and more important, to trace (if only in outline) the basic landmarks in the formation and development of the theory of limit theorems, to trace the chronological order and evolution of the methods, and to discuss the strengths and weaknesses of the various methods and the boundaries of their applicability.

§2. Different Types of Convergence. Skorokhod Topology

2.1. All our stochastic processes $X = (X_t)_{t \geq 0}$, $X^n = (X^n_t)_{t \geq 0}$, $n \geq 1$, will be stochastic processes with a time index $t \in \mathbb{R}_+$, taking values on the real line E^1 (for simplicity of exposition; more generally, taking values in E^d, $d < \infty$). Our problem is to give conditions for the "convergence" of X^n to X, where convergence will be understood in an appropriate sense.

Let us recall that for random variables ξ, ξ^n, $n \geq 1$, the following three forms of convergence are fundamental in probability theory:

convergence in probability:

$$\xi^n \xrightarrow{\mathbf{P}} \xi \Leftrightarrow \mathbf{P}(|\xi^n - \xi| \geq \epsilon) \to 0;$$

convergence with probability one, or *almost sure convergence*:

$$\xi^n \xrightarrow{\text{a.s.}} \xi, \ \xi^n \to \xi \ (\mathbf{P}\text{-a.s.}) \Leftrightarrow \mathbf{P}(\xi^n \not\to \xi) = 0$$
$$\Leftrightarrow \mathbf{P}(\sup_{m \geq n} |\xi^m - \xi| \geq \epsilon) \to 0, \ n \to \infty.$$

Convergence in distribution, or *in law*

$$\xi^n \xrightarrow{d} \xi, \ \xi^n \xrightarrow{\mathcal{L}} \xi \Leftrightarrow F^n(x) \to F(x), \ x \in \mathbf{C}(F),$$

where $\mathbf{C}(F)$ is the set of points of continuity of the limiting distribution function F ($F^n(x) = \mathbf{P}(\xi^n \leq x)$, $F(x) = \mathbf{P}(\xi \leq x)$).

One can express this last convergence in an equivalent manner as weak convergence

$$\int_{\mathbb{R}} g(x)dF^n(x) \to \int_{\mathbb{R}} g(x)dF(x), \ \forall g \in BC$$

(BC is the class of bounded continuous functions).

When we are dealing with random vectors $\xi^n = (\xi_1^n, \ldots, \xi_k^n)$, $\xi = (\xi_1, \ldots, \xi_k)$, then by their convergence $\xi^n \xrightarrow{d} \xi$, or $\xi^n \xrightarrow{\mathcal{L}} \xi$ we mean weak convergence of the corresponding finite-dimensional distributions $\mathbf{P}^n(\cdot) = \mathbf{P}(\xi^n \in \cdot)$, $\mathbf{P}(\cdot) = \mathbf{P}(\xi \in \cdot)$, that is,

$$\int_{\mathbb{R}^n} g(x)\mathbf{P}^n(dx) \to \int_{\mathbb{R}^n} g(x)\mathbf{P}(dx), \ g \in BC.$$

2.2. We can view (and define) a stochastic process (say, $X = (X_t)_{t \geq 0}$) in different ways:
a) as a family of random variables X_t, $t \geq 0$;
b) as a random function or a random element in some function space;
c) as a measure in a function space.

Our basic function space will be the space

$$\mathbf{D} = \{x : x = (x_t)_{t \geq 0} \text{ is cadlag}\},$$

which is the space of functions without second-order discontinuities, that is, the space of functions continuous to the right and having limits to the left ("cadlag" stands for "continue à droite avec des limites à gauche"). We denote by $\mathbf{C}$ the subspace of $\mathbf{D}$ consisting of the continuous functions: $\mathbb{R}_+ \to E^1$ (in the general case: $\mathbb{R}_+ \to E^d$).

We can introduce into the space $\mathbf{C}$ a locally uniform topology associated with the metric

$$\delta_{\text{lu}}(\alpha, \beta) = \sum_N 2^{-N}(1 \wedge \|\alpha - \beta\|_N), \tag{4.3}$$

where $\alpha = \alpha(t)$, $\beta = \beta(t)$, $t \in \mathbb{R}_+$ and $\|\alpha - \beta\|_N = \sup_{s \leq N} |\alpha(s) - \beta(s)|$. Relative to this locally uniform metric, the space $\mathbf{C}$ is complete and separable (in other words, Polish). It is basic in the study of *continuous* processes $X = (X_t)_{t \geq 0}$, which can be considered as random elements with values in $\mathbf{C}$. In this space, there are distinguished two σ-algebras:

$\mathcal{B}_{\text{cy}}(\mathbf{C})$ is the cylinder σ-algebra, that is, the σ-algebra generated by sets of the form $B = \{x : (x_{t_1}, \ldots, x_{t_k}) \in A^k\}$, where A^k are Borel sets in E^k;

$\mathcal{B}(\mathbf{C})$ is the Borel σ-algebra, that is, the σ-algebra generated by all open subsets (in the δ_{lu} metric).

It turns out that these σ-algebras are the same:

$$\mathcal{B}_{\text{cy}}(\mathbf{C}) = \mathcal{B}(\mathbf{C}).$$

It is understood that the metric δ_{lu} enables us to talk about the weak convergence ($\mathbf{P}^n \xrightarrow{w} \mathbf{P}$) of the probability distributions $\mathbf{P}^n$ to $\mathbf{P}$, where

$$\mathbf{P}^n(B) = \mathbf{P}\{\omega; X^n \in B\},$$
$$\mathbf{P}(B) = \mathbf{P}\{\omega : X \in B\}, \ B \in \mathcal{B}(C),$$

defined as the convergence of

$$\int_C g(x)\mathbf{P}^n(dx) \to \int_C g(x)\mathbf{P}(dx)$$

for each continuous bounded function $g = g(x)$ (in the δ_{lu} metric).

We will also denote the weak convergence $\mathbf{P}^n \xrightarrow{w} \mathbf{P}$ of the probability distributions of the processes X^n to X by $X^n \xrightarrow{\mathcal{L}} X$, saying that X^n converges in law (or in distribution) to X.

We need also to speak of the convergence

$$X^n \xrightarrow{\mathcal{L}(S)} X,$$

meaning the weak convergence of all possible finite-dimensional distributions of vectors $(X^n_{t_1}, \ldots, X^n_{t_k})$ to $(X_{t_1}, \ldots, X_{t_k})$, where $t_1, \ldots, t_k \in S$. Thus, if $S = \mathbb{R}_+$, then the notation

$$X^n \xrightarrow{\mathcal{L}(\mathbb{R}_+)} X$$

will denote the weak convergence of all finite-dimensional distributions of the process X^n to X.

The convergence $X^n \xrightarrow{\mathcal{L}(\mathbb{R}_+)} X$ follows from $X^n \xrightarrow{\mathcal{L}} X$ (we recall that X^n, X are continuous processes). The converse is not true in general, as is immediately apparent from the simple example in which $X_t \equiv 0$ and X^n is concentrated on the one trajectory depicted in the figure:

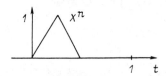

Clearly $X^n \xrightarrow{\mathcal{L}(\mathbb{R}_+)} X$, but $X^n \not\xrightarrow{\mathcal{L}} X$, since for $g(x) = \sup |x_s|$ we have $\int g(x) d\mathbf{P}^n = 1$, while $\int g(x) d\mathbf{P} = 0$. (Let us recall that if the family of measures $\{\mathbf{P}^n\}$ is relatively compact, then $X^n \xrightarrow{\mathcal{L}(\mathbb{R}_+)} X$ implies $X^n \xrightarrow{\mathcal{L}} X$.)

2.3. If we now turn to the space $\mathbf{D}$ endowed with the locally uniform topology, then it is easily seen that although it is, in fact, complete, it is not separable. For example, the family of functions $\alpha_s(t) = I_{[s,\infty)}(t)$, where the parameter $s \in [0, 1)$ is uncountable but $\delta_{\text{lu}}(\alpha_s, \alpha'_s) = 1/2$ for $s \neq s'$. This circumstance, in particular, leads to the general situation that the Borel σ-algebra $\mathcal{D}_{\text{lu}}$ turns out to be strictly wider than the cylinder σ-algebra $\mathcal{B}_{\text{cy}}(\mathbf{D})$,

$$\mathcal{D}_{\text{lu}} \supset \mathcal{B}_{\text{cy}}(\mathbf{D}),$$

so that if we consider sets of the form

$$B = \{\omega : X.(\omega) \in A\} = X^{-1}(A)$$

for $A \in \mathcal{D}_{\text{lu}}$, then it is by no means clear that $B \in \mathcal{F}$; hence it is also not clear whether one can speak of the probability of the set B in general, since it need not be an "event", that is, it need not belong to $\mathcal{F}$. As we have seen, this does not occur in the space **C**: if $A \in \mathcal{B}(\mathbf{C})$, then $B \in \mathcal{F}$, that is, B is an event for which the probability $\mathbf{P}(B)$ is defined. Let us note, by the way, that the σ-algebra $\mathcal{D}_{\text{lu}}[0]$ generated by the closed balls, coincides with $\mathcal{B}_{\text{cy}}(\mathbf{D})$, but $\mathcal{D}_{\text{lu}}[0]$ is strictly smaller than $\mathcal{D}_{\text{lu}}$:

$$\mathcal{B}_{\text{cy}}(\mathbf{D}) = \mathcal{D}_{\text{lu}}[0] \subset \mathcal{D}_{\text{lu}}.$$

In the space **D**, however, there is a metrizable topology, called the Skorokhod topology, which makes **D** a complete separable metric space. This topology is characterized by the property that a sequence of functions (α_n) converges to a function α if and only if there exists a sequence of changes of time $(\lambda_n) \in \Lambda$ (Λ is the set of continuous strictly increasing functions with $\lambda(0) = 0$, $\lambda(t) \uparrow \infty$, $t \uparrow \infty$) such that

$$\sup_s |\lambda_n(s) - s| \to 0, \tag{4.4}$$

$$\sup_{s \leq N} |\alpha_n \circ \lambda_n(s) - \alpha(s)| \to 0, \tag{4.5}$$

for all $N \geq 1$.

Let us observe that if $\delta_{\text{lu}}(\alpha_n, \alpha) \to 0$, then (4.4) and (4.5) are satisfied with $\lambda_n(s) = s$. Thus the Skorokhod topology is weaker than the locally uniform topology.

When α is a continuous function, $\alpha_n \to \alpha$ in the Skorokhod topology if and only if $\delta_{\text{lu}}(\alpha_n, \alpha) \to 0$.

We set

$$K_N(t) = \begin{cases} 1, & t \leq N, \\ N+1-t, & N < t < N+1, \\ 0, & t \geq N+1, \end{cases}$$

$$|||\lambda||| = \sup_{s<t} \left|\log \frac{\lambda(t) - \lambda(s)}{t - s}\right|,$$

and for $\alpha, \beta \in \mathbf{D}$

$$\delta_N(\alpha,\beta) = \inf_{\lambda \in \Lambda}(|||\lambda||| + ||(K_N \cdot \alpha) \circ \lambda - K_N \cdot \beta||_\infty),$$

$$\delta(\alpha,\beta) = \sum_N 2^{-N}(1 \wedge \delta_N(\alpha,\beta)). \tag{4.6}$$

It can be shown that δ is a distance in the space $\mathbf{D}$ (that is, $\delta \geq 0$, $\delta(\alpha,\beta) = \delta(\beta,\alpha)$, it satisfies the triangle inequality and if $\delta(\alpha,\beta) = 0$, then $\alpha = \beta$).

An important property of the distance δ is that the Borel σ-algebra $\mathcal{B}_\delta(\mathbf{D})$ coincides with the cylinder algebra $\mathcal{B}_{\text{cy}}(\mathbf{D})$:

$$\mathcal{B}_\delta(\mathbf{D}) = \mathcal{B}_{\text{cy}}(\mathbf{D}).$$

We give two examples illustrating the convergence $\alpha_n \to \alpha$ of the functions α_n to α in the Skorokhod topology.

Example 1. Let $\alpha_n(s) = x_n I(t_n \leq s)$:

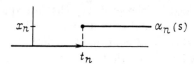

Then $\alpha_n \to \alpha$ if and only if
1) either $t_n \to \infty$ and then $\alpha \equiv 0$,
2) or alternatively, $t_n \to t < \infty$, $x_n \to x$ and then $\alpha(s) = xI(t \leq s)$:

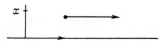

Example 2. Let

$$\alpha_n(s) = x_n I(t_n \leq s) + y_n I(r_n \leq s), \quad t_n < r_n$$

Then $\alpha_n \to \alpha$ if and only if:
1) $t_n \to \infty$ (hence, $r_n \to \infty$) and then $\alpha = 0$,
2) $t_n \to t < \infty$, $r_n \to \infty$, $x_n \to x$ and then $\alpha(s) = xI(t \leq s)$,
3) $t_n \to t < \infty$, $r_n \to r < \infty$, $x_n \to x$, $y_n \to y$ and $t < r$ if $x \neq 0 \neq y$; then $\alpha(s) = xI(t \leq s) + yI(r \leq s)$.

Note that in the Skorokhod topology it is by no means the case that $\alpha_n \to \alpha$, $\beta_n \to \beta$ imply that $\alpha_n + \beta_n \to \alpha + \beta$. Thus, the space $\mathbf{D}$ with the Skorokhod topology is not a topological *vector* space. However, if $\alpha_n \to \alpha$, $\beta_n \to \beta$ and β is continuous, then $\alpha_n + \beta_n \to \alpha + \beta$.

To understand convergence in the Skorokhod topology, we give a useful result (Theorem 1) in the case when the functions α under consideration belong to the set $\mathcal{V}^+$ of non-negative right-continuous non-decreasing functions equal to zero at zero. We denote by $\mathcal{V}^{+,1}$ the subset of $\mathcal{V}^+$ consisting of the *counting* functions

$$\alpha_s = \sum_{n \geq 1} I(t_n \geq s),$$

where $t_1 > 0$, $t_n < t_{n+1}$ if $t_n < \infty$, $t_n \uparrow \infty$.

Theorem 1. *Let* $\alpha_n, \alpha \in \mathcal{V}^+$.

a) *We have* $\alpha_n \to \alpha$ *in the Skorokhod topology if and only if there exists a dense subset* S *of* $\mathbb{R}^+$ *such that*

$$\alpha_n(s) \to \alpha(s), \tag{4.7}$$

$$\sum_{0 < u \leq s} |\Delta \alpha_n(u)|^2 \to \sum_{0 < u \leq s} |\Delta \alpha(u)|^2 \tag{4.8}$$

for $s \in S$.

b) *If* α *is continuous or if* $\alpha_n, \alpha \in \mathcal{V}^{+,1}$, *then only* (4.7) *guarantees the convergence* $\alpha_n \to \alpha$.

Thus, if we endow the space $\mathbf{D}$ with the metric δ, then we can speak of the continuity of the functions $g = g(x)$, $x \in \mathbf{D}$, and the notion of weak convergence $\mathbf{P}^n \xrightarrow{w} \mathbf{P}$ of the measures $\mathbf{P}^n$ to $\mathbf{P}$ defined on $(\mathbf{D}, \mathcal{B}(\mathbf{D}))$:

$$\int_{\mathbf{D}} g(x) \mathbf{P}^n(dx) \to \int_{\mathbf{D}} g(x) \mathbf{P}(dx)$$

makes sense.

In connection with the weak convergence of $\mathbf{P}^n \xrightarrow{w} \mathbf{P}$, we note that this does not imply (as in the case of the space $\mathbf{C}$) the weak convergence of *all* finite-dimensional distributions

$$\mathbf{P}^n_{t_1,\ldots,t_k} \xrightarrow{w} \mathbf{P}_{t_1,\ldots,t_k}, \quad t_i \in \mathbb{R}_+.$$

Namely, if $X = X_t(\omega))_{t \in \mathbb{R}_+}$ is a process with a distribution $\mathbf{P}$ in $(\mathbf{D}, \mathcal{B}(\mathbf{D}))$ and

$$J(X) = \{t \in \mathbb{R}_+ : \mathbf{P}(\Delta X_t \neq 0) > 0\}, \quad \Delta X_t = X_t - X_{t-},$$

then we can be sure only that

$$\mathbf{P}^n \xrightarrow{w} \mathbf{P} \Rightarrow \mathbf{P}^n_{t_1,\ldots,t_k} \xrightarrow{w} \mathbf{P}_{t_1,\ldots,t_k}$$

for all those t_j lying in the set

$$\mathbb{R}_+ \backslash J(X).$$

(This follows from the fact that the functions $\alpha \to \alpha(t)$, (as well as $\alpha \to \alpha(t-)$) are continuous on $\mathbf{D}$ at each point of α such that $t \in \mathbb{R}_+ \backslash J(\alpha)$, that is, such that $\Delta \alpha(t) = 0$, $t \in \mathbb{R}_+ \backslash J(\alpha)$.)

As was said above, one of our problems is the exposition of results obtained by martingale methods in the theory of the convergence $X^n \xrightarrow{\mathcal{L}} X$ and $X^n \xrightarrow{\mathcal{L}(S)} X$ of stochastic processes, that is the convergence $\mathbf{P}^n \xrightarrow{w} \mathbf{P}$ and $\mathbf{P}^n \xrightarrow{w(S)} \mathbf{P}$ (w is weak convergence, that is, weak convergence of finite-dimensional distributions at times belonging to the set $S \subseteq \mathbb{R}_+$).

We begin our account with a number of basic results in the area of classical limit theorems, placing particular emphasis on their methods of proof, which enable us to understand more fully the essence of "martingale" methods.

§3. Brief Review of a Number of Classical Limit Theorems of Probability Theory

3.1. The first limit theorem of probability theory was the law of large numbers of Jacob Bernoulli (1654–1705), who in his "Ars Conjectandi" (published in Latin in 1713) proved completely rigorously the following result.

Let $\xi_1, \xi_2, \ldots$ be a sequence of independent indentically distributed random variables assuming two values

$$\mathbf{P}(\xi_1 = 1) = p, \quad \mathbf{P}(\xi_1 = 0) = q \, (= 1 - p)$$

(this is called the Bernoulli scheme). If $S_n = \xi_1 + \ldots + \xi_n$ is the sum of "successes" after n trials, then the frequency

$$\frac{S_n}{n} \to p$$

in the sense that

$$\mathbf{P}\left(\left|\frac{S_n}{n} - p\right| > \epsilon\right) \to 0, \quad \forall \epsilon > 0. \tag{4.9}$$

This assertion is equivalent to

$$\frac{S_n}{n} \xrightarrow{\mathbf{P}} p,$$

or

$$F_{S_n/n}(X) \to F(x) = I(x \geq p), \quad x \in E^1 \setminus \{p\}. \tag{4.10}$$

Bernoulli's proof was based on a *direct* analysis of the distribution functions $F^n(x) = F_{S_n/n}(x)$, for which, clearly,

$$F^n(x) = \mathbf{P}(S_n \leq nx) = \sum_{k \leq nx} C_n^k p^k q^{n-k}, \tag{4.11}$$

which enabled him to prove the convergence of (4.10).

3.2. Historically, the next limit theorem, also proved by a *direct* analysis of the functions $F^n(x)$ was the de Moivre-Laplace Theorem (A. de Moivre: Miscellania Analytica, 1733 in Latin, Doctrine of Chances, 1738, 1756 in English, P.S. Laplace: Théorie analytique, 1812).

This theorem, which in modern terminology relates to the "central limit theorem", asserts that if

$$\Phi^n(x) = \mathbf{P}\left\{\frac{S_n - np}{\sqrt{npq}} \le x\right\},$$

$$\Phi(x) = \frac{1}{\sqrt{2\pi}} \int_{-\infty}^{x} e^{-y^2/2} dy,$$

then

$$\Phi^n(x) \to \Phi(x), \quad x \in E^1. \tag{4.12}$$

Since $\Phi(x)$ is a continuous function, the convergence to (4.12) is uniform with respect to x, that is,

$$\sup_x |\Phi^n(x) - \Phi(x)| \to 0, \quad n \to \infty. \tag{4.13}$$

De Moivre proved (1721) this theorem for $p = \frac{1}{2}$ (with a brief indication as to how everything is also done for $p \ne \frac{1}{2}$) by a direct study of the probabilities

$$\mathbf{P}_k^n = \mathbf{P}(S_n = k) = C_n^k p^k q^{n-k} \left(= \frac{n!}{k!(n-k)!} p^k q^{n-k}\right) \tag{4.14}$$

using for the analysis of the factorials a precursor of the "Stirling formula" (1730) which, incidentally, stimulated Stirling to discover the formula

$$n! \sim \sqrt{2\pi n}\, e^{-n} n^n, \tag{4.15}$$

which now bears his name. In a final exposition (in 1733) de Moivre now uses the Stirling formula, noting that although, in fact, not needed for his derivation, this formula nevertheless "has spread a singular Elegancy of the solution".

Formula (4.14) for even n, $k = \frac{n}{2}$ and $p = \frac{1}{2}$ has the form

$$\mathbf{P}_{n/2}^n = \mathbf{P}\left(S_n = \frac{n}{2}\right) = C_n^{n/2} 2^{-n}.$$

De Moivre shows that

$$C_n^{n/2} 2^{-n} \sim A \frac{\left(1 - \frac{1}{n}\right)^n}{\sqrt{n-1}}, \tag{4.16}$$

where the constant A ($\approx 2\frac{21}{125}$) is defined by him from the representation

$$\ln \frac{A}{2} = \frac{1}{12} - \frac{1}{360} + \frac{1}{1260} - \frac{1}{1680} + \dots \tag{4.17}$$

Very soon after this Stirling discovered that

$$\ln\sqrt{2\pi} = 1 - \left[\frac{1}{12} - \frac{1}{360} + \frac{1}{1260} - \frac{1}{1680} + \ldots\right]$$

and it thereby became clear that the de Moivre constant A is none other than $2e/\sqrt{2\pi}$.

Having obtained in this manner the approximation

$$\mathbf{P}^n_{n/2} \sim \frac{1}{\sqrt{\pi n/2}},$$

de Moivre then moves on to probabilities of the form $\mathbf{P}^n_{n/2+l}$ and for "not very large l" obtains the formula

$$\mathbf{P}^n_{n/2+l} = \mathbf{P}(S_n = n/2 + l) \sim \mathbf{P}^n_{n/2} \cdot e^{-2l^2/n} \sim \frac{1}{\sqrt{\pi n/2}} e^{-\frac{l^2}{n/2}}.$$

This route is embodied in contemporary presentations in that one says that for $k = k(n)$ that increases with n in such a way that

$$|k - np| = o(npq)^{2/3},$$

the "local theorem":

$$\mathbf{P}(S_n = k) \sim \frac{1}{\sqrt{2\pi npq}} e^{-\frac{(k-np)^2}{2npq}} \qquad (4.18)$$

holds. It is now not difficult to go over to the "integral theorem"

$$\Phi^n(x) = \mathbf{P}\left(\frac{S_n - np}{\sqrt{npq}} \le x\right) \to \Phi(x), \; n \to \infty. \qquad (4.19)$$

The Laplace method of obtaining an asymptotic formula for $\mathbf{P}^n_{n/2}$ was based on another remarkable idea which, in essence, was the beginning of the method of characteristic functions.

Laplace's idea was as follows.

Since (for even n)

$$(\alpha + \beta)^n = \alpha^n + C^1_n \alpha^{n-1}\beta + \ldots + C^{n/2}_n \alpha^{n/2}\beta^{n/2} + \ldots + \beta^n, \qquad (4.20)$$

it follows, on setting $\alpha = e^{-it}$, $\beta = e^{it}$, that

$$(e^{-it} + e^{it})^n = e^{-int}a_{-in} + e^{-i(n-1)t}a_{-(n-1)} + \ldots + a_0 + \ldots + a_n e^{int}, \qquad (4.21)$$

where $a_{-n} = 1$, $a_{-(n-1)} = C^1_n, \ldots, a_0 = C^{n/2}_n, \ldots, a_n = 1$. In seeking "suitable" formulas for $\mathbf{P}^n_{n/2} = C^{n/2}_n p^{n/2} q^{n/2}$ the basic difficulty lies in obtaining a "good" approximation for the number of combinations $C^{n/2}_n$ which, in accordance with (4.21), reduces to analyzing the coefficient a_0. The integration of (4.21) with respect to t immediately gives

Chapter 4.I. Weak Convergence of Measures on Metric Spaces

$$a_0 = \frac{1}{2\pi} \int_0^{2\pi} (e^{-it} + e^{it})^n dt = \frac{2^n}{2\pi} \int_0^{2\pi} \cos^n t \, dt$$

and hence ($p = q = \frac{1}{2}$),

$$\mathbf{P}^n_{n/2} = \frac{1}{2\pi} \int_0^{2\pi} \cos^n t \, dt.$$

In other words, the question of the asymptotic behavior of $\mathbf{P}^n_{n/2}$ reduces to studying the integrals $\int_0^{2\pi} \cos^n t \, dt$, the asymptotic analysis of which for Laplace-analytics did not present any difficulties.

3.3. Let us also recall that by the direct methods of Poisson (in the book "Recherches sur la probabilité des jugements en matière criminelle et en matière civile", 1837) in the "Bernoulli scheme", the "Poisson approximation" of the probabilities of rare events was obtained.

Namely, for each $n \geq 1$ let

$$\xi_{n1}, \ldots, \xi_{nn}$$

be a sequence of independent identically distributed random variables with

$$\mathbf{P}(\xi_{ni} = 1) = p_n, \quad \mathbf{P}(\xi_{ni} = 0) = q_n \, (= 1 - p_n),$$

where $np_n \to \lambda > 0$. Then a direct analysis of the probabilities

$$\mathbf{P}^n_k = \mathbf{P}(S_n = k) \, (= C_n^k p^k q^{n-k}),$$

where $S_n = \xi_{n1} + \ldots + \xi_{nn}$ leads to the "Poisson" approximation

$$\mathbf{P}^n_k \to \pi_k = \frac{e^{-\lambda}\lambda^k}{k!}, \quad k = 0, 1, \ldots. \tag{4.22}$$

3.4. In 1887 Chebyshev published a paper "Two Theorems regarding probabilities" in which he proposed a new method, called the method of moments, for a proof of the central limit theorem for sums of independent random variables, now not necessarily assuming only two values. (Let us recall, by the way, that twenty years earlier, in 1867 Chebyshev published an article "On mean variables" in which he went over from the consideration of random events and their probabilities to the study of random variables and their expectations. Thus it was Chebyshev who first recognized and used all the power of the concept of random variable and expectation.)

The statement of the problem considered by Chebyshev was as follows.

Let $F^n = F^n(x)$ be the distribution functions of the random variables ξ^n, $n \geq 1$, and

$$m_k^n = \int_{-\infty}^{\infty} x^k dF^n(x), \, k \geq 1,$$

be their moments, which are assumed to exist. Let

$$T^n = \{m_k^n, \ k \geq 1\}$$

be the set of these moments.

Also let $F = F(x)$ be some distribution function with set of moments

$$T = \{m_k, \ k \geq 1\}.$$

If $T^n \to T$ (in the sense that $m_k^n \to m_k$, $n \to \infty$, $k \geq 1$), then it is natural to expect also that $F^n \xrightarrow{w} F$.

Since $m_2^n \to m_2$, it follows from Chebyshev's inequality that

$$\lim_{c \uparrow \infty} \limsup_n \mathbf{P}(|\xi^n| \geq c) \lim_{c \uparrow \infty} \limsup_n \frac{\mathbf{E}|\xi^n|^2}{c^2} = 0 \qquad (4.23)$$

and hence for each $\epsilon > 0$ we can find a compact set $K_\epsilon = [-c_\epsilon, c_\epsilon]$ such that all the measures $\mathbf{P}^n$ corresponding to F^n, to within ϵ, "sit" on K_ϵ:

$$\sup_n \mathbf{P}(\xi^n \in K_\epsilon) = \sup_n \mathbf{P}(K_\epsilon) \geq 1 - \epsilon. \qquad (4.24)$$

This condition, called the tightness of the family of measures $\{\mathbf{P}^n\}$ (or the tightness of the family of random variables $\{\xi^n\}$ with respect to the "basic" measure $\mathbf{P}$) is equivalent (in a very general situation — this is Prokhorov's theorem) to the family of measures $\{\mathbf{P}^n\}$ being relatively compact, that is, each subsequence $\{\mathbf{P}^{\tilde{n}}\} \subseteq \{\mathbf{P}^n\}$ contains a weakly converging subsequence $\{\mathbf{P}^{n'}\} \subseteq \{\mathbf{P}^{\tilde{n}}\}$

$$\mathbf{P}^{n'} \to \mathbf{P}',$$

where $\mathbf{P}'$ is a probability measure that possibly does not belong to the initial family $\{\mathbf{P}^n\}$.

The following remark is important for the remainder of our discussion: if *any* weakly converging subsequence $\{\mathbf{P}^{n'}\}$ converges to one and the same limit $\mathbf{P}'$, then, in fact, the entire sequence $\{\mathbf{P}^n\}$ converges weakly to this limit $\mathbf{P}'$. (The proof is by contradiction.)

This remark suggests a general method of proving the implication

$$(T^n \to T) \Rightarrow (\mathbf{P}^n \xrightarrow{w} \mathbf{P}). \qquad (4.25)$$

I. Establish the tightness (relative compactness) of the family $\{\mathbf{P}^n\}$.
II. Characterize all weak limits $\mathbf{P}'$.
III. Identify $\mathbf{P}'$ with $\mathbf{P}$.

All this is illustrated as follows:

$$
(T^n \to T) \Rightarrow \boxed{\begin{array}{c}\text{I}\\ \text{Establishment of tightness} \\ \text{of the family } \{\mathbf{P}^n\}\end{array}} \oplus \boxed{\begin{array}{c}\text{II}\\ \text{Characterization of all} \\ \text{weak limits } \mathbf{P}'\end{array}}
$$

$$
\oplus \boxed{\begin{array}{c}\text{Establishment of tightness} \\ \text{of the family } \{\mathbf{P}^n\}\end{array}} \to \mathbf{P}^n \xRightarrow{w} (\mathbf{P}). \qquad (4.26)
$$

III

As applied to the method of moments under consideration:

(I) The establishment of the tightness of the family $\{\mathbf{P}\}$ follows from the convergence of $m_2^n \to m$ and the Chebyshev inequality (see (4.23));

(II) It is natural here to base the characterization of all the weak limits $\mathbf{P}'$ on the requirement that the convergence $\mathbf{P}^{n'} \stackrel{w}{\to} \mathbf{P}'$ must imply that

$$\int x^k \mathbf{P}^{n'}(dx) \to \int x^k \mathbf{P}'(dx), \quad k = 1, 2, \ldots. \tag{4.27}$$

This, of course, is so if all the measures $\mathbf{P}^{n'}$ and $\mathbf{P}'$ are concentrated on a finite interval $[a, b]$. In the general case this is, of course, not always true. Nevertheless, if it is so, then since along with (4.27) we also have $m_k^n \to m_k$, $k \geq 1$, we will be able to give this characterization of the weak limits $\mathbf{P}'$:

The measure $\mathbf{P}'$ is such that its moments $\int x^k \mathbf{P}'(dx)$ coincide with the moments m_k, $k \geq 1$.

Finally, the concluding stage of the *identification* of all weak limits $\mathbf{P}'$ with $\mathbf{P}$ is carried out starting from the usual assumption that the moments $T = \{m_k\}$ uniquely determine the distribution $\mathbf{P}$. For uniqueness, it is sufficient, for example, that the following Carleman condition be satisfied:

$$\sum \frac{1}{(m_{2n})^{1/2n}} = \infty. \tag{4.28}$$

Thus if, say, it is known *a priori* that all the distributions under consideration are concentrated on the same finite interval $[a, b]$, then (since (4.27) is evidently satisfied) the characterization and identification entail that all the weak limits $\mathbf{P}'$ coincide with $\mathbf{P}$.

Thus, in this case

$$T^n \to T \Rightarrow \boxed{\text{I} \oplus \text{II} \oplus \text{III}} \Longrightarrow \mathbf{P}^n \stackrel{w}{\to} \mathbf{P}.$$

As is well known, Chebyshev had a gap in the implementation of the method of moments for the proof of the central limit theorem (he considered that all the moments had to exist, but required convergence only for the first two), which Markov then corrected.

3.5. The next essential step in establishing the boundaries of applicability of the central limit theorem was made (in 1900, 1901) by Lyapunov who applied to his proof the method of characteristic functions, dating from Laplace. First Lyapunov assumed that the variables in the summation had third-order absolute moments, and then weakened this assumption to the existence of moments of order $2 + \delta$, $\delta > 0$.

If η_1 and η_2 are independent random variables with distribution functions F_{η_1} and F_{η_2}, then the distribution function of their sum is the convolution of F_{η_1} and F_{η_2}:

$$F_{\eta_1 + \eta_2} = F_{\eta_1} * F_{\eta_2},$$

that is,
$$F_{\eta_1+\eta_2}(x) = \int F_{\eta_1}(x-y) dF_{\eta_2}(y).$$

Convolution is not a simple formation and a *direct* analysis of the convolution $F^n = F_{\eta_1+\ldots+\eta_n}$ of the sum of independent random variables is in practice not possible in any general situation.

However, the characteristic function $f^n(t) = f_{\eta_1+\ldots+\eta_n}(t)$ of the sum of independent random variables is the *product* of the characteristic functions,
$$f^n(t) = f_{\eta_1}(t) \ldots f_{\eta_n}(t),$$
which yields more simply to analysis than the convolution F^n.

Let us consider how the method of characteristic functions "works" in analyzing the question of the weak convergence of the distributions F^n of the random variables ξ^n to the distribution F of the random variable ξ.

Let
$$f^n(t) = \mathbf{E} e^{it\xi^n} \left(= \int e^{itx} dF^n(x) = \int e^{itx} \mathbf{P}^n(dx) \right),$$
$$f(t) = \mathbf{E} e^{it\xi} \left(= \int e^{itx} dF(x) = \int e^{itx} \mathbf{P}(dx) \right)$$
be the corresponding characteristic functions. We set
$$T^n = \{f^n(t),\ t \in \mathbb{R}\},\ T = \{f(t),\ t \in \mathbb{R}\}.$$

It is well known that the characteristic functions f and the distribution functions F, and hence also the corresponding probability distributions $\mathbf{P}$ are in one-to-one correspondence. This gives some hope that the convergence of the characteristic functions will imply the weak convergence of the corresponding probability distributions:
$$T^n \to T \Rightarrow \mathbf{P}^n \xrightarrow{w} \mathbf{P}. \tag{4.29}$$

This is indeed the case. It is proved by invoking the intermediate stages I$\oplus$II$\oplus$III ("tightness"$\oplus$"characterization"$\oplus$"identification") in the following manner.

Since we have the estimate
$$\mathbf{P}^n\left(\mathbf{E}^1 \setminus \left[-\frac{1}{a}, \frac{1}{a}\right]\right) \le \frac{K}{a} \int_0^a [1 - \operatorname{Re} f^n(t)] dt, \tag{4.30}$$
the tightness of the family $\{\mathbf{P}^n\}$ follows from the convergence of $f^n(t) \to f(t)$, $t \in \mathbb{R}$, and the continuity of $f(t)$ at zero.

If, further, $\mathbf{P}^{n'}$ is a weakly converging subsequence, $\mathbf{P}^{n'} \xrightarrow{w} \mathbf{P}'$, then it follows from the very definition of weak convergence that $f^{n'}(t) \to f'(t)$ for each $t \in \mathbb{R}$, where $f' = f'(t)$ is the characteristic function of $\mathbf{P}'$. But since, by hypothesis, $f^n(t) \to f(t)$, it follows that $f' = f$, and hence we obtain the following *characterization* of the limit points of $\mathbf{P}'$:

the function f is the characteristic function of the distributions $\mathbf{P}'$.

But, as was already noted, a distribution is uniquely defined by the characteristic function and, consequently, the *identification* of $\mathbf{P}'$ with $\mathbf{P}$ is brought about, so that the required implication holds. (Note that the reverse implication is trivial and that in (4.29) there is no need to stipulate that $f = f(t)$ is a characteristic function: only its continuity at zero is needed; it then automatically follows from the considerations presented above that it is the characteristic function of some distribution $\mathbf{P}$ and $\mathbf{P}^n \xrightarrow{w} \mathbf{P}$).

Thus, in the method of characteristic functions, all the intermediate stages I, II, III are verified and all this can be depicted in the following manner:

$$
\begin{array}{ccc}
& \text{characteristic function } f' & \text{distribution uniquely} \\
\text{from estimate} & \text{of } \mathbf{P}' \text{ is equal to } f & \text{recovered from} \\
& & \text{characteristic function} \\
\downarrow & \downarrow & \downarrow
\end{array}
$$

$T^n \to T \Rightarrow \boxed{\text{Density}} \oplus \boxed{\text{Characterization of } \mathbf{P}'} \oplus \boxed{\text{Identification}}$

$\Rightarrow \mathbf{P}^n \to \mathbf{P}$.

3.6. The method of characteristic functions has proved to be very powerful in proving limit theorems of probability theory for sums of independent terms. Applied to the central limit theorem (c.l.t.) this method led Lyapunov (1901) and Lindeberg (1922) to the following conditions of validity of the c.l.t.

Theorem 4.1. *Let $S_n = \xi_{n1} + \ldots + \xi_{nn}$, where for each $n \geq 1$*

$$\xi_{n1}, \ldots, \xi_{nn}$$

are independent random variables, $\mathbf{E}\xi_{ni} = 0$, $D_n = \sum_{k=1}^{n} \mathbf{E}\xi_{nk}^2 = 1$. Then

$$(Lya) \quad \sum_{k=1}^{n} \mathbf{E}|\xi_{nk}|^{2+\delta} \to 0 \Longrightarrow \mathbf{P}(S_n \leq x) \to \Phi(x), \quad x \in \mathbb{R},$$

(Lyapunov condition),

$$(L) \quad \sum_{k=1}^{n} \mathbf{E}[|\xi_{nk}|^2; |\xi_{nk}| > \epsilon] \to 0 \Longrightarrow \mathbf{P}(S_n < x) \to \Phi(x), \quad x \in \mathbb{R},$$

(Lindeberg condition).

Clearly (Lya) $\Rightarrow$ (L). It is appropriate to note straight away that the Lindeberg condition implies that the above random variables $\xi_{n1}, \ldots, \xi_{nn}$ are *asymptotically negligible* in the sense that

$$\max_{k \leq n} \mathbf{P}(|\xi_{nk}| > \epsilon) \to 0, \ n \to \infty, \tag{4.31}$$

and, *a fortiori*,

$$\max_{k \leq n} D\xi_{nk} \to 0, \ n \to \infty. \tag{4.32}$$

However, the simplest examples show that the c.l.t. can, in fact, hold even without the satisfaction of the Lindeberg condition or even, say, the satisfaction of condition (4.32). For example, if

$$\xi_{nk} = \frac{\xi_k}{\sqrt{\sum_{i=1}^n D\xi_i}},$$

where $\xi_1, \xi_2, \ldots$ is a sequence of independent normally distributed random variables with $\mathbf{E}\xi_k = 0$, $D\xi_1 = 1$, $D\xi_k = 2^{k-2}$, $k \geq 2$, then it is clear that $\mathbf{P}(S_n \leq x) = \Phi(x)$. However, condition (4.32) is not satisfied here:

$$\max_{k \leq n} D\xi_k = \frac{1}{2}.$$

In this connection, let us recall a result of Zolotarev, who gave a necessary and sufficient condition for the validity of the central limit theorem for sums of independent random variables without the assumption of asymptotic negligibility. In the modified form proposed by Rotar' (Rotar' 1978), Zolotarev's result (Zolotarev 1967) is as follows.

Let $\xi_{n_1}, \ldots, \xi_{nn}$ be independent random variables with

$$\mathbf{E}\xi_{nk} = 0, \quad \mathbf{E}\xi_{nk}^2 \equiv \sigma_{nk}^2 \leq \infty, \quad \sum_{k=1}^n \sigma_{nk}^2 = 1, \quad F_{nk}(x) = \mathbf{P}(\xi_{nk} \leq x).$$

Then the condition

(L) $$\sum_{k=1}^n \int_{|x|>\epsilon} |x| \left| F_{n^k}(x) - \Phi\left(\frac{x}{\sigma_{n^k}}\right) \right| dx \to 0, \quad \epsilon > 0,$$

is necessary and sufficient for the validity of the central limit theorem.

We also note the following. As far back as 1926 in a course on limit theorems, Khinchin posed the problem on whether there is a connection between the law of large numbers and the central limit theorem. The answer, found in 1938 by Raikov and Bobrov states, in particular, the following.

Let $\xi_{n1}, \ldots, \xi_{nn}$ be a sequence of independent identically distributed random variables,

$$\mathbf{E}\xi_{nk} = 0, \quad \sum_{k=1}^n D\xi_{nk} = 1,$$

satisfying the condition (4.31) of asymptotic negligibility. Then in order that $\mathbf{P}(S_n \leq x) \to \Phi(x)$, $x \in E^1$, it is necessary and sufficient that (in the terminology of Khinchin) the following relative stability condition should hold:

$$\sum_{k=1}^n \xi_{nk}^2 \xrightarrow{\mathbf{P}} 1. \qquad (4.33)$$

It turns out, and this becomes clear in what follows, that one can formulate the conditions of validity of the central limit theorem in a similar form, for the case of not necessarily independent terms. The probabilistic sense of the sum of squares in (4.33) will also be explained below.

3.7. Convergence conditions in many limit theorems have been found by the method of characteristic functions. For the sake of definiteness, and also in keeping with the subsequent presentation, we present a number of formulations of limit theorems in the case of convergence to so-called infinitely divisible distributions.

We recall that a random variable ξ is called infinitely divisible if for any $n \geq 1$ it can be represented in the form

$$\xi \stackrel{d}{=} \xi_{n1} + \ldots + \xi_{nn},$$

where $\xi_{n1}, \ldots, \xi_{nn}$ are independent identically distributed random variables, and the equality "$\stackrel{d}{=}$" is understood in the sense of the coincidence of their distributions. In terms of characteristic functions, the condition of infinite divisibility of ξ means that its characteristic function $f(\lambda)$ can for any $n \geq 1$ be represented in the form

$$f(\lambda) = [f_n(\lambda)]^n, \ \lambda \in E^1, \tag{4.34}$$

where $f_n(\lambda)$ is some characteristic function.

First, Kolmogorov (1933) for the case of variables with finite variance, and then Lévy (Lévy 1939) and Khinchin (Khinchin 1938) in the general case, established that the characteristic function $f(\lambda)$ of an infinitely divisible random variable admits a representation in the form

$$f(\lambda) = e^{\psi(\lambda)}, \tag{4.35}$$

where (the cumulant) $\psi(\lambda) = \psi_{b,C,F}(\lambda)$ is determined by the triple of parameters (b, C, F) and has the following structure:

$$\psi_{b,C,F}(\lambda) = i\lambda b - \frac{\lambda^2}{2}C + \int_{E^1} (e^{i\lambda x} - 1 - i\lambda h(x))F(dx), \tag{4.36}$$

where the "truncation" function $h = h(x)$ is any bounded function with compact support such that $h(x) = x$ in some neighborhood of zero (in many cases it is convenient to choose the function $h = h(x)$ as continuous); $b \in E^1, C \geq 0$, $F(dx)$ is a measure on E^1 with

$$F(\{0\}) = 0 \text{ and } \int_{E^1} (x^2 \wedge 1)F(dx) < \infty.$$

Let us observe that the characteristics C and F are "internal" characteristics of the distribution in the sense that their values do not depend on the choice of h. As regards the characteristic b, it depends on h and for two different truncation functions h and h', the corresponding characteristics b and b' are connected by the relation

$$b' - b = \int_{E^1} [h'(x) - h(x)] F(dx). \tag{4.37}$$

The triple of characteristics $T = (b, C, F)$ uniquely determines (for a chosen function $h = h(x)$) the characteristic function. Therefore, if we pose the question on the convergence conditions of distributions of random variables ξ^n with triples $T^n = (b^n, C^n, F^n)$ to the distribution of a random variable ξ with triple $T = (b, C, F)$, then we can expect that natural convergence conditions can be expressed in terms of the convergence of the components of these triples.

For the formulation of the corresponding classical result, it is convenient to introduce the following notation:

$$\widetilde{C} = C + \int_{E^1} h^2(x) F(dx), \quad \widetilde{C}^n = C^n + \int_{E^1} h^2(x) F^n(dx), \tag{4.38}$$

and

$$F(g) = \int_{E^1} g(x) F(dx), \quad F^n(g) = \int_{E^1} g(x) F^n(dx). \tag{4.39}$$

Then we have the following

Theorem 4.2 (Gnedenko 1939). *Let $h = h(x)$ be a continuous truncation function. Then for the convergence of the infinitely divisible random variables $\xi^n \xrightarrow{d} \xi$, it is necessary and sufficient that we have convergence $\widetilde{T}^n \to \widetilde{T}$ of the triples $\widetilde{T}^n = (b^n, \widetilde{C}^n, F^n)$ to $\widetilde{T} = (b, \widetilde{C}, F)$ in the sense that*

$$\begin{aligned} b^n &\to b, \\ \widetilde{C}^n &\to \widetilde{C}, \\ F_n(g) &\to F(g), \ g \in C_2, \end{aligned} \tag{4.40}$$

where C_2 is the set of bounded continuous functions having limits at infinity and equal to zero in a neighborhood of zero.

Remark 1. If conditions (4.40) are satisfied, then $F_n(g) \to F(g)$ also for functions g in the class $C_3 \supset C_2$, consisting of all bounded continuous functions satisfying the condition $g(x) = o(x^2)$, $x \to 0$.

Remark 2. In (4.40) it is sufficient to require that the functions g belong only to the class C_1, defined as the subclass of C_2 containing all functions of the form $g_a(x) = (a|x| - 1)^+ \wedge 1$ for all positive rational a. Thus

$$C_1 \subset C_2 \subset C_3.$$

The true meaning of the importance of the class of infinitely divisible distributions is that they (and only they) act as limiting distributions for the sums of independent *asymptotically negligible (small)* terms $\xi_{n1}, \ldots, \xi_{nn}$.

Namely, we have the following

Theorem 4.3 (see, for example, Gnedenko and Kolmogorov 1949).

1) *If* $S^n = \xi_{n1} + \ldots + \xi_{nn} \xrightarrow{d} \xi$, *then* ξ *is infinitely divisible;*

2) *In order that* $S^n \xrightarrow{d} \xi$, *where* ξ *is an infinitely divisible random variable with triple* $T = (b, C, F)$, *it is necessary and sufficient that the following conditions hold:*

$$\sum_{k=1}^{n} \mathbf{E}h(\xi_{nk}) \to b,$$

$$\sum_{k=1}^{n} [\mathbf{E}h^2(\xi_{nk}) - (\mathbf{E}h(\xi_{nk}))^2] \to \widetilde{C}, \qquad (4.41)$$

$$\sum_{k=1}^{n} \mathbf{E}g(\xi_{nk}) \to F(g), \quad g \in C_2 \text{ or } g \in C_1.$$

In the case when S^n, $n \geq 1$, *are infinitely divisible random variables, conditions* (4.40) *and* (4.41) *are equivalent, as can be shown.*

3.8. In connection with this theorem, it would now be natural to put the question whether it is possible to generalize it to the case of dependent random variables, preserving as far as possible the structure of conditions (4.41). It turns out that this can be done, as we now show.

We will assume that our entire discussion takes place in some probability space $(\Omega, \mathcal{F}, \mathbf{P})$ with a flow of σ-algebras $\mathbf{F}^n = (\mathcal{F}_k^n)_{k \geq 0}$, $n \geq 1$ distinguished on it, such that

$$\mathcal{F}_0^n \subseteq \mathcal{F}_1^n \subseteq \ldots \subseteq \mathcal{F}.$$

In the simplest cases $\mathcal{F}_k^n = \sigma\{\eta_{n0}, \eta_{n1}, \ldots, \eta_{nk}\}$ is the σ-algebra generated by certain random variables $\eta_{n0}, \eta_{n1}, \ldots, \eta_{nk}$. This means that $\mathcal{F}_k^n$ is the collection of events (in the nth series) observable up to time k inclusively.

Now let us assume that for each $n \geq 1$

$$\xi_{n1}, \ldots, \xi_{nn}$$

is a sequence of (generally speaking, dependent) random variables, such that

$$\xi_{nk} \text{ are } \mathcal{F}_{nk}\text{-measurable.}$$

Also let $S^n = \xi_{n1} + \ldots + \xi_{nn}$. One can ask under what conditions $S^n \xrightarrow{d} \xi$, where ξ is an infinitely divisible random variable with triple $T = (b, C, F)$.

The next theorem (Jacod, Mémin 1980); (Liptser, Shiryaev 1982) gives an answer to this question.

Theorem 4.4. *Suppose that the asymptotic negligibility conditions*

$$\sup_{k \leq n} \mathbf{P}(|\xi_{nk}| > \epsilon | \mathcal{F}_{k-1}^n) \xrightarrow{\mathbf{P}} 0$$

hold. Suppose further that

(β)
$$\sum_{k=1}^{n} \mathbf{E}[h(\xi_{nk})|\mathcal{F}_{k-1}^{n}] \xrightarrow{\text{P}} b,$$

(γ)
$$\sum (\mathbf{E}[h^2(\xi_{nk})|\mathcal{F}_{k-1}^{n}] - [\mathbf{E}h(\xi_{nk}|\mathcal{F}_{k-1}^{n})]^2) \xrightarrow{\text{P}} \widetilde{C}, \qquad (4.42)$$

(δ)
$$\sum \mathbf{E}[g(\xi_{nk})|\mathcal{F}_{k-1}^{n}] \xrightarrow{\text{P}} F(g), \ g \in C_1.$$

Then $S^n \xrightarrow{\text{d}} \xi$.

It is more convenient to obtain the proof of this theorem as a particular case of the weak convergence of the finite-dimensional distributions of the stochastic processes

$$X_t^n = \sum_{k}^{[nt]} \xi_{nk}, \ t \geq 0,$$

to some process with independent increments X_t, $t \geq 0$. Furthermore, such an imbedding enables one to clarify the sense of the expressions entering in conditions (β)–(δ).

§4. Convergence of Processes with Independent Increments

4.1. Processes with independent increments X are defined as processes for which any increments

$$X_{t_k} - X_{t_{k-1}}, \ \ldots, \ X_{t_2} - X_{t_1}, \ X_{t_1} - X_{t_0}$$

with $t_0 < t_1 < \ldots < t_k$, $k \geq 2$, form a collection of independent random variables.

Any process X with independent increments can be represented in the form

$$X_t = A_t + Y_t,$$

where $A = (A_t)_{t \geq 0}$ is a deterministic function ($A \in \mathbf{D}$) that is of locally unbounded variation in general, and $Y = (Y_t)_{t \geq 0}$ is a process with independent increments for which for any $\lambda \in \mathbb{R}$ the function

$$g_t^Y(\lambda) = \mathbf{E}e^{i\lambda Y_t}, \ t \geq 0, \qquad (4.43)$$

has locally bounded variation as a function of t. Since the deterministic component A is not of probabilistic interest, we will assume in what follows that $A_t \equiv 0$, so that our process X with independent increments will be such that the corresponding function $g_t(\lambda) = g_t^Y(\lambda)$, $t \geq 0$, has locally bounded variation.

Under this assumption it can be established that the characteristic function

$$g_t(\lambda) = \mathbf{E}e^{i\lambda X_t}$$

is a solution (in the class of functions in **D**) of the equation (λ is fixed)

$$dg_t(\lambda) = g_{t-}(\lambda)dG_t(\lambda) \qquad (4.44)$$

with

$$G_t(\lambda) = i\lambda B_t - \frac{\lambda^2}{2}C_t + \int_0^t \int_{E^1} (e^{i\lambda x} - 1 - i\lambda h(x))\nu(ds,dx), \qquad (4.45)$$

where
$B = (B_t)$ is a function of locally bounded variation, $B_0 = 0$;
$C = (C_t)$ is a continuous non-decreasing function, $C_0 = 0$;

$$\nu([0,t] \times A) = \mathbf{E}\mu([0,t] \times A),$$

and μ is the jump measure of the process X, that is,

$$\mu([0,t] \times A) = \sum_{0<s\leq t} I(\Delta X_s \in A)I(\Delta X_s \neq 0), \quad \Delta X_s = X_s - X_{s-}, \qquad (4.46)$$

where $\int_0^t \int_{E^1}(x^2 \wedge 1)\nu(ds,dx) < \infty$.

The solution of equation (4.44) is the so-called *generalized exponential* $\mathcal{E}(G(\lambda))_t$, that is,

$$g_t(\lambda) \equiv \mathcal{E}(G(\lambda))_t = e^{G_t(\lambda)} \prod_{0<s\leq t} (1 + \Delta G_s(\lambda))e^{-\Delta G_s(\lambda)}. \qquad (4.47)$$

In the case when the initial process is continuous in probability (equivalently, there are no discontinuities at fixed times as, say, for the process $X_t = \sum_{k \leq t} \xi_k$), then B, C, ν are continuous in t and the characteristic function $g_t(\lambda)$ takes the form

$$g_t(\lambda) = e^{G_t(\lambda)}, \qquad (4.48)$$

since then $\Delta G_t(\lambda) = 0$.

Finally, if our process with independent increments is, in addition, a process with homogeneous increments, then

$$B_t = bt, \quad C_t = ct, \quad \nu(dt,dx) = dtF(dx),$$

where $c \geq 0$, $\int_{E^1}(x^2 \wedge 1)F(dx)$, and then

$$g_t(\lambda) = \exp\left(t[i\lambda b - \frac{\lambda^2}{2}c + \int(e^{i\lambda x} - 1 - i\lambda h(x)F(dx)]\right), \qquad (4.49)$$

is none other than the Lévy-Khinchin representation for the characteristic function of a uniform process with independent increments.

In analytic probability theory the study of the properties of stochastic processes is realized by a direct analysis of their distributions. Moreover, the method of characteristic functions is a powerful tool for analysis, especially for such processes as, say, processes with independent increments. However, as has already been noted, the transition to more complicated processes does not provide the possibility of effective use of the method of characteristic functions.

4.2. This circumstance compels one to seek other paths of analysis, one of which is based on ideas of the pathwise (i.e., trajectorywise) study of stochastic processes based on stochastic calculus.

Let us clarify the fundamental idea used in what follows with an example of processes with independent increments. It turns out that any process with independent increments can be represented in the following form:

$$X_t = X_0 + B_t + X_t^c + \int_0^t \int_{E^1} h(x) d(\mu - \nu) + \int_0^t \int_{E^1} (x - h(x)) d\mu, \quad (4.50)$$

where $B = (B_t)$ is the function of locally bounded variation entering into the formula for the characteristic function, $X^c = (X_t^c)$ is a continuous Gaussian process with independent increments, μ is the jump measure and $\nu(dt, dx) = \mathbf{E}\mu(dt, dx)$.

The representation (4.50), called *canonical*, in the case $h(x) = xI(|x| \leq 1)$ assumes the form

$$X_t = X_0 + B_t + X_t^c + \int_0^t \int_{|x| \leq 1} x d(\mu - \nu) + \int_0^t \int_{0|x| > 1} x d\mu, \quad (4.51)$$

from which it follows that it is natural to call $B = (B_t)$ the drift of X, the term $\int_0^t \int_{|x|>1} x d\mu$ characterizes the "large" jumps of the process X; X^c is the continuous martingale part (since X^c is a continuous martingale), and $\int_0^t \int_{|x| \leq 1} x d(\mu - \nu)$ is the jump martingale part (since this process is a purely discontinuous martingale).

We set $C_t = DX_t^c$ and call the collection of objects

$$T = (B, C, \nu)$$

the triple of characteristics of a process with independent increments.

This triple consists of deterministic objects. Here, $C = (C_t)_{t \geq 0}$ is a nondecreasing continuous function, $C_0 = 0$; $B = (B_t)_{t \geq 0}$ is a function of locally bounded variation with the property that $B_0 = 0$,

$$\Delta B_t = \int h(x) \nu(\{t\}, dx);$$

$\nu = \nu(dt, dx)$ is a measure on $\mathbb{R}_+ \times E^1$ such that

Chapter 4.I. Weak Convergence of Measures on Metric Spaces

$$\nu(\mathbb{R}_+ \times \{0\}) = 0, \quad \nu(\{0\} \times E^1) = 0,$$

$$\int_{(0,t] \times E^1} (|x|^2 \wedge 1)\nu(ds, dx) < \infty, \quad t > 0.$$

It is very remarkable that any triple $T = (B, C, \nu)$ of objects with the above properties uniquely determines on the space $(\mathbf{D}, \mathcal{D})$ a probability measure $\mathbf{P}$ such that the canonical process $X_t(\alpha) = \alpha(t)$, $\alpha \in \mathbf{D}$, is (with respect to this measure $\mathbf{P}$) a process with independent increments, for which the triple T coincides precisely with the initial ones. (Clearly, this result generalizes the situation that an infinitely divisible distribution is uniquely determined by its characteristics (b, C, F).)

The proof of this is based on the fact that the above problem is equivalent to the solution of the so-called martingale problem, consisting in the following.

We form the process

$$X(h)_t = X_t - \int_0^t \int_{E^1} (x - h(x)) d\mu,$$

where μ is the jump measure of the canonical process (or, equivalently, let $X(h) = X - \check{X}(h)$, where $\check{X}(h)_t = \sum_{s \leq t}[\Delta X_s - h(\Delta X_s)]$ and let

$$M(h)_t = X(h)_t - X_0 - B_t(h).$$

(If we turn to (4.50), then we see that $M(h)$ is a martingale.) We also set

$$\widetilde{C}_t = C_t + \int_0^t \int_{E^1} h^2(x)\nu(ds, dx) - \sum_{0 < s \leq t} \left(\int_{E^1} h(x)\nu(\{s\}, dx) \right)^2$$

$$= C_t + \int_0^t \int_{E^1} h^2(x)\nu(ds, dx) - \sum_{0 < s \leq t} (\Delta B_s)^2.$$

It turns out that the probability measure $\mathbf{P}$ on $(\mathbf{D}, \mathcal{D})$ is the measure of the (canonical) process with independent increments $X_t(\alpha) = \alpha(t)$, $t \geq 0$, with triple $T = (B, C, \nu)$ if and only if (with respect to this measure $\mathbf{P}$) the following processes
1) $M(h) = (M(h)_t)_{t \geq 0}$,
2) $M^2(h) - \widetilde{C}$,
3) $g * \mu - g * \nu$, $g \in \mathcal{G}^+$

are martingales. Here $\mathcal{G}$ is defined as the class of bounded Borel functions on E^1 equal to zero in a neighborhood of zero and such that if for two measures η and $\tilde{\eta}$ with the property

$$h(\{0\}) = \tilde{\eta}(\{0\}) = 0, \quad \eta(x : |x| > \epsilon) < \infty, \quad \eta'(x : |x| > \epsilon) < \infty$$

we have $\eta(f) = \tilde{\eta}(f)$ for $f \in \mathcal{G}^+$, then $\eta = \tilde{\eta}$; this class is non-empty and there exist families $\mathcal{G}^+$ that are, for example, countable and contain only continuous functions. Particular cases of this theorem are well known.

Example 1. If $X_0 = 0$, C_t is a continuous non-decreasing function and $B = 0$, $\nu = 0$, then the martingale problem is formulated as:
1) $X = (X_t)_{t \geq 0}$ is a martingale,
2) $X^2 - C = (X_t^2 - C_t)_{t \geq 0}$ is a martingale. The corresponding measure **P** (which is unique) with respect to this (in the space of continuous functions on $[0, \infty)$) is none other than Wiener measure. This assertion is called the "Lévy Theorem".

Example 2. Let $X = (X_t)_{t \geq 0}$ be cadlag functions that are piecewise constant and with jumps of size $+1$, $X_0 = 0$. If
$$B_t(h) = h(1)A_t,$$
$$C(t) = 0,$$
$$\nu(dt, dx) = dA_t \otimes \epsilon_1(dx),$$
where $A = (A_t)_{t \geq 0}$ is a non-negative continuous non-decreasing function (say, $A_t = \lambda t$, $\lambda > 0$), and $\epsilon_1(dx)$ is the Dirac measure concentrated at the point $x = 1$, then (in the canonical space of these functions), there exists a unique probability measure, with respect to which the process X is a Poisson process with $\mathbf{E} X_t = A_t$. This assertion is a theorem of Watanabe.

Example 3. Let
$$X_t = \sum_{k \leq t} \xi_k, \quad \xi_0 = 0,$$
be a stochastic process generated by a sum of independent random variables. Here the corresponding triple $T = (B, C, \nu)$ has the following form:
$$B_t = \sum_{k \leq t} \mathbf{E} h(\xi_k),$$
$$C_t = 0,$$
$$\nu([0, t] \times g) = \sum_{k \leq t} \mathbf{E}[g(\xi_k) I(\xi_k) \neq 0)].$$

In this case the "discrete characteristic" ν completely determines the triple T. The measures $\nu(\{k\} \times dx)$ ($= \mathbf{P}(\xi_k \in dx)$), $k \geq 1$, uniquely define the probability distribution of the process X (in the coordinate representation).

4.3. We now turn to conditions ensuring the convergence $X^n \xrightarrow{\mathcal{L}(S)} X$ of finite-dimensional distributions of processes X^n with independent increments to X (for times belonging to the set S).

Let $T^n = (B^n, C^n, \nu^n)$ and $T = (B, C, \nu)$ be their triples. We introduce the conditions
$(\beta - S)$ $\quad B_s^n \to B_s, \ s \in S,$
$(\gamma - S)$ $\quad \tilde{C}_s^n \to C_s, \ s \in S,$
$(\delta_i - S)$ $\quad g * \nu_s^n \to g * \nu_s, \ g \in C_i.$

Chapter 4.I. Weak Convergence of Measures on Metric Spaces

Theorem 4.5. *Suppose that the process with independent increments X does not have fixed jump times (equivalently, it is continuous in probability). Suppose further that the following asymptotic negligibility condition of the jumps for the predetermined processes X^n holds:*

$$\limsup_n \sum_{s \le t} \nu^n(\{s\} \times \{|x| > \epsilon\}) = 0, \ \epsilon > 0, \ t \in S. \qquad (4.52)$$

Then conditions $(\beta - S)$, $(\gamma - S)$, $(\delta_i - S)$ for $i = 1$ or $i = 2$ are necessary and sufficient for the weak convergence $X^n \xrightarrow{\mathcal{L}(S)} X$.

If, in addition, the set S is everywhere dense in $\mathbb{R}_+$, then the condition $(\delta_1 - S)$ implies (4.52) and hence

$$(\beta - S), \ (\gamma - S), \ (\delta_1 - S) \Rightarrow X^n \xrightarrow{\mathcal{L}(S)} X. \qquad (4.53)$$

Remark 1. In (4.53) the reverse implication is not true in general. Here is an example.

Consider the deterministic function

$$X_t^n = \begin{cases} 1, & \frac{1}{n} \le t \le \frac{2}{n}, \\ 0, & \text{otherwise} \end{cases}$$

and $X_t \equiv 0$. Then $X^n \xrightarrow{\mathcal{L}(S)} X$, but condition (4.52) is not satisfied.

Remark 2. Theorem 4.2 (on the convergence of infinitely divisible distributions under conditions (4.40)) and Theorem 4.3 (on the convergence of distributions of the sum of asymptotically negligible independent random variables to an infinitely divisible one) follow from Theorem 4.5. In order to convince oneself of this it is sufficient to set $X_t^n = \sum^{[nt]} \xi_{nk}$ and to take as X a homogeneous process with independent increments and characteristic function (4.49) and then take $S = \{1\}$.

Although Theorem 4.2 is a particular case of Theorem 4.3, it is convenient to combine them in the following form.

Let ζ^n, $n \ge 1$, be infinitely divisible random variables with triples (b^n, C^n, ν^n), and let $(\xi_{nk})_{1 \le k \le n}$, $n \ge 1$, be a triangular scheme of asymptotically negligible independent (for each n) random variables that are also independent of ζ^n, $n \ge 1$. Let $h = h(x)$ be a continuous truncation function and suppose that the following conditions hold:

$$b^n + \sum \mathbf{E}h(\xi_{nk}) \to b,$$
$$\tilde{C}^n + \sum [\mathbf{E}h^2(\xi_{nk}) - (\mathbf{E}h(\xi_{nk}))^2] \to \tilde{C}, \qquad (4.54)$$
$$F^n(g) + \sum \mathbf{E}g(\xi_{nk}) \to F(g), \ g \in C_1 \text{ or } g \in C_2,$$

where $T = (b, C, F)$ is the triple of the infinitely divisible random variable ζ.

Then
$$\zeta^n - \sum \xi_{nk} \xrightarrow{d} \zeta. \qquad (4.55)$$

To derive this result from Theorem 4.5 it is sufficient to define, for example, the processes
$$X_t^n = \zeta^n + \sum_{k=1}^n \xi_{nk} I\left(t \geq \frac{1}{k}\right),$$
$$X_t \equiv \zeta$$

and to take as S the set consisting of one point $S = \{1\}$.

The proof of Theorem 4.5 is carried out by the method of characteristic functions and is based on the fact that the characteristic functions
$$g_t^n(\lambda) = \mathbf{E} e^{i\lambda X_t^n} = \mathcal{E}(G^n(\lambda))_t,$$
$$g_t(\lambda) = \mathbf{E} e^{i\lambda X_t} = \mathcal{E}(G(\lambda))_t$$

are explicitly defined in terms of the triples T^n and T, since
$$\mathcal{E}(G^n(\lambda))_t = e^{G_t^n(\lambda)} \prod_{0 < s \leq t} (1 + \Delta G_s^n(\lambda)) e^{-\Delta G_s^n(\lambda)}$$

with
$$G_t^n(\lambda) = i\lambda B_t^n - \frac{\lambda^2}{2} C_t^n + \int_0^t \int_{E^1} (e^{i\lambda X} - 1 - i\lambda h(x)) \nu^n(ds, dx),$$

and (for a process X that is continuous in probability)
$$\mathcal{E}(G(\lambda))_t = e^{G_t(\lambda)}$$

with
$$G_t(\lambda) = i\lambda B_t - \frac{\lambda^2}{2} C_t + \int_0^t \int_{E^1} (e^{i\lambda x} - 1 - i\lambda h(x)) \nu(ds, dx).$$

Using these explicit representations, it is shown (the same as in the classical Theorem 4.2) that the conditions $(\beta - S)$, $(\gamma - S)$, $(\delta_i - S)$ ($i = 1$ or 2) ensure the convergence of the characterstic functions of the corresponding finite-dimensional distributions $\mathbf{P}_{s_1,\ldots,s_k}^n$ to $\mathbf{P}_{s_1,\ldots,s_k}$, $s_1, \ldots, s_k \in S$.

Now that we have the convergence conditions of the finite-dimensional distributions $X^n \xrightarrow{\mathcal{L}} X$ with S dense in $\mathbb{R}_+$, we can pose the natural question of the validity of the functional convergence $X^n \xrightarrow{\mathcal{L}} X$. For this, it is sufficient to find conditions for the tightness of the family of distributions $\{\mathbf{P}^n\}$ of the processes X^n, $n \geq 1$. In fact, here (to establish the convergence $X^n \to X$) we can again use the method based on the three stages I $\oplus$ II $\oplus$ III ("tightness"$\oplus$"characterization"$\oplus$"identification"), since the tightness of the family $\{\mathbf{P}^n\}$ along with the convergence of the finite-dimensional distributions ensures that all the weak limits $\mathbf{P}' =: w - \lim \mathbf{P}^{n'}$ are characterized by the property that the finite-dimensional distributions for $\mathbf{P}'$ coincide with the finite-dimensional distributions for $\mathbf{P}$. But in a Skorokhod space the σ-algebra of cylinder sets $\mathcal{B}_{cy}(\mathbf{D})$ coincides with the Borel σ-algebra $\mathcal{B}(\mathbf{D})$, and the measures on $\mathcal{B}_{cy}(\mathbf{D})$ are completely determined by their finite-dimensional distributions. In this way, the fact of convergence of the finite-dimensional distributions identifies all the weak limits $\mathbf{P}'$ as $\mathbf{P}$. Hence, w-$\lim \mathbf{P}^n = \mathbf{P}$.

Theorem 4.6. *Suppose that the condition*

$$(\sup \text{-}\beta) \quad \sup_{s \leq t} |B_s^n - B_s| \to 0, \; t > 0$$

holds and the conditions $(\gamma - S)$, $(\delta_1 - S)$ *hold for some everywhere dense set S in $\mathbb{R}_+$. Then the sequence $\{\mathbf{P}^n\}$ is tight.*

We will return later to the question of establishing (by martingale methods) the tightness of the family $\{\mathbf{P}^n\}$ of probability distributions of a sufficiently wide class of processes X^n, $n \geq 1$ (semimartingales in the context below). The next result on the convergence $X^n \xrightarrow{\mathcal{L}} X$ also follows from Theorems 4.5 and 4.6.

Theorem 4.7. *Let X^n be processes with independent increments with triples $T^n = (B^n, C^n, \nu^n)$. Let X be a process with independent increments, without fixed times of jumps and with triple $T = (B, C, \nu)$.*
Then the conditions
$(\sup \text{-}\beta) \quad \sup_{s \leq t} |B_s^n - B_s| \to 0, \; t > 0,$
$(\gamma) \quad \tilde{C}_t^n \to C_t, \; t > 0,$
$(\delta) \quad g * \nu_t^n \to g * \nu_t, \; t > 0, \; g \in C_1 \text{ or } C_2,$
are sufficient (and also necessary) in order that $X^n \xrightarrow{\mathcal{L}} X$.

One can deduce several useful corollaries from this theorem.

Corollary 1. *Let X^n and X be processes without fixed times of jumps. Then $X^n \xrightarrow{\mathcal{L}} X$ if and only if the convergence of all finite-dimensional distributions $X^n \xrightarrow{\mathcal{L}(\mathbb{R}_+)} X$ holds, and condition ($\sup \text{-}\beta$) is satisfied.*

Corollary 2. *Let X^n and X be processes with stationary independent increments. Then $B_t^n = b^n \circ t$, $C_t^n = c^n \circ t$, $\nu^n(dt, dx) = dt F(dx)$, $B_t = b \circ t$, $C_t = c \circ t$, $\nu(dt, dx) = dt F(dx)$ and the following conditions are equivalent:*
a) $X^n \xrightarrow{\mathcal{L}} X$,
b) $X_1^n \xrightarrow{\mathcal{L}} X_1$,
c) $b^n \to b, \quad \tilde{C}^n \to \tilde{C}, \quad F_n(g) \to F(g), \; g \in G_1.$

Corollary 3 (Donsker 1951). *Let $(\xi_k)_{k \geq 1}$ be a sequence of independent identically distributed random variables with $\mathbf{E}\xi_k = 0$, $\mathbf{E}\xi_k^2 = 1$. Then the processes*

$$X_t^n = \frac{1}{\sqrt{n}} \sum_{1 \leq k \leq [nt]} \xi_k, \; t \geq 0,$$

converge in distribution to a Wiener process.

In fact, since here we are dealing with the square-integrable case, we can simply take as the function $h = h(x)$ the function $h(x) = x$. Then

$$B_t^n = \sum_{k \leq [n']} \mathbf{E}\left(\frac{\xi_k}{\sqrt{n}}\right) = 0,$$

$$\tilde{C}_t^n = \sum_{k \leq [nt]} \mathbf{E}\left(\frac{\xi_k}{\sqrt{n}}\right)^2 = \frac{[nt]}{n} \to t,$$

$$g * \nu_t^n = \sum_{k \leq [nt]} \mathbf{E}g\left(\frac{\xi_k}{\sqrt{n}}\right) = [nt]\mathbf{E}g\left(\frac{\xi_1}{\sqrt{n}}\right).$$

Note that for $g \in C$ we have

$$|g(x)| \leq c|x|^2 I\{|x| > a\}$$

for some constant c, and since

$$|x|^2 I(|x| > a) * \nu_t^n = [nt]\frac{1}{n}\mathbf{E}(|\xi_1|^2 I(|\xi_1| > a\sqrt{n})) \to 0,$$

it follows that condition (δ) is satisfied.

We now turn to the proof of Theorem 4.7.

As has been noted above, the sufficiency is based on verifying the tightness and the convergence of the finite-dimensional distributions, which in turn relies, in essence, on a classical result, namely, Theorem 4.6. The necessity is a more complicated matter. It turns out, however, that the method of proof (see Jacod and Shiryaev 1987) of the necessity of conditions of the type in Theorem 4.7 "works" even without the assumption that the process X does not have fixed times of jumps. Furthermore, the "method of necessary conditions" proves to be useful for the proof of the sufficiency in the general situation of processes with independent increments X and X^n, $n \geq 1$.

We now give the essence of the "method of necessary conditions".

A crucial result is the following.

Theorem 4.8. *Assume that Y and Y^n, $n \geq 1$ are processes with independent increments, $Y^n \xrightarrow{\mathcal{L}} Y$ and that for each $t \geq 0$ the family of random variables $\{\sup_{s \leq t} |Y_s^n|\}_{n \geq 1}$ is uniformly integrable. If $\mathcal{L}_n(t) = \mathbf{E}Y_t^n$ and $\mathcal{L}(t) = \mathbf{E}Y_t$, then $\mathcal{L}_n \to \mathcal{L}$ in the Skorokhod topology.*

The following general fact is based on this result.

Theorem 4.9. *In order that $X^n \xrightarrow{\mathcal{L}} X$, where X and X^n are processes with independent increments, it is necessary and sufficient that the following conditions hold:*

$(S_k\text{-}\beta)\quad S_k(B^n, B) \to 0,$
$(S_k\text{-}\gamma)\quad S_k(\tilde{C}^n, \tilde{C}) \to 0,$ \hfill (4.56)
$(S_k\text{-}\delta)\quad S_k(g*\nu^n, g*\nu) \to 0,$

where $S_k(\alpha, \beta)$ is a distance compatible with the Skorokhod topology, for example, the distance $\delta(\alpha, \beta)$ introduced in Sect. 2.

The proof of the theorem is along the following lines. First, we establish the necessity of conditions $(S_k\text{-}\beta)$, $(S_k\text{-}\gamma)$, $(S_k\text{-}\delta)$ (on the basis of Theorem 8). We then prove the sufficiency as follows.

Our first task is to establish that these conditions ensure the tightness of the family $\{X^n\}$. (We will return later to questions of this type.) It therefore remains to prove that all the weak limits $\mathbf{P}' = \text{w-}\lim \mathbf{P}^{n'}$ coincide with the distribution $\mathbf{P}$ of the process X. Let the corresponding limiting process X' have triple $T' = (B', C', \nu')$. Then, by the necessity, $S_k(B^{n'}, B') \to 0$, $S_k(\tilde{C}^{n'}, \tilde{C}) \to 0$, $S_k(g*\nu^{n'}, g*\nu) \to 0$, $g \in C_1$, and hence by conditions (4.56), $B' = B$, $C' = C$, $\nu' = \nu$.

4.4. As a specific instance of the results presented above, let us consider the question of the conditions of convergence to a continuous Gaussian process X with triple $(B, C, 0)$.

Theorem 4.10. *Let X^n be processes with independent increments for which $|x|^2 * \nu_t^n < \infty$, and suppose that the Lindeberg condition*

$(L)\quad |x|^2 I(|x| > \epsilon) * \nu_t^n \to 0,\ \epsilon > 0,\ t \in S \subseteq \mathbb{R}_+$

holds.
Then
a) $X^n \overset{\mathcal{L}(S)}{\to} X$ if and only if

$$B_t^n \to B_t,\ t \in S,$$
$$\tilde{C}_t^n \to C_t,\ t \in S,$$

(with $h(x) = x$)
b) $X^n \overset{\mathcal{L}}{\to} X$ if and only if

$$\sup_{s \leq t} |B_s^n - B_s| \to 0,\ t \geq 0,$$

and

$$\tilde{C}_t^n \to C_t,\ t \geq 0.$$

§5. The Convergence of Semimartingales to Processes with Independent Increments

5.1. Having studied the questions of the convergence of finite-dimensional distributions $X^n \xrightarrow{\mathcal{L}(S)} X$ and the convergence $X^n \xrightarrow{\mathcal{L}} X$ for processes with independent increments (where for the convergence $X^n \xrightarrow{\mathcal{L}(S)} X$ we proceed via the method of characteristic functions, while for the convergence $X^n \xrightarrow{\mathcal{L}} X$ we proceed by a method based on the verification of the tightness of the family $\{X^n\}$ and establish the convergence of the finite-dimensional distributions), it is natural to move on to the study of a convergence in a more general situation. To this end, we first assume that the "limit" process X is again a process with independent increments. On the other hand, the "prelimit" processes X^n will be processes of a more general nature.

The following question is difficult but of fundamental importance. How broad can the supply of processes X^n be in order to be able to introduce for them an analogue of the triples T^n considered above, and in order that the form of the answer be, for example:

$$\text{``}T^n \to T \Rightarrow X^n \xrightarrow{\mathcal{L}} X\text{''},$$

where the "convergence $T_n \to T$" is understood in an appropriate sense? Let us note straight away that an analysis of the proof of the preceding theorems for processes X with independent increments shows that the appearance of the triple $T = (B, C, \nu)$ was suggested by the canonical representation (4.50):

$$X_t = X_0 + B_t + X_t^c + \int_0^t \int_{E^1} h(x) d(\mu - \nu) + \int_0^t \int_{E^1} (x - h(x)) d\mu. \quad (4.57)$$

Let us also note that here the process

$$M_t = X_t^c + \int_0^t \int_{E^1} h(x) d(\mu - \nu) \quad (4.58)$$

is a martingale, while the process

$$A_t = B_t + \int_0^t \int_{E^1} (x - h(x)) d\mu \quad (4.59)$$

is a process of locally bounded variation and $X_t = X_0 + M_t + A_t$.

In this connection, we now consider stochastic processes X that can be represented in the form

$$X_t = X_0 + M_t + A_t, \quad (4.60)$$

where $M = (M_t)_{t \geq 0}$ is an arbitrary process of "martingale" type and $A = (A_t)_{t \geq 0}$ is an arbitrary process of local bounded variation.

The development of the general theory of stochastic processes has shown that the appropriate formalization presupposes the introduction of the concept

Chapter 4.I. Weak Convergence of Measures on Metric Spaces

of a "stochastic basis" and the concept of a "local martingale", defined in the following manner.

5.2. The axioms of probability theory proposed by Kolmogorov imply that there is a given triple $(\Omega, \mathcal{F}, \mathbf{P})$ where the set Ω is treated as the set of elementary outcomes ω, $\mathcal{F}$ is a σ-algebra of subsets $A \subseteq \Omega$ called events, and $\mathbf{P} = \mathbf{P}(A)$, $A \in \mathcal{F}$, is a countably additive set function such that $0 \leq \mathbf{P} \leq 1$ and $\mathbf{P}(\Omega) = 1$.

For stochastic calculus (and for everything that follows) it is important that there be further distinguished in the measure space $(\Omega, \mathcal{F})$ a flow of σ-algebras $\mathbf{F} = (\mathcal{F}_t)_{t \geq 0}$, where $\mathcal{F}_s \subseteq \mathcal{F}_t \subseteq \mathcal{F}$, $s \leq t$. We treat $\mathcal{F}_t$ as the collection of events observable up to and including time t and we call the set of objects $\mathcal{B} = (\Omega, \mathcal{F}, \mathbf{F} = (\mathcal{F}_t)_{t \geq 0}, \mathbf{P})$ the *stochastic basis* (with additional assumptions of a purely technical character, namely, that the family $\mathbf{F}$ be right-continuous, that is, $\mathcal{F}_t = \mathcal{F}_{t+} \equiv \bigcap_{s > t} \mathcal{F}_s$, and that $\mathcal{F}_0$ be completed by the $\mathbf{P}$-null sets of $\mathcal{F}$).

The assumption regarding the presence of the additional structure $\mathbf{F} = (\mathcal{F}_t)_{t \geq 0}$ on the probability space $(\Omega, \mathcal{F}, \mathbf{P})$ makes possible the introduction and exploitation of new concepts such as, for example, stopping times, predictability, local martingale, etc.

Of immediate importance for us is the general definition of a martingale $M = (M_t, \mathcal{F}_t)_{t \geq 0}$ as a process $(M_t)_{t \geq 0}$ given on a stochastic basis $\mathcal{B}$ such that
1) the M_t are $\mathcal{F}_t$-measurable, $t \geq 0$,
2) $\mathbf{E}(M_t | \mathcal{F}_s) = M_s$, $s < t$.

We say that a mapping $\tau = \tau(\omega)$; $\Omega \to \bar{\mathbb{R}}_+$ is a stopping time if it is such that $\{\omega : \tau \leq t\} \in \mathcal{F}_t$ for any $t \geq 0$.

The concepts "martingale" and "stopping time" lead to the new concept of a "local martingale" as a stochastic process $X = (X_t)_{t \geq 0}$ such that X_t are $\mathcal{F}_t$-measurable, $t \geq 0$, there exists an increasing sequence $(\tau_n)_{n \geq 1}$ of Markov times with the property $\lim_n \tau_n = \infty$ ($\mathbf{P}$-a.s.), and that for each n the "stopped" process $X^{\tau_n} = (X_{t \wedge \tau_n})$ is a martingale. If $\mathcal{M}$ is the class of martingales and $\mathcal{M}_{\mathrm{loc}}$ is the class of local martingales, then

$$\mathcal{M} \subseteq \mathcal{M}_{\mathrm{loc}},$$

since each martingale is a local martingale (it is necessary to take $\tau_n \equiv n$).

We will also say that a stochastic process $A = (A_t)_{t \geq 0}$ is a process of class $\mathcal{V}$, or a process of locally bounded variation, if on each interval $[0, t]$ the trajectories $A_s(\omega)$, $s \in [0, t]$, have bounded variation for any $\omega \in \Omega$ and the A_t are $\mathcal{F}_t$-measurable, $t \geq 0$.

5.3. We are now in a position to give the definition of a "semimartingale" which is an important notion for us.

We say that a process $X = (X_t)_{t \geq 0}$, given on a stochastic basis $\mathcal{B} = (\Omega, \mathbf{F}, \mathbf{P})$, is a semimartingale if the X_t are $\mathcal{F}_t$-measurable, $t \geq 0$, and there is a (generally speaking, non-unique) representation of the form

$$X_t = X_0 + M_t + A_t, \tag{4.61}$$

where M is a local martingale ($M \in \mathcal{M}_{\text{loc}}$) and A is a process of locally bounded variation ($A \in \mathcal{V}$).

For what follows, we require the important concepts of the predictable σ-algebra $\mathcal{P}$ and a predictable process.

By definition a σ-algebra $\mathcal{P}$ of subsets in the space $\Omega \times \mathbb{R}_+ = \{(\omega, t) : \omega \in \Omega, \ t \in \mathbb{R}_+\}$ is called the σ-algebra of predictable sets if it is generated by all left-continuous processes $Y = Y(\omega, t)$, considered as mappings of $\Omega \times \mathbb{R}_+$ to $\mathbb{R}$ that are $\mathcal{F}_t$-measurable for each t. (In an equivalent manner, one can define the σ-algebra $\mathcal{P}$ as the system generated by sets of the form $A \times \{0\}$, $A \in \mathcal{F}_0$, and $A \times (s, t]$, $s < t$, $A \in \mathcal{F}_s$ or as the system generated by sets of the form $A \times \{0\}$, $A \in \mathcal{F}_0$, and the stochastic intervals $[0, \tau] = \{(\omega, t) : 0 \leq t \leq \tau(\omega)\}$, where $\tau = \tau(\omega)$ are stopping times.)

A stochastic process $Y = (Y_t(\omega))_{t \geq 0}$ is called predictable if it can also be considered as a mapping of $\Omega \times \mathbb{R}_+$ to E^1 that is measurable with respect to the σ-algebra $\mathcal{P}$:

$$\{(\omega, t) : Y_t(\omega) \in B\} \in \mathcal{P}$$

for each Borel set B in E^1.

The terminology "predictable" process is appropriate, since a process for which the trajectories are left-continuous, possesses the property that one can predict the value of Y_t from the values of Y_s, $s < t$. Every deterministic process is, of course, predictable. Indeed, "predictability" can be interpreted as "stochastic determinism".

It is very remarkable that each semimartingale $X = (X_t)_{t \geq 0}$ given on a stochastic basis $\mathcal{B}$ admits a (canonical) representation

$$X_t(\omega) = X_0 + B_t(\omega) + X_t^c(\omega)$$
$$+ \int_0^t \int_{E^1} h(x) d(\mu - \nu) + \int_0^t \int_{E^1} (x - h(x)) d\mu \tag{4.62}$$

outwardly resembling the representation (4.57) for processes with independent increments and differing from it only in that instead of deterministic B, ν and a continuous process X with independent increments, in (4.62) we have:

1) $B = (B_t(\omega))_{t \geq 0}$ is a predictable process of class $\mathcal{V}$,
2) $X^c = (X_t^c(\omega))_{t \geq 0}$ is the continuous martingale component of X,
3) ν is the compensator of the jump measure μ of X; the process $(x^2 \wedge 1) * \nu \equiv \int_0^t \int_{E^1} (x^2 \wedge 1) d\nu$ is locally integrable and

$$\Delta B_t = \int_{E^1} h(x) \nu(\{t\} \times dx).$$

Let us recall that every continuous martingale is locally square-integrable and by the "Doob-Meyer decomposition", there exists an increasing predictable process $\langle X^c \rangle$ such that $(X^c)^2 - \langle X^c \rangle$ is a local martingale. We denote this process $\langle X^c \rangle$ by $C = (C_t(\omega))_{t \geq 0}$; it is called the *quadratic characteristic* of

Chapter 4.I. Weak Convergence of Measures on Metric Spaces

X^c. We also recall that the compensator $\nu = \nu(\omega; dx, dt)$ of the jump measure $\mu = \mu(\omega; dx, dt)$ of the process X is defined as the non-negative predictable random measure such that

$$g * \mu - g * \nu \left(= \int_0^t \int_{E^1} g(x)\mu(\omega; dx, ds) - \int_0^t \int_{E^1} g(x)\nu(\omega; dx, ds) \right) \quad (4.63)$$

is a local martingale (for a sufficiently broad class of bounded non-negative Borel functions $g = g(x)$ vanishing in a neighborhood of zero).

The collection

$$T = (B, C, \nu) \quad (4.64)$$

is called the *triple of predictable characteristics* of the semimartingale X.

For a semimartingale X the compensator ν of its jump measure satisfies the property that the process

$$(|x|^2 \wedge 1) * \nu \left(= \int_0^t \int_{E^1} (|x|^2 \wedge 1) d\nu \right) \quad (4.65)$$

is locally integrable, and

$$\Delta B_t = \int h(x)\nu(\{t\} \times dx), \ t > 0. \quad (4.66)$$

To explain how the representation (4.62) arises for semimartingales, we set

$$\check{X}(h)_t = \sum_{0 < s \leq t} [\Delta X_s - h(\Delta X_s)] = \int_0^t \int_{E^1} (x - h(x)) d\mu$$

and let

$$X(h)_t = X_t - \check{X}_t(h) = X_t - \int_0^t \int_{E^1} (x - h(x)) d\mu \quad (4.67)$$

be a process with bounded jumps. It is known that each such process $X(h)$ admits a unique representation

$$X(h) = X_0 + M(h) + B(h) \quad (4.68)$$

with $M(h) \in \mathcal{M}_{\text{loc}}$ and the predictable process $B(h) \in \mathcal{V}$. In its turn, the local martingale $M(h)$ can be written in the form

$$M(h)_t = M^c(h)_t + M^d(h)_t, \quad (4.69)$$

where $M^c(h)$ is a continuous local martingale, while $M^d(h)$ is a purely discontinuous local martingale representable in the form

$$M^d(h)_t = \int_0^t \int_{E^1} h(x) d(\mu - \nu). \quad (4.70)$$

From (4.67)–(4.70) we obtain the above representation (4.62):

$$X_t = X_0 + B(h)_t + M^c(h)_t + \int_0^t\!\!\int_{E^1} h(x)d(\mu - \nu) + \int_0^t\!\!\int_{E^1} (x - h(x))d\mu.$$

Here the process $M^c(h)$ is a continuous local martingale that is independent of the choice of the "truncation" function h and is therefore an "internal" characteristic of the process X, so to speak. This process is denoted by X^c and is called the continuous martingale component of the semimartingale X. The compensator ν also does not depend on h and also is an "internal" characteristic of X.

5.3. It is now appropriate to note the connection between the semimartingale X with triple $T = (B, C, \nu)$ and a certain "martingale problem". Namely, we have the following result.

Theorem 4.11. *The condition*
a) X *is a semimartingale with triple* $T = (B, C, \nu)$ *is equivalent to the condition:*
b) *the processes*

$$M(h) \equiv X(h) - X_0 - B, \quad M^2(h) - C, \quad g*\mu - g*\nu$$

are local martingales $(g \in \mathcal{G} + (\mathbb{R}_+))$.

5.4. We now move on to an account of the results on the convergence $X^n \xrightarrow{\mathcal{L}(S)} X$ and the convergence $X^n \xrightarrow{\mathcal{L}} X$ of the semimartingales X^n to a process with independent increments. Moreover, we will approach it in order to give the answer in the symbolic form

$$\text{``}T^n \xrightarrow{(S)} T \Rightarrow X^n \xrightarrow{\mathcal{L}(S)} X\text{''}, \tag{4.71}$$

$$\text{``}T^n \to T \Rightarrow X^n \xrightarrow{\mathcal{L}} X\text{''}, \tag{4.72}$$

understanding by the convergence of the triples, the appropriate convergence of their components.

Concerning the finite-dimensional convergence (4.71), here again it is natural to attempt to apply the method of characteristic functions. However, the immediate verification that, say, the one-dimensional distributions for X_t^n converge to the one-dimensional distributions X_t is now not so simple, since it is difficult to express in terms of the triples the characteristic function

$$g_t^n(\lambda) = \mathbf{E}e^{i\lambda X_t^n}.$$

For a process X with independent increments the characteristic function

$$g_t(\lambda) = \mathbf{E}e^{i\lambda X_t},$$

as was indicated in (4.47), has the simple form:

$$g_t(\lambda) = \mathcal{E}(G(\lambda))_t = e^{G_t(\lambda)} \prod_{0 < s \leq t} (1 + \Delta G_s(\lambda)) e^{-\Delta G_s(\lambda)}, \tag{4.73}$$

where

$$G_t(\lambda) = i\lambda B_t - \frac{\lambda^2}{2} G_t + \int_0^t \int_{E^1} (e^{i\lambda x} - 1 - i\lambda h(x)) \nu(ds, dx). \tag{4.74}$$

For an arbitrary semimartingale X^n one can also introduce the corresponding cumulants

$$G_t^n(\lambda) = i\lambda B_t^n - \frac{\lambda^2}{2} C_t^n + \int_0^t \int_{E^1} (e^{i\lambda x} - 1 - i\lambda h(x)) \nu^n(ds, dx) \tag{4.75}$$

which exist, since the process $(x^2 \wedge 1) * \nu$ is locally integrable.

How are the stochastic exponentials $\mathcal{E}(G^n(\lambda))_t$ and the characteristic functions $g_t^n(\lambda) = \mathbf{E} e^{i\lambda X_t^n}$ of the process X^n related? This question is quite legitimate if one bears in mind that for processes X with independent increments $g_t(\lambda) = \mathbf{E} e^{i\lambda X_t} = \mathcal{E}(G(\lambda))_t$.

It might be thought that $g_t^n(\lambda)$ is $\mathbf{E}\mathcal{E}(G^n(\lambda))_t$, but this is not so! Indeed, we have the following result.

Theorem 4.12. *Let $\mathcal{E}(G(\lambda))$ be non-vanishing for all $\lambda \in E^1$. The process X^n is a semimartingale with triple $T^n = (B^n, C^n, \nu^n)$ if and only if for each $\lambda \in E^1$ the process*

$$\frac{e^{i\lambda X^n}}{\mathcal{E}(G^n(\lambda))} \tag{4.76}$$

is a local martingale.

Thus (in the conditions of the theorem) one can assert that

$$e^{i\lambda X_t^n} = \mathcal{E}(G^n(\lambda))_t m^n(\lambda)_t,$$

where $m^n(\lambda)$ is a local martingale, and hence the characteristic function

$$g^n(\lambda)_t = \mathbf{E}\mathcal{E}(G^n(\lambda))_t m^n(\lambda)_t.$$

Locally the martingale $m^n(\lambda)$ is "badly controlled" from the point of view of its properties. However, it is only the property that $m^n(\lambda)$ be a martingale that provides to be fully adequate for the following (and many other) considerations. (In the general case, when $\mathcal{E}(G^n(\lambda))$ vanishes, there is an analogue of Theorem 4.12, but the formulation is more complicated.)

Theorem 4.13. *Let the "limit" process X be a process with independent increments without fixed times of jumps, that is $g(\lambda)_t \neq 0$ for all λ and t. If*

$$\mathcal{E}(G^n(\lambda))_t \xrightarrow{\mathbf{P}} g(\lambda)_t = \mathcal{E}(G(\lambda))_t, \ t \in S, \tag{4.77}$$

then $X^n \xrightarrow{\mathcal{L}(S)} X$, that is, weak convergence of the finite-dimensional distributions occurs at the moments of time belonging to the set S.

The idea of the proof of this theorem can be explained as follows (we restrict ourselves to the one-dimensional case $S = \{t\}$).

In accordance with the method of characteristic functions, all we need to prove is the convergence of the characteristic functions

$$g^n(\lambda)_t \to g(\lambda)_t, \ \lambda \in E^1,$$

or that

$$\frac{g^n(\lambda)_t}{g(\lambda)_t} - 1 \to 0, \ \lambda \in E^1, \tag{4.78}$$

since $g(\lambda)_t \neq 0$. But by (4.76), taking $X_0^n \equiv 0$, we find that

$$1 = \mathbf{E}\frac{e^{i\lambda X_t^n}}{\mathcal{E}(G(\lambda))_t};$$

therefore (4.78) is equivalent to

$$\mathbf{E}\left[\frac{e^{i\lambda x_t^n}}{\mathcal{E}(G(\lambda))_t} - \frac{e^{i\lambda X_t^n}}{\mathcal{E}(G^n(\lambda))_t}\right] \to 0,$$

for which, of course, it is sufficient that

$$\mathbf{E}\left[\frac{1}{\mathcal{E}(G(\lambda))_t} - \frac{1}{\mathcal{E}(G^n(\lambda))_t}\right] \to 0, \ n \to \infty. \tag{4.79}$$

If

$$\frac{1}{\mathcal{E}(G(\lambda))_t} + \frac{1}{\mathcal{E}(G^n(\lambda))_t} \leq C(\lambda, t),$$

then (4.79) follows from (4.77) and the Lebesgue dominated convergence theorem.

The general case reduces to the considered one by introducing suitable stopping times.

The convergence of the stochastic exponentials $\mathcal{E}(G^n(\lambda))_t$ to $\mathcal{E}(G(\lambda))_t$, ensuring the convergence of the corresponding characteristic functions $g^n(\lambda)_t$ to $g(\lambda)_t$, gave rise to a name "the method of stochastic exponential" for the implication (4.77) $\Rightarrow X^n \xrightarrow{\mathcal{L}(S)} X$.

In the case when, in addition, the processes X^n are processes with independent increments, the convergence of (4.77) simply becomes the convergence of the characteristic functions. So in this case, the method of stochastic exponentials is the same as the method of characteristc functions.

The fact that the cumulants $G^n(\lambda)$ and $G(\lambda)$ are expressed in terms of the triples T^n and T leads to the hope that the convergence (4.77) can be expressed by means of a suitable convergence of the triples T^n to T. The next theorem makes this idea more specific.

Theorem 4.14. *Suppose that the process X has no fixed times of discontinuity (equivalently, $\nu(\{t\} \times E^1) = 0$ for all $t \in \mathbb{R}_+$, so that B and $\tilde{C}$ are continuous). Suppose further that*

$$\sup_{s \le t} \nu^n(\{s\} \times \{|x| \ge \epsilon\}) \xrightarrow{P} 0, \ t \in S, \ \epsilon > 0, \qquad (4.80)$$

and that the following conditions are satisfied:

(β) $\quad B_t^n \xrightarrow{P} B_t, \quad t \in S,$
(γ) $\quad \tilde{C}_t^n \xrightarrow{P} \tilde{C}_t, \quad t \in S,$
(δ) $\quad g * \nu_t^n \xrightarrow{P} g * \nu, \quad t \in S, g \in C_1.$

Then $X^n \xrightarrow{\mathcal{L}(S)} X$.

If S is everywhere dense in $\mathbb{R}_+$, then condition (δ) ensures condition (4.80) (so that in this case for convergence, the satisfaction of conditions (β), (γ), (δ) is sufficient).

5.5. It is natural now to pass to the functional case. Since there are already conditions for the convergence of finite-dimensional distributions, the simplest way would be to explain whether the conditions (β), (γ), (δ) or their strengthenings guarantee the relative compactness (tightness) of the family of distributions of the processes X^n, $n \ge 1$. It turns out that it suffices merely to strengthen condition (β):

Theorem 4.15. *Let X be a process with independent increments without fixed times of discontinuity. If S is an everywhere dense set in $\mathbb{R}_+$, then the condition*

$(\sup \beta)$ $\qquad\qquad \sup_{s \le t} |B_s^n - B_s| \to 0, \ t \in S, \qquad (4.81)$

and the conditions (γ), (δ) ensure the tightness of the family of distributions of the processes X^n, $n \ge 1$, and the functional convergence $X^n \xrightarrow{\mathcal{L}} X$.

It is now appropriate to see what this theorem gives for the triangular scheme of the series

$$X_t^n = \sum_{1 \le k \le [nt]} \xi_n^k, \ 0 \le t \le 1. \qquad (4.82)$$

First we note that in this case

$$G_t^n(\lambda) = \sum_{k \le [nt]} \int_{E^1} (e^{i\lambda x} - 1)\nu(\{k\} \times dx) = \sum \mathbf{E}[e^{i\lambda \xi_k^n} - 1|\mathcal{F}_{k-1}^n],$$

$$\mathcal{E}(G^n(\lambda))_t = \prod_{k \le [nt]} \mathbf{E}[e^{i\lambda \xi_k^n}|\mathcal{F}_{k-1}^n].$$

Further,

$$B_t^n = \sum_{k \le [nt]} \mathbf{E}[h(\xi_k^n)|\mathcal{F}_{k-1}^n],$$

$$C_t^n = 0,$$

$$\nu([0,t] \times g) = \sum_{k \le [nt]} \mathbf{E}[g(\xi_k^n)I(\xi_k^n \ne 0)|\mathcal{F}_{k-1}^n].$$

Then if $\mu = \mu(a, b, F)$ is an infinitely divisible distribution with parameters (b, C, F) and if the conditions

$$\sup_{k \leq n} \mathbf{P}(|\xi_k^n| > \epsilon | \mathcal{F}_{k-1}^n) \xrightarrow{\mathbf{P}} 0, \qquad (4.80')$$

(β') $\qquad\qquad \sum_{k \leq n} \mathbf{E}[h(\xi_k^n) | \mathcal{F}_{k-1}^n] \xrightarrow{\mathbf{P}} b,$

(γ') $\sum \{\mathbf{E}[h^2(\xi_k^n) | \mathcal{F}_{k-1}^n] - (\mathbf{E}[h(\xi_k^n) | \mathcal{F}_{k-1}^n])^2 \xrightarrow{\mathbf{P}} \tilde{C} = C + \int h^2(x) F(dx),$

(δ') $\qquad\qquad \sum \mathbf{E}[g(\xi_k^n) | \mathcal{F}_{k-1}^n] \xrightarrow{\mathbf{P}} F(g), \; g \in C_1,$

are satisfied, then $\mathcal{L}\left(\sum_{k \leq n} \xi_k^n\right) \to \mu$.

But if X is a process with independent increments without fixed times of jumps and the triple $T = (B, C, \nu)$, then for the functional convergence $X^n \xrightarrow{\mathcal{L}} X$ it is sufficient that

(sup β') $\qquad \sup_{s \leq t} \left| \sum_{k \leq [ns]} \mathbf{E}[h(\xi_k^n) | \mathcal{F}_{k-1}^n] - B_s \right| \xrightarrow{\mathbf{P}} 0,$

(γ') $\qquad \sum_{k \leq [nt]} \{\mathbf{E}[h^2(\xi_k^n) | \mathcal{F}_{k-1}^n] - (\mathbf{E}(h(\xi_k^n) | \mathcal{F}_{k-1}^n))^2\} \xrightarrow{\mathbf{P}} \tilde{C}_s, \; s \leq 1,$

(δ') $\qquad \sum_{k \leq [nt]} \mathbf{E}[g(\xi_k^n) | \mathcal{F}_{k-1}^n] \xrightarrow{\mathbf{P}} (g * \nu)_s, \; s \leq 1, \; g \in C_1.$

5.6. We present a number of other conditions (also expressed in predictable terms) for the convergences $X^n \xrightarrow{\mathcal{L}(S)} X$ and $X^n \xrightarrow{\mathcal{L}} X$.

Firstly, if $g_t(\lambda) \neq 0$, then as we already know from Theorem 4.13 the condition

$$\mathcal{E}(G^n(\lambda)_t) \to g(\lambda)_t, \; t \in S, \; \lambda \in E^1 \implies X^n \xrightarrow{\mathcal{L}(S)} X.$$

Thus there arises the natural question of how to strengthen the convergence of the stochastic exponentials here in order to have functional convergence.

Theorem 4.16. *Suppose that the process X with independent increments has no fixed times of jumps ($g_t(\lambda) \neq 0$). If*

$$\sup_{|\lambda| \leq \theta} \sup_{s \leq t} |\mathcal{E}(G^n(\lambda))_s - g(\lambda)_s| \xrightarrow{\mathbf{P}} 0, \; t > 0, \; \theta > 0, \qquad (4.83)$$

then $X^n \xrightarrow{\mathcal{L}} X$.

Secondly, one can go even further and namely try to give conditions for the convergences $X^n \xrightarrow{\mathcal{L}(S)} X$ and $X^n \xrightarrow{\mathcal{L}} X$ in terms of the convergence of the corresponding cumulants $G^n(\lambda)$ to $G(\lambda)$. No general result is known here, but if both X and X^n are left quasicontinuous (that is, if $\tau^k \uparrow \tau$, then $X^n_{\tau_k} \to X^n_\tau$ a.s.), then

$$G^n(\lambda)_t \xrightarrow{P} G(\lambda)_t,\ t \in \mathbb{R}_+,\ \lambda \in E^1 \implies X^n \xrightarrow{\mathcal{L}(\mathbb{R}_+)} X. \qquad (4.84)$$

Condition (4.84) combined with condition (sup β), entails the functional convergence $X^n \xrightarrow{\mathcal{L}} X$. If, in addition,

$$\sup_{\substack{s \leq t \\ \lambda \leq \theta}} |G^n(\lambda)_s - G(\lambda)_s| \xrightarrow{P} 0,\ \lambda \in E^1,\ \theta \in E^1, \qquad (4.85)$$

then $X^n \xrightarrow{\mathcal{L}} X$.

The results presented above are implemented (aside from the case of "discrete time" (4.82) already considered) in many useful and interesting cases, among which we note the following:

I. The central limit theorem (X is a continuous Gaussian process with independent increments and triple $(B, C, 0)$).

II. The central limit theorem, when X^n are local martingales, and X is a continuous Gaussian process with independent increments and triple $(0, C, 0)$. If, say, $|\Delta X^n| \leq K$, then

$$X^n \xrightarrow{\mathcal{L}} X \iff [X^n, X^n]_t \xrightarrow{P} C_t,\ t \in \mathbb{R}_+;$$

$$X^n \xrightarrow{\mathcal{L}} X \iff \langle X^n \rangle_t \xrightarrow{P} C_t,$$

$$\nu^n([0,t] \times \{|x| > \epsilon\}) \xrightarrow{P} 0,\ t \in \mathbb{R}_+,\ \epsilon > 0,$$

where $[X^n, X^n]$ and $\langle X^n \rangle$ are the quadratic variations and the characteristic of X^n.

This result remains true if the condition

$$\lim_{a \uparrow \infty} \limsup_n \mathbf{P}(|x|I(|x| > a) * \nu^n_t > \eta) = 0,\ \eta > 0,\ t > 0 \qquad (4.86)$$

holds, which is true, for example, if the "family $\{\sup_{s \leq t} |\Delta X^n_s|\}$ is uniformly integrable for all $t > 0$."

These results interpret the relative stability condition as the convergence of the quadratic variation $[X^n, X^n]_1 \equiv \sum \xi^2_{nk} \xrightarrow{P} C_1 = 1$.

It is useful in this connection to note that all the theorems presented above on the convergence $X^n \xrightarrow{\mathcal{L}} X$, $X^n \xrightarrow{\mathcal{L}(S)} X$ contain as sufficient (and frequently, also necessary) conditions, conditions of the type $T^n \xrightarrow{P} T$, where the convergence "$\xrightarrow{P}$" is understood as convergence in probability of the corresponding components of the triples, that is, as the satisfiability for them of the appropriate ergodic theorems. One can therefore say that our theorems have the nature of reducing weak convergence to the satisfiability of the laws of large numbers for the components of the triples; this is in accordance with the spirit of Khinchin's question on the connection between the central limit theorem and the law of large numbers.

III. The convergence of point processes X^n with compensators A^n, $n \geq 1$, to a Poisson process with a continuous deterministic compensator A:

$$A_t^n \xrightarrow{\mathbf{P}} A_t, t \in S \Longrightarrow X^n \xrightarrow{\mathcal{L}(S)} X,$$

$$A_t^n \xrightarrow{\mathbf{P}} A_t, t \in \mathbb{R}_+ \Longrightarrow X^n \xrightarrow{\mathcal{L}} X.$$

IV. The convergence of normalized sums of identically distributed local martingales $Y^k = (Y_t^k)_{t \geq 0}$, $k \geq 1$. Let Y be a continuous Gaussian local martingale with a (continuous) function $C_t = \mathbf{E} Y_t^2$, $t \geq 0$, $C_0 = 0$. If

$$X^n = \frac{1}{\sqrt{n}} \sum_{k \leq n} Y^k,$$

then $X^n \xrightarrow{\mathcal{L}} X$, where X is a Wiener process with characteristics $(0, C, 0)$.

V. Limit theorems for Markov processes. Let $(\Omega, \mathcal{F}, \mathcal{F}_t, \theta_t, Y_t, \mathbf{P}_x)$ be a Markov process with right-continuous trajectories and with values in some topological space E. Here we suppose that there exists a probability measure μ and the σ-algebra of invariant sets with respect to the semigroup $(\theta_t)_{t \geq 0}$ is $\mathbf{P}_\mu$-trivial ($\mathbf{P}_\mu = \int \mathbf{P}_x \mu(dx)$), that is, the ergodicity condition is satisfied.

Then if $f = f(x)$, $x \in E$, is a bounded Borel function of the form

$$f = \mathcal{A}g,$$

where $\mathcal{A}$ is a weak infinitesimal operator, g and g^2 belong to the domain of definition of $\mathcal{A}$, then the processes

$$X_t^n = \frac{1}{\sqrt{n}} \int_0^{nt} f(Y_s) ds, \ t \geq 0,$$

with respect to the measure $\mathbf{P}_\mu$ converge ($X^n \xrightarrow{\mathcal{L}} X$) to the process $X = \sqrt{\beta} W$, where W is the standard Wiener process and

$$\beta = -2 \int g(x) \mathcal{A} g(x) \mu(dx).$$

VI. Limit theorems for stationary processes. Let $Y = (Y_t)_{t \geq 0}$ be an ergodic stationary process, $\mathcal{F}_t = \sigma(Y_s, s \leq t)$,

$$p \in [2, \infty], \ q \in [1, 2], \ \frac{1}{p} + \frac{1}{q} = 1$$

and

$$\|Y_0\|_p < \infty,$$
$$\int_0^\infty \|\mathbf{E}(Y_t|\mathcal{F}_0)\|_q dt < \infty.$$

Then the processes
$$X_t^n = \frac{1}{\sqrt{n}} \int_0^{nt} Y_s ds, \ t \geq 0,$$

converge in the sense of the convergence of finite-dimensional distributions ($X^n \xrightarrow{\mathcal{L}(\mathbb{R}_+)} X$) to the process $X = \sqrt{c}W$, where W is the standard Wiener process and $c = 2\int_0^\infty \mathbf{E}(Y_0 Y_t) dt$.

Furthermore, if $p = 2$, then $X^n \xrightarrow{\mathcal{L}} X$.

5.7. What is known with regard to the convergence of semimartingales X^n to an *arbitrary* process with independent increments X?

If one assumes that $g(\lambda)_t$ does not vanish, then the method of stochastic exponentials is again applicable and the following result is proved.

Theorem 4.17. *Suppose that the following conditions hold:*

(β) $\qquad\qquad\qquad B_t^n \xrightarrow{\mathbf{P}} B_t, \ t \in S,$

(γ) $\qquad\qquad\qquad \tilde{C}_t^n \xrightarrow{\mathbf{P}} \tilde{C}_t, \ t \in S,$

(δ) $\qquad\qquad\qquad g*\nu_t^n \xrightarrow{\mathbf{P}} g*\nu_t, \ t \in S, \ g \in C_1,$

$(S_k\text{-}\delta)$ $\qquad\qquad\qquad g*\nu^n \xrightarrow{\mathbf{P}} g*\nu,$

where S is an everywhere dense subset of $\mathbb{R}_+$.

Then $X^n \xrightarrow{\mathcal{L}(S)} X$.

As regards functional convergence, we have the following result.

Theorem 4.18. *Suppose that the following conditions hold:*

$(S_k\text{-}\beta)$ $\qquad\qquad B^n \xrightarrow{\mathbf{P}} B$ *in the Skorokhod topology,*

(γ) $\qquad\qquad\qquad \tilde{C}_t^n \xrightarrow{\mathbf{P}} \tilde{C}_t, \ t \in \mathbb{R}_+,$

$(S_k\text{-}\delta)$ $\qquad\qquad g*\nu^n \xrightarrow{\mathbf{P}} g*\nu$ *in the Skorokhod topology for $g \in C_1$.*

Then $X^n \xrightarrow{\mathcal{L}} X$.

The application of these theorems to point processes X^n and X with compensators A^n and a deterministic compensator A leads to the next assertion.

Theorem 4.19. a) *If*

$$A_t^n \xrightarrow{P} A_t, \ t \in \mathbb{R}_+,$$

$$\sum_{s \leq t}(\Delta A_s^n)^2 \xrightarrow{P} \sum_{s \leq t}(\Delta A_s)^2, \ t \in \mathbb{R}_+,$$

then $X^n \xrightarrow{\mathcal{L}(\mathbb{R}_+)} X$ *and* $X^n \xrightarrow{\mathcal{L}} X$.

b) *If* $A^n \xrightarrow{P} A$ *in the Skorokhod topology, then* $X^n \xrightarrow{\mathcal{L}} X$.

§6. Relative Compactness and Tightness of Families of Distributions of Semimartingales

6.1. A vital step in establishing results of the type

$$\text{``} T^n \to T \text{''} \implies \text{``} X^n \xrightarrow{\mathcal{L}} X \text{''}$$

was the proof of the tightness of the family $\{\mathbf{P}^n\}$ of probability distributions of semimartingales $\{X^n\}$, expressed in terms of the triples $T^n = (B^n, C^n, \nu^n)$, $n \geq 1$.

It is natural to start the exposition of the corresponding results with the following well known general result, which lies at the basis of the search for simple tightness criteria of the family $\{\mathbf{P}^n\}$.

Theorem 4.20. *A family* $\{\mathbf{P}^n\}$ *of probability distributions of stochastic processes* $X^n = (X_t^n)_{t \geq 0}$ *with values in the space* $(\mathbf{D}, \mathcal{D})$ *is tight if and only if*

(a) *for any* $N \in \mathbb{N}^* = \{1, 2, \ldots\}$ *and* $\epsilon > 0$ *there exist* $n_0 = n_0(N, \epsilon)$ *and* $K = K(N, \epsilon)$ *such that*

$$n \geq n_0 \Rightarrow \mathbf{P}(\sup_{t \leq N}|X_t^n| > K) \leq \epsilon;$$

(b) *for all* $N \in \mathbb{N}^*$, $\epsilon > 0$, $\eta > 0$ *there exist* $n_0 = n_0(N, \epsilon, \eta)$ *and* $\theta = \theta(N, \epsilon, \eta) > 0$ *such that*

$$n \geq n_0 \Rightarrow \mathbf{P}(w_N'(X^n, \theta) \geq \eta) \leq \epsilon,$$

where the modulus of continuity

$$w_N'(\alpha; \theta) = \inf\{\max_{i \leq r} w(\alpha; [t_{i-1}, t_i)) : 0 = t_0 < \ldots < t_r = N, \inf_{i < r}(t_i - t_{i-1}) > \theta\}$$

and $w(\alpha, I) = \sup_{s,t \in I}|\alpha(s) - \alpha(t)|$, I *is an interval in* $\mathbb{R}_+$.

(Let us recall that a subset A of $\mathbf{D}$ is relatively compact in the Skorokhod topology if and only if
(1) $\sup_{\alpha \in A} \sup_{s \leq N} |\alpha(s)| < \infty$ for all $N \in \mathbb{N}^*$;
(2) $\lim_{\theta \downarrow 0} \sup_{\alpha \in A} w_N'(\alpha, \theta) = 0$ for all $N \in \mathbb{N}^*$;

this criterion is also used to prove the preceding Theorem 4.20.)

Clearly Theorem 4.20 is too general, and (in the case of the space $(\mathbf{D}, \mathcal{D})$) the following sufficient condition (Kolmogorov-Chentsov), which is derivable from this theorem, is frequently used.

Theorem 4.21. *Suppose that*
(a) *the sequence* $\{X_0^n\}$ *is tight (in* $\mathbb{R}$*);*
(b) $\lim_{\delta \downarrow 0} \limsup_n \mathbf{P}(|X_\delta^n - X_0^n| > \epsilon) = 0, \ \epsilon > 0;$
(c) *there exist a continuous function* F *on* $\mathbb{R}_+$ *and two constants* $\gamma \geq 0$, $\alpha > 1$ *such that*

$$\forall \lambda > 0, \quad \forall s < r < t, \quad \forall n \in N^*$$
$$\mathbf{P}(|X_r^n - X_s^n| \geq \lambda, \ |X_t^n - X_r^n| \geq \lambda) \leq \lambda^{-\delta}[F(t) - F(s)]^\alpha. \qquad (4.87)$$

Then the sequence $\{X^n\}$ *is dense.*

This criterion is highly applicable, for example, in the case of diffusion processes (even with jumps), when all the coefficients (parameters) are bounded uniformly with respect to n. But this criterion *no longer works* in the diffusion case, when the coefficients are not bounded; this is related in this case to the fact that "deterministic control" of the increments, as in (4.87), is now insufficient.

6.2. Apparently, the first criterion in which the limitation was taken into account of the applicability of Theorem 4.19 was caused by the "uniformity in n" and the "deterministc nature of the control" of the increments in (4.87) was the Aldous criterion which is applicable for adapted stochastic processes X^n given on stochastic bases $\mathcal{B}^n = (\Omega^n, \mathcal{F}^n, \mathbf{F}^n = (\mathcal{F}_t^n)_{t \geq 0}, \mathbf{P}^n)$.

Theorem 4.22. *Suppose that for all* $N \in N^*$

$$\lim_{\theta \downarrow 0} \limsup_n \sup_{\substack{S, T \in \mathcal{F}_N^n \\ S \leq T \leq S+\theta}} \mathbf{P}^n(|X_T^n - X_S^n| \geq \epsilon) = 0, \qquad (4.88)$$

where $\mathcal{F}_N^n$ *is the set of all* $\mathbf{F}^n$-*stopping times. Suppose also that condition* (a) *of Theorem 4.21 is satisfied. Then the family* $\{X^n\}$ *is tight.*

We make a number of remarks with regard to this criterion before proceeding to its applications.

1. Clearly, the tightness of the family $\{X^n\}$ is in no way connected with the filtrations $\mathbf{F}^n$, $n \geq 1$. It is also clear that it is necessary to approach this, so that the class $\mathcal{F}_N^n$ can be as "small" as possible (but, of course, so that $\mathcal{F}_t^n \supseteq \sigma(X_s^n, \ s \leq t)$).

2. The Aldous criterion "works" when the processes X^n, $n \geq 1$, are "asymptotically left quasicontinuous". One can obtain some idea about this if one simply assumes that all the $X^n \equiv X$. Then, clearly, condition (a) of Theorem 4.21 is satisfied, while (4.88) is satisfied if and only if the process X

is left quasicontinuous (that is, $\mathbf{P}(\Delta X_T \neq 0, T < \infty) = 0$ for any predictable time T).

3. If $\mathbf{F}^n$ is the maximum possible filtration, that is, $\mathcal{F}_t^n = \mathcal{F}^n$ for all $t \geq 0$, then condition (4.88) will ensure the C-tightness of the family $\{X^n\}$.

The following corollaries are deduced from Theorem 4.22.

Corollary 1. *If X^n are locally square-integrable martingales and $G^n = \langle X^n \rangle$, $n \geq 1$, then for the tightness of the family $\{X^n\}$, it is sufficient that*
(a) *the family $\{X_0^n\}$ be tight (in E^1),*
(b) *the family $\{\langle X^n \rangle\}$ be C-tight (in $\mathbf{D}$), that is, the family $\{\langle X^n \rangle\}$ be tight and all the weak limits be continuous processes.*

Thus, if it is known a priori that $\langle X^n \rangle_t \xrightarrow{\mathbf{P}} C_t$, $t \geq 0$, where $C = (C_t)_{t \geq 0}$ is a continuous process, then the family $\{X^n - X_0^n\}$ is tight.

Corollary 2. *Let X^n be semimartingales, $n \geq 1$, with triples $T^n = (B^n, B^n(h), C^n, \nu^n)$, where $h = h(x)$ is a bounded function with compact support and satisfying the property $h(x) = x$ in a neighborhood of zero. Let*

$$\tilde{C}^n = C^n + h^2 * \nu - \sum_{s \leq \cdot} (\Delta B_s^n)^2$$

be a modified second characteristic of X^n. Suppose that the following conditions are satisfied:
(1) *the sequence (X_0^n) is tight (in E^1);*
(2) *for all $N > 0$, $\epsilon > 0$*

$$\lim_{a \uparrow \infty} \limsup_n \mathbf{P}^n \{\nu^n([0, N] \times \{x : |x| > a\}) > \epsilon\} = 0;$$

(3) *each of the sequences (B^n), $(\tilde{C}^n)$, $(g_p * \nu^n)$ with $g_p(x) = (p|x| - 1)^+ \wedge 1$, $p \in N^*$, is C-tight.*

Then the sequence $\{X^n\}$ is tight.

These criteria "serve well" the cases of convergence to quasicontinuous processes. In the general case, however, it is necessary to resort to more complex criteria. It is important, moreover, to emphasize that they are expressed in predictable terms.

§7. Convergence of Semimartingales to a Semimartingale

7.1. Let us now move on to the results on the weak convergence of semimartingales X^n to a semimartingale X. It should be noted straight away that here the principle " $T^n \to T$ " $\implies$ " $X_n \xrightarrow{\mathcal{L}} X$ " in terms of "tightness"$\oplus$"convergence of finite-dimensional distributions" does not work for the reason that it is difficult to establish the weak convergence $X^n \xrightarrow{\mathcal{L}(\mathbb{R}_+)} X$, when the process X is not a process with independent increments, but is a

Chapter 4.I. Weak Convergence of Measures on Metric Spaces

semimartingale of some general structure (even, say, a process of diffusion type).

However, the scheme of proving "$T^n \to T$" $\Longrightarrow$ "$X_n \xrightarrow{\mathcal{L}} X$" presented above with the use of the three intermediate stages

I: establishing the tightness of the family $\{X^n\}$,

II: characterizing all the weak limits $\mathbf{P'}$ ($= (\text{w-}\lim \mathbf{P}^{n'})$),

III: identifying the limit points $\mathbf{P'}$ with $\mathbf{P}$

works in the general situation as well, but requires drawing upon new "martingale" ideas. Let us clarify this with the example when $X^n (= M^n)$ are martingales (more generally, local martingales).

Thus, let M^n, $n \geq 1$, be martingales, where $|M^n| \leq b$ for all n and the distributions $\mathcal{L}(M^n)$ converge (weakly) to the distribution $\mathcal{L}(M)$ of some process M. It is not hard to show that then the process M is a martingale (with respect to the natural filtration and the law of $\mathcal{L}(M)$). A similar result holds when the M^n, $n \geq 1$, are local martingales with uniformly bounded jumps, $|\Delta M^n| \leq b$, $n \geq 1$: if $\mathcal{L}(M^n) \xrightarrow{w} \mathcal{L}(M)$, then M is a local martingale.

In other words, the class of local martingales (with uniformly bounded jumps) is stable in the sense of weak convergence.

Let us now move on to the question of the *characterization* of processes that are weak limits of semimartingales X^n with triples $T^n = (B^n, C^n, \nu^n)$, $n \geq 1$.

To this end, we introduce a number of conditions and some notation.

First of all, we will assume that the "limit" process X is canonical: $X_t(\alpha) = \alpha(t)$, where $\alpha = \alpha(t)$, $t \geq 0$, are functions in the space $\mathbf{D}$. We also assume that we are given objects $T = (B, C, \nu)$, playing the role of the triple of the "limit" process X in what follows and defined in the following manner:

$B = (B_t(\alpha))_{t \geq 0}$ is a predictable process of locally bounded variation, $B_0 = 0$;

$C = (C_t(\alpha))_{t \geq 0}$ is a continuous non-decreasing adapted process with $C_0 = 0$;

$\nu = \nu(\alpha; dt, dx)$ is a predictable random measure on $\mathbb{R}_+ \times E^1$ such that

$$\nu(\mathbb{R}_+ \times \{0\}) = \nu(\{0\} \times E^1) = 0, \quad (1 \wedge |x|^2) * \nu_t < \infty,$$

$$\int_{E^1} \nu(\{t\} \times dx) h(x) = \Delta B_t, \quad \nu(\{t\} \times E^1) \leq 1, \ t \geq 0.$$

We also set

$$\tilde{C} = C + h^2 * \nu - \sum_{s \leq \cdot}(\Delta B_s)^2 \tag{4.89}$$

and introduce the following (at first glance, possibly, somewhat strange) conditions

$(\beta\text{-}S)$ $\qquad\qquad B_t^n - B_t \circ X^n \xrightarrow{P} 0, \quad t \in S,$

$(\gamma\text{-}S)$ $\qquad\qquad \tilde{C}_t^n - \tilde{C}_t \circ X^n \xrightarrow{P} 0, \quad t \in S,$

$(\delta\text{-}S)$ $\qquad\qquad g * \nu^n - (g * \nu_t) \circ X^n \xrightarrow{n} 0, \quad t \in S,$

which reduce to the conditions already considered earlier, when B, C, ν are simply deterministic functions.

We also require the following conditions (majorizability and continuity):

$$\sup_\alpha |\tilde{C}_t(\alpha)| < \infty, \ \sup_\alpha |g * \nu_t(\alpha)| \text{ for all } t \geq 0, \ g \in G_1 \qquad (4.90)$$

and for all $t \geq 0$ and $g \in G_1$ the functions $\alpha \to B_t(\alpha), \tilde{C}_t(\alpha), g * \nu_t(\alpha)$ are **P**-a.s. continuous in the Skorokhod topology, where

$$\mathbf{P} \text{ is the weak limit of the laws of } \alpha(X^n). \qquad (4.91)$$

We can now state the basic result which "services" stage II of the characterization of weak limits of $\mathcal{L}(X^n)$.

Theorem 4.23. *Let $\mathcal{L}(X^n)$ weakly converge to some probability distribution* **P**. *Suppose that the conditions $(\beta$-$S)$, $(\gamma$-$S)$, $(\delta$-$S)$ hold for some everywhere dense subset in $\mathbb{R}_+$, lying in $\mathbb{R}_+ \setminus \{t > 0 : \mathbf{P}(\Delta X_t \neq 0) > 0\}$ and suppose further that the conditions of majorizability (4.90) and continuity (4.91) hold.*

Then X is a semimartingale on $(\mathbf{D}, \mathcal{D}, \mathbf{P})$ with triple of predictable characteristics $T = (B, C, \nu)$.

The idea of the proof of this theorem is based on the fact that in accordance with Theorem 4.11, the verification of the "semimartingaleness" is equivalent to verifying that the three processes (see Theorem 4.11 b) are local martingales. Therefore, the processes $M^n(h) \equiv X^n(h) - X_0^n - B^n$, $(M^n(h))^2 - C^n$, $g * \mu^n - g * \nu^n$ are local martingales (with respect to the measures $\mathcal{L}(X^n)$) and the proof of the theorem reduces to verifying that the conditions of the theorem guarantee that the corresponding processes $M(h) \equiv X(h) - X_0 - B$, $(M(h))^2 - C$, $g * \mu - g * \nu$ on the stochastic basis $(\mathbf{D}, \mathcal{D}, (\mathcal{D}_t), \mathbf{P})$ are also local martingales.

There are also other results of "characterization" type. For example, let

$$(X^n, B^n, \tilde{C}^n) \xrightarrow{\mathcal{L}} (X, B, C) \qquad (4.92)$$

and

$$(X^n, g * \nu^n) \xrightarrow{\mathcal{L}} (X, g * \nu), \ g \in C_1. \qquad (4.93)$$

Then the process X is a semimartingale with triple (B, C, ν) (see Jacod and Shiryaev 1987, Chap. IX, Theorem 2.4.).

Or suppose, for example, that X is a canonical process, $S_{k_3}[\cdot], S_{k_2}[\cdot]$ are distances compatible with the Skorokhod topology in $\mathbf{D}(E^3), \mathbf{D}(E^2)$. Then if $\mathcal{L}(X^n) \xrightarrow{w} \mathbf{P}$,

$$S_{k_3}[(X^n, B^n, \tilde{C}^n), (X, B, \tilde{C}) \circ X^n] \xrightarrow{\mathbf{P}} 0,$$
$$S_{k_2}[(X^n, g * \nu^n), (X, g * \nu) \circ X^n] \xrightarrow{\mathbf{P}} 0, \ g \in C_1,$$

and the functions $\alpha \to (\alpha, B(\alpha), C(\alpha))$ and $\alpha \to (\alpha, g * \nu(\alpha))$, $g \in C_1$, are **P**-a.s. continuous, then X will be a semimartingale on the stochastic basis $(\mathbf{D}, \mathcal{D}, (\mathcal{D}_t), \mathbf{P})$ with triple $T = (B, C, \nu)$. (See Jacod and Shiryaev 1987, Chap. IX, Theorem 2.22.)

7.2. Let us now turn to our basic problem of obtaining results for semimartingales of the type "$T^n \to T$" $\Longrightarrow$ "$X^n \xrightarrow{\mathcal{L}} X$". We will again follow the scheme

$$
(T^n \to T) \Rightarrow \underbrace{\boxed{\begin{array}{c}\text{Establishment of tightness}\\ \text{of the laws } \mathbf{P}^n = \mathcal{L}(X^n)\end{array}}}_{\text{I}} \oplus \underbrace{\boxed{\begin{array}{c}\text{Characterization of all}\\ \text{weak limits } \mathbf{P}' = \text{w-lim}\,\mathbf{P}^{n'}\end{array}}}_{\text{II}}
$$

$$
\oplus \underbrace{\boxed{\begin{array}{c}\text{Identification}\\ \mathbf{P}' \sim \mathbf{P}\end{array}}}_{\text{III}} \Rightarrow X^n \xrightarrow{\mathcal{L}} X.
$$

The results presented above "service" only stage II ("characterization") for the present. Let us now consider results with respect to the character of the convergence "$T^n \to T$", which ensure simultaneously both stage I ("tightness") and stage II ("characterization"). As regards stage III, we will simply assume that the triple $T = (B, C, \nu)$ is such that it uniquely determines the measure $\mathbf{P}$ so that the canonical process X with respect to this measure $\mathbf{P}$ has the set $T = (B, C, \nu)$ as its triple. (We will return to this question of the existence and uniqueness of the "semimartingale" problem in Sect. 8.)

We now give the precise (if lengthy) statements of the theorems concerning convergence, assuming that $X_0^n \equiv 0$, $X_0 \equiv 0$.

We start with the case when the "limit" process X is left quasicontinuous, that is $\nu(\{t\} \times E^1) = 0$, $t \geq 0$.

Theorem 4.24. *Let S be an everywhere dense subset of $\mathbb{R}_+$ and suppose that the following conditions on $T = (B, C, \nu)$ hold:*

(1) strict majorizability in the sense that there exists a continuous deterministic increasing function $t \to F_t$ that strictly majorizes[1] the functions $\text{Var}(B(\alpha))$ *and* $C(\alpha) + (|x|^2 \wedge 1) * \nu(\alpha)$ *for all $\alpha \in \mathbf{D}$;*

(2) condition on "large jumps":

$$\limsup_{a\uparrow\infty}\,\sup_{\alpha \in \mathbf{D}} \nu(\alpha; [0,t] \times \{x : |x| > a\}) = 0;$$

(3) continuity condition: for all $t \in S$, $g \in C_1$ the functions $\alpha \to B_t(\alpha)$, $\tilde{C}_t(\alpha)$, $g * \nu_t(\alpha)$ are continuous in the Skorokhod topology.

---
[1] We say that a function $t \to F_t$ strictly majorizes a function $t \to G_t$ (written $G \prec F$) if the function $t \to F_t - G_t$ is non-decreasing.

Suppose further that the convergence "$T^n \to T$" satisfies the following conditions:

(sup -β) $$\sup_{s \leq t} |B_t^n - B_t \circ X^n| \xrightarrow{\mathbf{P}} 0, \quad t > 0,$$

(γ-S) $$\tilde{C}_t^n - \tilde{C}_t \circ X^n \xrightarrow{\mathbf{P}} 0, \quad t \in S,$$

(δ-S) $$g * \nu_t^n - (g * \nu_t) \circ X^n \to 0, \quad t \in S,$$

where S is some everywhere dense subset of $\mathbb{R}_+$.

Then $\mathcal{L}(X^n) \xrightarrow{w} \mathbf{P}$, that is, $X^n \xrightarrow{\mathcal{L}} X$.

Remark. In the case when the limit process X is a process with independent increments without fixed times of jumps, Theorem 4.7 follows from this theorem.

The rejection of the assumption of "left quasicontinuity" for the limit process X leads to the next general result.

Theorem 4.25. *Suppose that the following conditions hold:*

(1) *strict majorizability in the sense that there exist deterministic increasing functions $t \to F_t$, $t \to F_t^g$ such that for all $\alpha \in \mathbf{D}$, $g \in C_1$*

$$\mathrm{Var}(B(\alpha)) \prec F, \quad \tilde{C}(\alpha) \prec F, \quad g * \nu(\alpha) \prec F^g;$$

(2) *condition on "large jumps"*

$$\lim_{a \uparrow \infty} \sup_\alpha \nu(\alpha; [0,t] \times \{x : |x| > a\}) = 0;$$

(3) *continuity condition: if S is an everywhere dense subset of $\mathbb{R}_+$ that lies in the set of those moments of time t for which $\Delta F_t = 0$ and $\Delta F_t^g = 0$ for all $g \in C_1$ (F and F^g are defined in (1)), then the functions $\alpha \to B_t(\alpha)$, $\tilde{C}_t(\alpha)$, $g * \nu_t(\alpha)$ are continuous in the Skorokhod topology for all $t \in S$, $g \in C_1$.*

Suppose further that the convergence "$T^n \to T$" satisfies the following conditions:

(S_k-β) $$S_k(B^n, B \circ X^n) \xrightarrow{\mathbf{P}} 0,$$

(S_k-γ) $$S_k(\tilde{C}^n, \tilde{C} \circ X^n) \xrightarrow{\mathbf{P}} 0,$$

(S_k-δ) $$S_k(g * \nu^n, (g * \nu) \circ X^n) \xrightarrow{\mathbf{P}} 0, \quad g \in C_1,$$

where S_k is a metric compatible with the Skorokhod topology.

Then (again under the assumption that the triple T uniquely determines the measure $\mathbf{P}$) $X^n \xrightarrow{\mathcal{L}} X$, that is $\mathcal{L}(X^n) \xrightarrow{w} \mathbf{P}$.

Remark. If the limit process X is a process with independent increments, then we obtain the result of Theorem 4.9 from the above theorem.

7.3. Let us present a number of examples illustrating limit theorems on the weak convergence of probability distributions of semimartingales.

Chapter 4.I. Weak Convergence of Measures on Metric Spaces

Example 1. The convergence of diffusion processes with jumps.

We will assume that the processes X and X^n, $n \geq 1$, have the following structure.

The process X given on a canonical space is a uniform diffusion process with jumps, that is,

$$B_t = \int_0^t b(X_s)ds, \quad C_t = \int_0^t c(X_s)ds, \quad c(x) \geq 0,$$

$$\nu(dt, dx) = dt K(X_t, dx), \quad \int_{E^1} K(x, dy)(|y|^2 \wedge 1) < \infty.$$

We will also set

$$\tilde{C}_t = \int_0^t \tilde{c}(X_s)ds, \quad \tilde{c}(x) = c(x) + \int_{E^1} h^2(y) K(x, dy)$$

and we assume that the corresponding martingale problem has a unique solution $\mathbf{P}_x$ (for each initial state $X_0 = x$).

Under this assumption the process X is Markov. Moreover, if $f = f(x)$ is of class C^2 and

$$\mathcal{A}f(x) = b(x)\frac{\partial f}{\partial x} + \frac{1}{2}c(x)\frac{\partial^2 f}{\partial x^2}$$
$$+ \int K(x, dy)\left[f(x+y) - f(y) - h(y)\frac{\partial f(y)}{\partial y}\right],$$

then

$$f(X_t) - f(X_0) - \int_0^t \mathcal{A}f(X_s)ds$$

is a local martingale (with respect to any of the measures $\mathbf{P}_x$, $x \in E^1$) and hence $\mathcal{A}$ is an extended infinitesimal operator of the process X.

Concerning the processes X^n, $n \geq 1$, we will assume that they have the same structure as the process X with

$$B_t^n = \int_0^t b^n(X_s^n)ds, \quad C_t^n = \int_0^t c^n(X_s^n)ds,$$

$$\nu^n(dt, dx) = dt K^n(X_t^n, dx), \quad \tilde{C}_t^n = \int_0^t \tilde{c}^n(X_s^n)ds.$$

However, we do not assume the uniqueness of the solution of the corresponding martingale problem, so that X^n is not a Markov process in general.

Let us assume that the set (b, c, K) is such that

$$\lim_{b\uparrow\infty} \sup_{|x| \leq a} K(x, \{y : |y| > b\}) = 0, \quad a > 0, \tag{4.94}$$

and the functions

$$x \to b(x), \; \tilde{c}(x), \; \int K(x,dy)g(y) \text{ are continuous for } g \in C_1. \tag{4.95}$$

Also let

$$b^n \to b, \quad \tilde{c}^n \to c, \quad \int_{E^1} K^n(\cdot,dy)g(y) \to \int_{E^1} K(\cdot,dy)g(y) \tag{4.96}$$

be locally uniform, $g \in C_1$,

$$\eta^n \xrightarrow{d} \eta, \text{ where } \eta^n = \mathcal{L}(X_0^n), \; \eta = \mathcal{L}(X_0). \tag{4.97}$$

Theorem 4.26. *If conditions* (4.94)–(4.97) *are satisfied, then*

$$\mathcal{L}(X^n) \xrightarrow{w} \mathbf{P} = \int_{E^1} \mathbf{P}_x \eta(dx).$$

Remark. Let us assume that X^n is a Markov process, $n \geq 1$, with an extended infinitesimal operator $\mathcal{A}^n$. Then conditions (4.96) are equivalent to

$$\mathcal{A}^n f \to \mathcal{A} f \text{ locally uniformly for all } f \in C^2. \tag{4.98}$$

Thus in the Markov case, we obtain from Theorem 26, the Trotter-Kato theorem: if conditions (4.95), (4.96), (4.98) are satisfied, then $\mathcal{L}(X^n) \xrightarrow{w} \mathbf{P}$.

Example 2. The convergence of pure jump Markov processes to a diffusion process.

Let X^n be a pure jump Markov process with

$$\mathcal{A}^n f(x) = \int_{E^1} K^n(x,dy)[f(x+y) - f(x)], \tag{4.99}$$

where K^n is a finite transition measure on E^1.

Then

$$b^n(x) = \int_{E^1} K^n(x,dy)h(y), \quad c^n(x) = 0, \quad \tilde{c}^n = \int_{E^1} h^2(y) K^n(x,dy).$$

For simplicity we will assume that $\int |y|^2 K^n(x,dy) < \infty$. Under this assumption one can take $h(y) = y^2$. Also let X be a diffusion process (with $K = 0$) and let

$$b^n \to b, \; \tilde{c}^n \to c \text{ locally uniformly} \tag{4.100}$$

$$\sup_{x:|x| \leq a} \int_{E^1} K^n(x,dy)|y|^2 I(|y| > \epsilon) \to 0, \; n \uparrow \infty, \; \epsilon > 0. \tag{4.101}$$

$$\eta^n \xrightarrow{d} \eta \tag{4.102}$$

Theorem 4.27. *Under assumptions* (4.99)–(4.102) *the distributions* $\mathcal{L}(X^n)$ *converge weakly to* $\mathbf{P} = \int \eta(dx) \mathbf{P}_x$ — *the distribution of a diffusion process with coefficients* b, c *and initial distribution* η.

Example 3. In the previous example let $\eta^n = \eta = \epsilon_x$ for some $x \in \mathbb{R}$ (ϵ_x is the distribution concentrated at the point x). Suppose that (4.101) holds, and also the conditions

$$b^n \to b, \quad c^n \to 0 \text{ locally uniformly} \tag{4.103}$$
$$b = b(x) \text{ satisfies a Lipschitz condition.} \tag{4.104}$$

Under these assumptions the limit process X is degenerate, that is, the measure $\mathbf{P}_x$ "sits" on the solution of the equation

$$dx_t(x) = b(x_t(x))dt, \quad x_0(x) = x$$

and hence

$$\sup_{s \leq t}|X_s^n - x_s(x)| \xrightarrow{\mathbf{P}} 0, \; t \geq 0 \tag{4.105}$$

since Skorokhod convergence to a continuous function is the same as locally uniform convergence. It turns out that it is easy to obtain from the preceding results a refinement of the speed of convergence in (4.105).

Theorem 4.28. *Let $(a_n)_{n \geq 1}$ be a sequence of positive numbers such that $a_n \uparrow \infty$ and*

$$a_n^2 \tilde{c}^n \to \hat{c} \text{ locally uniformly,} \tag{4.106}$$

$\hat{c} = \hat{c}(x)$ is a continuous function;

$$\limsup_n a_n^2 \int_{|x| \leq a} \int_{E^1} K^n(x, dy)|y|^2 I\left\{|y| > \frac{\epsilon}{u_n}\right\} = 0, \; a > 0, \; \epsilon > 0. \tag{4.107}$$

Then the process

$$Y_t^n = a_n\left(X_t^n - X_0^n - \int_0^t b^n(X_s^n)\right)ds$$

converges in distribution to a continuous Gaussian process with independent increments and triple $T = (0, \hat{C}(x), 0)$, where $\hat{C}(x)_t = \int_0^t \hat{c}_s(x_s(x))ds$.

Thus we can write symbolically:

$$X_t^n = X_0^n + \int_0^t b^n(x_s(x))ds + \frac{1}{a_n}Y_t. \tag{4.108}$$

§8. The Martingale Problem

8.1. Let us briefly dwell on results relating to the "martingale problem" that have a direct bearing on stage III ("identification"), in which all the weak limits $\mathbf{P}' = \text{w-}\lim \mathbf{P}^{n'}$ are identified with the probability distribution $\mathbf{P}$ of a semimartingale X with triple $T = (B, C, \nu)$. In Theorem 4.23 as stated above, the limit process turned out to be a semimartingale with a given triple (B, C, ν). Thus it is necessary to occupy ourselves with the question of when the triple of a semimartingale uniquely determines its distribution. Without additional assumptions, of course, a triple does not uniquely determine a distribution. It is sufficient, for example, to consider the equation $dx_t = b(x_t)dt$, for which a solution exists, but is not unique. So, in this case the triple

$$T = (B, 0, 0), \quad B_t = \int_0^t b(x_s)ds$$

does not determine the "distribution" of a degenerate semimartingale uniquely.

Let us formulate the "semimartingale" problem in its full generality.

The ingredients of this problem are:

(1) a measure space $(\Omega, \mathcal{F})$ with a flow of σ-algebras $\mathbf{F} = (\mathcal{F}_t)_{t \geq 0}$, where $\mathcal{F}_0 \subseteq \mathcal{F}_s \subseteq \mathcal{F}_t \subset \mathcal{F}$, $s \leq t$;

(2) $X = (X_t(\omega))_{t \geq 0}$, which is an adapted process with values in the space $\mathbb{D}$ (the "candidate" for being a semimartingale);

(3) a triple $T = (B, C, \nu)$ (the "candidate" for being the triple of the semimartingale X), where B is an $\mathbf{F}$-predictable process of locally bounded variation, $B_0 = 0$; C is an $\mathbf{F}$-predictable continuous non-decreasing process, $C_0 = 0$; ν is an $\mathbf{F}$-predictable random measure on $\mathbb{R}_+ \times E^1$ such that $\nu(\mathbb{R}_+ \times \{0\}) = \nu(\{0\} \times E^1) = 0$, $(x^2 \wedge 1) * \nu_t < \infty$, $\int \nu(\omega; \{t\} \times dx)h(x) = \Delta B_t(\omega)$, $\nu(\omega; \{t\} \times E^1) \leq 1$; $h = h(x)$ is a truncation function.

Definition. By a solution of the semimartingale problem connected with $(\mathcal{F}_0, X)$ and $\mathbf{P}_0$; B, C, ν is meant a probability measure $\mathbf{P}$ on $(\Omega, \mathcal{F})$ such that

(a) $\mathbf{P}|\mathcal{F}_0 = \mathbf{P}_0$

(b) X is a semimartingale on the stochastic basis $(\Omega, \mathcal{F}, \mathbf{F}, \mathbf{P})$ with characteristics $T = (B, C, \nu)$ (with respect to a given truncation function $h = h(x)$).

We denote by $S(\mathcal{F}_0, X | \mathbf{P}_0; B, C, \nu)$ the set of all solutions (that is, measures $\mathbf{P}$). If we set

$$X(h) = X - \sum_{s \leq \cdot} [\Delta X_s - h(\Delta X_s)],$$
$$M(h) = X(h) - X_0 - B,$$
$$\tilde{C} = C + h^2 * \nu - \sum_{s \leq \cdot} (\Delta B_s)^2,$$

then, as was noted in Theorem 4.11, the probability measure $\mathbf{P} \in S(\mathcal{F}_0, X | \mathbf{P}_0; B, C, \nu)$ if and only if $\mathbf{P}|\mathcal{F}_0 = \mathbf{P}_0$ and each of the processes

Chapter 4.I. Weak Convergence of Measures on Metric Spaces

$$M(h), \quad M^2(h) - \tilde{C}, \quad g * \mu^x - g * \nu$$

is a local martingale $g \in \mathcal{G}^+(\mathbb{R}_+)$.

It is easy to deduce from this equivalent formulation of the "semimartingale" problem that $S(\mathcal{F}_0, X | \mathbf{P}_0; B, C, \nu)$ is a convex set.

We present a number of results relating to the existence and uniqueness of semimartingale problems.

1) Processes with independent increments. If (B, C, ν) is deterministic and η is a probability measure on $\mathbb{R}$, then in the canonical representation, the corresponding semimartingale problem $S(\sigma(X_0, X | \eta; B, C, \nu)$ has a solution (which is *unique*).

2) Diffusion with jumps. This is a semimartingale X on a stochastic basis $(\Omega, \mathcal{F}, \mathbf{F}, \mathbf{P})$ for which

$$B_t(\omega) = \int_0^t b(s, X_s(\omega)) ds,$$

$$C_t(\omega) = \int_0^t c(s, X_s(\omega)) ds,$$

$$\nu(\omega; dt \times dx) = dt \times K_t(X_t(\omega); dx),$$

where $b = b(s, x)$ is a Borel function, $c(s, x)$ is a non-negative Borel function and $K_s(x, dy)$ is a Borel transition kernel with $K_s(x, \{0\}) = 0$.

If $\nu = 0$, then X is called a diffusion; if b, c, K do not depend on s, then X is called a homogeneous diffusion with jumps.

The construction of a diffusion with jumps can be realized if one turns to the consideration of the solutions of stochastic differential equations.

Suppose that we are given on the stochastic basis $\mathcal{B}' = (\Omega', \mathcal{F}', \mathbf{F}', \mathbf{P}')$:

$W = (W_t)_{t \geq 0}$ — the standard Wiener process,

$\pi = (\pi(dt, dx))_{t \in \mathbb{R}_+, \, x \in E^1}$ — a Poisson random measure on $\mathbb{R}_+ \times E^1$ with intensity $q(dt, dx) = dt \otimes F(dx)$, where F is a positive σ-finite measure on $(E^1, \mathcal{B}(E^1))$. Suppose further that we are given the (Borel) coefficients:

$$\beta = \beta(t, x), \ \gamma = \gamma(t, x), \ \delta = \delta(t, x, z), \ t \in \mathbb{R}_+, \ x \in F, \ z \in E^1,$$

and an initial random $\mathcal{F}_0'$-measurable variable ξ_0.

Let us consider the stochastic differential equation

$$\begin{aligned} dY_t &= \beta(t, Y_t) dt + \gamma(t, Y_t) dW_t + h \circ \delta(t, Y_{t-}, z) \pi(dt, dz) \\ &\quad - q(dt, dz) + h' \circ \delta(t, Y_{t-}, z) \pi(dt, dz), \end{aligned} \quad (4.109)$$

where $h = h(x)$ is some cut-off function, $h'(x) = x - h(x)$ and $Y_0 = \xi$. (Note that if the measure π has a jump at the point (t, z), then $\Delta Y_t = \delta(t, Y_{t-}, z)$.)

For equations (4.109) we can consider two types of solution: strong solutions (or solutions that are processes) and weak solutions (or solutions that are measures).

It turns out that the set of all weak solutions of equation (4.109) with $\alpha(\xi_0) = \eta$ is the same as the set $S(\sigma(Y_0), Y|\eta; B, C, \nu)$, where

$$B = \beta, \ C = \gamma^2, \ K_t(y, A) = \int_{E^1} I_{A\setminus\{0\}}(\delta(t, y, z)) F(dz). \tag{4.110}$$

(This result admits a certain converse, namely, from C and K one can construct γ and δ satisfying (4.110).)

3) Point processes. A point process $N = (N_t)_{t\geq 0}$ given on a stochastic basis $(\Omega, \mathcal{F}, \mathbf{F} = (\mathcal{F}_t)_{t\geq 0})$ is an adapted process of the form

$$N_t = \Sigma I(\tau_n \leq t), \tag{4.111}$$

where $(\tau_n)_{n\geq 1}$ are Markov times and $\tau_0 \equiv 0$, $\tau_n < \tau_{n+1}$ if $\tau_n < \infty$.

Let $N_t = A_t + m_t$ be the Doob-Meyer decomposition, where $A = (A_t)_{t\geq 0}$ is a predictable increasing process (or the compensator of the process N). If $h = h(x)$ is a truncation function, then

$$B(h) = h(1)A, \ C = 0, \ \nu(dt, dx) = dA_t \otimes \epsilon_1(dx).$$

It turns out that the case of point processes is remarkable in that the corresponding semimartingale problem has a solution, which is unique, if one takes as the space Ω the set of all counting functions, which may even be "exploding", that is, with $\lim \tau_n \leq \infty$.

II. Applications: The Invariance Principle and Diffusion Approximation

§1. The Invariance Principle for Stationary and Markov Processes

1.1. Stationary Processes. Let $\xi = (\xi_k)_{-\infty < k < \infty}$ be a sequence stationary in the strict sense or let $\xi = (\xi_t)_{t \in E^1}$ be a measurable process stationary in the strict sense (in the case of continuous time).

The invariance principle for ξ means that the sequence of stochastic processes $X^n = (X^n_t)_{t \geq 0}$, $n \geq 1$

$$X^n_t = \frac{1}{\sqrt{n}} \sum_{k=1}^{nt} \xi_k \tag{4.112}$$

or, in the case of continuous time,

$$X^n_t = \frac{1}{\sqrt{n}} \int_0^{nt} \xi_s ds \tag{4.113}$$

weakly converges in the Skorokhod topology (see I. Sect. 2) to a homogeneous Gaussian process with independent increments, in particular, to a Wiener process $W = (W_t)_{t \geq 0}$.

We will assume that ξ is given on a complete probability space $(\Omega, \mathcal{F}, \mathbf{P})$ with a distinguished filtration on it

$$\mathbf{F}^\xi = (\mathcal{F}^\xi_t)_{t \in \mathbb{R}} :$$

in the case of discrete time

$$\mathcal{F}^\xi_t = \sigma\{\xi_k, -\infty < k \leq t\} \vee \mathcal{N},$$

while in the case of continuous time

$$\mathcal{F}^\xi_t = \bigcap_{\epsilon > 0} \sigma\{\xi_s, -\infty < s \leq t + \epsilon\} \vee \mathcal{N},$$

where $\mathcal{N}$ is the system of sets of $\mathcal{F}$ of zero $\mathbf{P}$ measure.

We denote by J^ξ the σ-algebra of invariant sets corresponding to ξ (recall that J^ξ is the collection of sets $A \in \mathcal{F}$, such that there exists a measurable set B in the space of trajectories ξ and for all t

$$A = \{\omega : (\xi_l(\omega))_{l \geq t} \in B\}$$

or, in the case of continuous time,

$$A = \{\omega : (\xi_s(\omega))_{s \geq t} \in B\}).$$

We will assume that the σ-algebra J^ξ is completed by the sets of $\mathcal{F}$ of zero $\mathbf{P}$ measure, and observe that for any t

$$J^\xi \subseteq \mathcal{F}_t^\xi.$$

If the σ-algebra J^ξ contains only sets of measure 0 or 1, then ξ is called an ergodic process.

In the non-ergodic case, one can give the following generalizations of the invariance principle: the sequence X^n, $n \geq 1$, weakly converges in the Skorokhod topology to a mixture of Wiener processes;

$$X^n \xrightarrow{\mathcal{L}} \eta W,$$

where W is a Wiener process not depending on η and is a J^ξ-measurable random variable. Moreover, for any function $f = f(X)$ bounded and continuous in the Skorokhod topology, we have convergence in probability

$$\mathbf{E}\{f(X^n)|J^\xi\} \xrightarrow{\mathbf{P}} \mathbf{E}\{f(\eta W)|J^\xi\}.$$

In what follows we will denote this form of convergence by

$$X^n \xrightarrow{\mathcal{L}} \eta W \ (J^\xi\text{-stable})$$

The modern approach to proving the invariance principle for stationary sequences and processes relies on the functional central limit theorem for square-integrable martingales, which is the functional analogue of the Lindeberg Theorem (see Theorem 4.1) with the Khinchin condition (4.33).

Let $(\Omega, \mathcal{F}, \mathbf{F}^n = (\mathcal{F}_t^n)_{t \geq 0}, \mathbf{P})$, $n \geq 1$, $(\Omega, \mathcal{F}, \mathbf{F} = (\mathcal{F}_t)_{t \geq 0}, \mathbf{P})$ the stochastic basis, and G some sub-σ-algebra of $\mathcal{F}_0$. A Wiener process $W = (W_t)_{t \geq 0}$ is given on $(\Omega, \mathcal{F}, \mathbf{F}, \mathbf{P})$. For each $n \geq 1$ on $(\Omega, \mathcal{F}, \mathbf{F}^n, \mathbf{P})$ we are given a square-integrable martingale $M^n = (M_t^n)_{t \geq 0}$ ($[M^n, M^n] = ([M^n, M^n]_t)_{t \geq 0}$ is the quadratic variation of M^n, $\Delta M^n = (\Delta M_t^n)_{t \geq 0}$ is the jump process of M^n).

Theorem 4.29. *Suppose that the following conditions hold:*

(O) $\qquad\qquad\qquad G \subseteq \cap_{n \geq 1} \mathcal{F}_0^n;$

(L) $\qquad \mathbf{E} \sum_{s \leq t} (\Delta M_s^n)^2 I(|\Delta M_s^n| > \epsilon) \to 0, \ n \to \infty, \ \forall t > 0, \ \epsilon > 0,$

(the Lindeberg condition);

(C) $\qquad\qquad\qquad [M^n, M^n]_t \xrightarrow{\mathbf{P}} \eta^2 t,$

where η^2 is a G-measurable random variable (the Khinchin condition). Then

$$M^n \xrightarrow{\mathcal{L}} \eta W \ (G\text{-stable}),$$

where $\eta = \sqrt{\eta^2}$, in fact, does not depend on W.

Chapter 4.II. The Invariance Principle and Diffusion Approximation

1.2. Donsker's Invariance Principle (Donsker 1951). This is one of the first results in this area. In this case in (4.113), ξ is a sequence of independent random variables with

$$\mathbf{E}\xi_0^2 = \sigma^2, \quad \mathbf{E}\xi_0 = 0.$$

The conditions of Theorem 4.29 are verified in obvious fashion.

The process X^n is a square-integrable martingale with respect to the filtration $\mathbf{F}^n = (\mathcal{F}_{nt}^\xi)_{t\geq 0}$. We take as G the σ-algebra J^ξ, which in the present instance consists only of sets of measure 0 or 1. Hence condition (O) is satisfied. The Lindeberg and Khinchin conditions

(L) $$\lim_n \mathbf{E}\frac{1}{n} \sum_{1\leq k\leq nt} \xi_k^2 I(|\xi_k| > \sqrt{n}\epsilon) = 0,$$

(C) $$[X^n, X^n]_t = \frac{1}{n}\sum_{1\leq k\leq nt} \xi_k^2 \xrightarrow{\mathbf{P}} t\mathbf{E}\xi_0^2 = t\sigma^2,\ t > 0,$$

are satisfied (condition (C) by virtue of the Birkhoff-Khinchin ergodic theorem). Consequently,

$$X^n \xrightarrow{\mathcal{L}} \sigma W, \qquad (4.114)$$

where $\sigma = \sqrt{\sigma^2}$.

1.3. A Generalization of Donsker's Invariance Principle. We now drop the assumption of independence of the random variables in the sequence ξ. We consider three cases.

(i) ξ is an ergodic (!) sequence that is stationary in the strict sense

$$\mathbf{E}\xi_0^2 = \sigma^2,$$

this being the martingale-difference: $\mathbf{E}(\xi_k|\mathcal{F}_{k-1}^\xi) = 0$, $-\infty < k < \infty$. In this case, the same proof as in Donsker's invariance principle gives (4.114).

(ii) Suppose that the condition in (i) holds without the assumption of ergodicity. In this case

$$X^n \xrightarrow{\mathcal{L}} \eta W\ (J^\xi\text{-stable}),$$

where $\eta = \sqrt{\mathbf{E}(\xi_0^2|J^\xi)}$ does not depend on W. Here we merely need to note that condition (O) of Theorem 4.29 is satisfied with $G = J^\xi$, since $J^\xi \subseteq \mathcal{F}_0^\xi$.

(iii) We now turn to the analogue of (ii) for continuous time. In this case, we will assume that

$$X_t^n = \frac{1}{\sqrt{n}} M_{nt}, \qquad (4.115)$$

where $M = (M_t)_{t\geq 0}$ is a square-integrable martingale that is a process with stationary increments in the strict sense. In order to simplify the situation, we will consider as given a probability space $(\Omega, \mathcal{F}, \mathbf{P})$ and a group of measure-preserving transformations $\theta = (\theta_t)_{t\in E^1}$ with the group operation $\theta_t\theta_s =$

θ_{t+s} ($\theta_0\omega = \omega$), where the mapping $t, \omega \to \theta_t\omega$ is $\mathcal{B}(E^1) \otimes \mathcal{F}/\mathcal{F}$-measurable. Let $\mathcal{F}^{\mathbf{P}}$ be the completion of $\mathcal{F}$ with respect to the measure $\mathbf{P}$, and $\mathcal{N}$ the collection of sets in $\mathcal{F}^{\mathbf{P}}$ of zero $\mathbf{P}$ measure. We define a filtration $\mathbf{F}^{\mathbf{P}} = (\mathcal{F}_t^{\mathbf{P}})_{t \in \mathbb{R}}$, by setting

$$\mathcal{F}_t^{\mathbf{P}} = \mathcal{F}_t^0 \vee \mathcal{N},$$

where $\mathcal{F}_t^0 = \theta_t^{-1}(G)$, G is a sub-σ-algebra of $\mathcal{F}$ such that $\theta_t^{-1}(G) \subset G$ for $t < 0$. The filtration so defined is right-continuous, while the σ-algebra of invariant sets J is defined in the following manner:

$$J = \{A \in \mathcal{F}^{\mathbf{P}} : \mathbf{P}(I_A(\omega) = I_A(\theta_t\omega), \forall t \in E^1\}.$$

We also assume that M is a martingale with respect to the given filtration $\mathbf{F}^{\mathbf{P}}$ and that for any $t, s, h \in \mathbb{R}$

$$M_{t+h}(\omega) - M_{s+h}(\omega) = M_t(\theta_h\omega) - M_s(\theta_h\omega) \quad \mathbf{P}\text{-a.s.} \tag{4.116}$$

where for $t < 0$, M_t is determined by the formula

$$M_{-t}(\omega) = -M_t(\theta_{-t}\omega), \ t > 0.$$

Remark. By passing to the appropriate version, the equation in (4.116) holds for all ω (see Sam Lazaro and Meyer 1975).

Theorem 4.30. *Let $M = (M_t)_{t \geq 0}$ be a square-integrable martingale satisfying property (4.116) on the stochastic basis $(\Omega, \mathcal{F}^{\mathbf{P}}, \mathbf{F}^{\mathbf{P}} = (\mathcal{F}_t^{\mathbf{P}})_{t \in E^1}, \mathbf{P})$ and suppose that the process X^n for each $n \geq 1$ is determined by formula (4.115). Then*

$$X^n \xrightarrow{\mathcal{L}} \eta W \ (J\text{-stable}),$$

where $W = (W_t)_{t \geq 0}$ is a Wiener process not depending on

$$\eta = \{\mathbf{E}(M_1^2 | J)\}^{1/2}.$$

Corollary. *If the σ-algebra J contains sets of measure 0 or 1, then $X^n \xrightarrow{\mathcal{L}} \sigma W$, $\sigma = \sqrt{\mathbf{E}M_1^2}$.*

The proof of this theorem also follows from Theorem 4.29 with $G = J$ and $\mathcal{F}_t^n = \mathcal{F}_{nt}^{\mathbf{P}}$. Condition (O) of Theorem 4.29 is satisfied, since $J \subseteq \mathcal{F}_0^{\mathbf{P}}$. The Lindeberg condition (L) is satisfied, since by (4.116),

$$\mathbf{E}\sum_{s \leq t}(\Delta X_s^n)^2 I(|\Delta X_s^n| > \epsilon) = \mathbf{E}\frac{1}{n}\sum_{s \leq nt}(\Delta M_s)^2 I(|\Delta M_s| > \sqrt{n}\epsilon)$$

$$= \mathbf{E}\sum_{s \leq t}(\Delta M_s)^2 I(|\Delta M_s| > \sqrt{n}\epsilon) \to 0, \ n \to \infty.$$

The verification of the Khinchin condition (C) is based on the fact that property (4.116) also holds for the quadratic variation $[M, M]$ of the martingale M:

Chapter 4.II. The Invariance Principle and Diffusion Approximation   217

$$[M,M]_{t+h}(\omega) - [M,M]_{s+h}(\omega) = [M,M]_t(\theta_h\omega) - [M,M]_s(\theta_h\omega) \text{ \textbf{P}-a.s.}$$

Then
$$[X^n, X^n]_t(\omega) = \frac{1}{n}[M,M]_{nt}(\omega)$$
$$= \frac{1}{n}\sum_{k=1}^{nt}([M,M]_k(\omega) - [M,M]_{k-1}(\omega)) = \frac{1}{n}\sum_{k=1}^{nt}[M,M]_1(\theta_{k-1}\omega).$$

By the Birkhoff-Khinchin theorem, therefore, as $n \to \infty$ **P**-a.s.
$$[X^n, X^n]_t \to t\mathbf{E}([M,M]_1|J) = t\mathbf{E}(M_1^2|J).$$

1.4. The results presented above show that success in proving the invariance principle is closely connected with X^n being a martingale. For a general process X^n this is not so. In connection with this, in the general situation, the method of proving the invariance principle uses the decomposition
$$X^n = M^n + U^n, \tag{4.117}$$
where M^n is a square-integrable martingale that is a process with stationary increments in the strict sense and U^n is an asymptotically negligible process.

Decomposition (4.117) holds under certain conditions of weak dependence of the variables ξ_t and $\{\xi_s,\ s \leq 0\}$ ($t > 0$). Here the following conditions are used:
$$\sum_{k=1}^{\infty}\|\mathbf{E}(\xi_k|\mathcal{F}_0^\xi)\|_2 < \infty \tag{4.118}$$
or, in the case of continuous time,
$$\int_0^\infty \|\mathbf{E}(\xi_t|\mathcal{F}_0^\xi)\|_2 dt < \infty, \tag{4.119}$$
where $\|\alpha\|_2 = \sqrt{\mathbf{E}\alpha^2}$.

We now present auxiliary results necessary for the validity of (4.117).

Lemma 4.1. *Let $\xi = (\xi_k)_{-\infty < k < \infty}$ be a sequence that is stationary in the strict sense with $\mathbf{E}\xi_0^2 = \sigma^2$, $\mathbf{E}\xi_0 = 0$, for which condition (4.118) is satisfied. Then for any $n \geq 1$*
$$\sum_{k=1}^n \xi_k = M_n + V_0 - V_{n+1},$$
where

1) $(M_n)_{n\geq 1}$ *is a square-integrable martingale,* $M_n = \sum_{k=1}^n m_k$, $(m_k)_{k\geq 1}$ *is a sequence*
$$m_k = \sum_{i=k}^\infty [\mathbf{E}(\xi_i|\mathcal{F}_k^\xi) - \mathbf{E}(\xi_i|\mathcal{F}_{k-1}^\xi)]$$

of martingale-differences ($\mathbf{E}(m_k|\mathcal{F}_{k-1}) = 0$ **P**-a.s.) that is stationary in the strict sense with

$$\mathbf{E}m_1^2 = \sigma^2 + 2\sum_{k=1}^{\infty} \mathbf{E}(\xi_k\xi_0),$$

$$\mathbf{E}(m_1^2|J^\xi) = \mathbf{E}(\xi_0^2|J^\xi) + 2\sum_{k=1}^{\infty} \mathbf{E}(\xi_k\xi_0|J^\xi);$$

2) $(V_n)_{n\geq 0}$ is a sequence that is stationary in the strict sense with

$$V_n = \sum_{i=1}^{\infty} \mathbf{E}(\xi_i|\mathcal{F}_n^\xi),$$

such that $(V_n, \xi_n)_{n\geq 0}$ is a stationary sequence in the strict sense and

$$\sup_{t\leq T} \frac{1}{\sqrt{n}} V_{[nt]}| \xrightarrow{\mathbf{P}} 0.$$

In the case of continuous time we assume that $\xi_t(\omega) = \xi(\theta_t\omega)$, where $\theta = (\theta_t)_{t\in E^1}$ is a measure-preserving transformation, $\xi(\omega)$ is some random variable and the stochastic basis $(\Omega, \mathcal{F}^{\mathbf{P}}, \mathbf{F}^{\mathbf{P}}(\mathcal{F}_t^{\mathbf{P}})_{t\in E^1}, \mathbf{P})$ is defined in the same manner as for Theorem 4.30.

Lemma 4.2. Let $\xi = (\xi_t)_{t\in E^1}$ be a process stationary in the strict sense with $\xi_t(\omega) = \xi(\theta_t\omega)$, $\mathbf{E}\xi^2(\omega) = \sigma^2$, $\mathbf{E}\xi(\omega) = 0$, for which condition (4.119) is satisfied. Then

$$\int_0^t \xi_s ds = M_t + V_0 - V_t,$$

where

1) $M = (M_t)_{t\geq 0}$ is a square-integrable martingale with property (4.116) that is a version of the stochastic process $M' = (M_t')_{t\geq 0}$ with

$$M_t' = \int_0^\infty [\pi_t(\xi_u) - \pi_0(\xi_u)]du,$$

$\pi_t(\xi_u)$ is the optional projection of the random variable ξ_u with respect to $\mathbf{F}^{\mathbf{P}}(\pi_t(\xi_u) = \mathbf{E}(\xi_u|\mathcal{F}_t\mathbf{P})$ **P**-a.s.), possessing the following properties:

$$\mathbf{E}M_1^2 = 2\int_0^\infty \mathbf{E}(\xi_t\xi_0)dt,$$

$$\mathbf{E}(M_1^2|J) = 2\int_0^\infty \mathbf{E}(\xi_t\xi_0|J)dt;$$

2) $V_t = (V_t)_{t\geq 0}$ is a stationary process in the strict sense that is a version of the stochastic process $V' = (V_t')_{t\geq 0}$ with

$$V_t' = \int_t^\infty \pi_t(\xi_u)du$$

such that $(V_t, \xi_t)_{t \geq 0}$ is stationary in the strict sense,

$$\sup_{t \leq T} \frac{1}{\sqrt{n}} |V_{nt}| \xrightarrow{P} 0, \; \forall T > 0.$$

The decompositions presented in Lemmas 4.1 and 4.2 allow one to establish that

$$X_t^n = \frac{1}{\sqrt{n}} M_{[nt]} + \frac{1}{\sqrt{n}} (V_0 - V_{[(n+1)t]})$$

($[a]$ is the integer part of a) and in the case of continuous time

$$X_t^n = \frac{1}{\sqrt{n}} M_{nt} + \frac{1}{\sqrt{n}} (V_0 - V_{nt}).$$

Next, using the properties of the sequences $(V_n)_{n \geq 0}$ and the processes $(V_t)_{t \geq 0}$ (see Lemmas 4.1 and 4.2), it is not hard to see that the weak limits of the sequences (X^n) and $\left(\frac{1}{\sqrt{n}} M_{[n \cdot]}\right)$ and, in the case of continuous time, (X^n) and $\left(\frac{1}{\sqrt{n}} M_{n \cdot}\right)$ coincide.

Theorem 4.31. *Let $\xi = (\xi_k)_{-\infty < k < \infty}$ be a stationary sequence in the strict sense with $\mathbf{E} \xi_0^2 < \infty$ and $\mathbf{E} \xi_0 = 0$ for which condition (4.118) is satisfied. Then*

$$X^n \xrightarrow{\mathcal{L}} \eta W \; (J^\xi\text{-stable}),$$

where W is a Wiener process that is independent of

$$\eta = \left\{ \mathbf{E}(\xi_0^2 | J^\xi) + 2 \sum_{k=1}^{\infty} \mathbf{E}(\xi_k \xi_0 | J^\xi) \right\}^{1/2}.$$

Theorem 4.32. *Let $\xi = (\xi_t)_{t \in \mathbb{R}}$ be a stationary process in the strict sense with $\mathbf{E} \xi_0^2 < \infty$ and $\mathbf{E} \xi_0 = 0$, for which condition (4.119) is satisfied. Then*

$$X^n \xrightarrow{\mathcal{L}} \eta W \; (J^\xi\text{-stable})$$

where W is a Wiener process independent of

$$\eta = \left\{ 2 \int_0^\infty \mathbf{E}(\xi_t \xi_0 | J^\xi) dt \right\}^{1/2}.$$

In the proof of these theorems, we may suppose that ξ is a coordinate process and use the decompositions in Lemmas 4.1 and 4.2. Then the proof of Theorems 4.31 and 4.32 is a simple consequence of the results of Sect. 3. The transition to the original formulation is carried out in the usual fashion. If the σ-algebra J^ξ contains the sets of measures 0 or 1, then

$$X^n \xrightarrow{\mathcal{L}} aW,$$

where $a = \left\{\sigma^2 + 2\sum_{k=1}^{\infty} \mathbf{E}\xi_k\xi_0\right\}^{1/2}$ or $a = \left\{2\int_0^{\infty} \mathbf{E}(\xi_t\xi_0)dt\right\}^{1/2}$

in the case of continuous time.

1.5. We now discuss conditions (4.118), (4.119). To prove the invariance principle, conditions expressed in terms of strong and uniformly strong mixing coefficients $\alpha(t)$ and $\phi(t)$ are traditionally used (let us recall that

$$\alpha(t) = \sup |\mathbf{P}(AB) - \mathbf{P}(A)\mathbf{P}(B)|,$$
$$\phi(t) = \sup |\mathbf{P}(B|A) - \mathbf{P}(B)|,$$

where the sup is taken over all the sets $A \in \mathcal{F}_0^\xi$ and $B \in \sigma\{\xi_u, \ u \geq t\}$).

The conditions expressed in terms of $\|\mathbf{E}(\xi_t|\mathcal{F}_0^\xi)\|_2$ turn out to be weaker in the sense that

$$\left\|\mathbf{E}\cdot\left(\xi_t|\mathcal{F}_0^\xi\right)\right\|_2 \leq \begin{cases} 4C\alpha^{1/2}(t), & |\xi_0| \leq C, \\ 2\mathbf{E}\|\xi_0\|_2\phi^{1/2}(t), & \|\xi_0\|_2 < \infty \end{cases}$$

(see (Ibragimov and Linnik 1965) and (Serfling 1968)). Therefore, conditions (4.118) and (4.119) are satisfied if

$$\sum_{k=1}^{\infty} \alpha^{1/2}(k) < \infty \text{ or } \sum_{k=1}^{\infty} \phi^{1/2}(k) < \infty,$$

and in the case of continuous time

$$\int_0^{\infty} \alpha^{1/2}(t)dt < \infty \text{ or } \int_0^{\infty} \phi^{1/2}(t)dt < \infty.$$

Since $\alpha(t)$ and $\phi(t)$ are non-increasing functions, the satisfaction of the above conditions implies that $\alpha(t) \to 0$, $\phi(t) \to 0$, $t \to \infty$. This means that J^ξ contains the sets of measure 0 or 1. Therefore, $X^n \xrightarrow{\mathcal{L}} aW$, where the constant a is presented at the end of Sect. 4.

1.6. Two Examples.

Example 1. Let $Z = (Z_t)_{t \in E^1}$ be a stochastic process with homogeneous and independent increments whose trajectories are right-continuous with left limits such that

$$\mathbf{E}(Z_t - Z_s)^2 = |t-s|, \ \mathbf{E}(Z_t - Z_s) = 0.$$

The stationary process $\xi = (\xi_t)_{t \in E^1}$ (in the strict sense) is defined by the stochastic integral (Wold decomposition)

Chapter 4.II. The Invariance Principle and Diffusion Approximation 221

$$\xi_t = \int_{-\infty}^{t} b(t-s) dZ_s$$

of the measurable function $b = b(t)$ such that

$$\int_0^\infty b^2(t) dt < \infty.$$

In this case the σ-algebra J^ξ contains sets of measure 0 or 1, condition (4.119) is equivalent to the condition

$$\int_0^\infty \left\{ \int_t^\infty b^2(s) ds \right\}^{1/2} dt < \infty \tag{4.120}$$

and

$$\int_0^\infty \mathbf{E}(\xi_t \xi_0) dt = \int_0^\infty b(s) \int_0^s b(u) du\, ds. \tag{4.121}$$

Therefore when (4.120) holds we have $X^n \xrightarrow{\mathcal{L}} aW$ with constant

$$a = \left\{ 2 \int_0^\infty b(s) \int_0^s b(u) du\, ds \right\}^{1/2}.$$

In particular if, in addition to (4.120), the condition

$$\int_0^\infty |b(s)| ds < \infty,$$

holds, then

$$a = \left| \int_0^\infty b(s) ds \right|.$$

Example 2. Let $\xi = (\xi_t)_{t \geq 0}$ be an ergodic process that is stationary in the strict sense with

$$|\xi_0| \leq C, \quad \mathbf{E}\xi_0 = 0,$$

for which condition (4.119) is satisfied.

Assume that for each $n \geq 1$ the equation

$$Y_t^n = \sqrt{n} \int_0^t \xi_{Y_s^n + ns} ds \tag{4.122}$$

with respect to $Y^n = (Y_t^n)_{t \geq 0}$ has a solution (a solution of equation (4.122) exists if ξ_t is a smooth function with a bounded derivative).

Let us establish the invariance principle for the sequence Y^n, $n \geq 1$.

For $n \geq 4c^2$ the change of variables $u = Y_s^n + ns$ is possible in the integral on the right hand side of (4.122). This change leads to the representation

$$Y_t^n = \frac{1}{\sqrt{n}} \int_0^{Y_t^n + nt} \frac{\xi_u}{\frac{1}{\sqrt{n}}\xi_u + 1} du = Z_{Y_t^n + nt}^n,$$

where
$$Z_t^n = \frac{1}{\sqrt{n}} \int_0^{nt} \frac{\xi_u}{\frac{1}{\sqrt{n}}\xi_u + 1} du.$$

Since $\sup_{t \leq T} \frac{|Y_t^n|}{n} \leq \frac{CT}{\sqrt{n}}$, the weak limits for the sequences Y^n, $n \geq 1$, and Z^n, $n \geq 1$, are the same. Let $X_t^n = \frac{1}{\sqrt{n}} \int_0^{nt} \xi_u du$. Then

$$Z_t^n = X_t^n - \frac{1}{n} \int_0^{nt} \xi_u^2 du + \frac{1}{n^{3/2}} \int_0^{nt} \frac{\xi_u^3}{\frac{1}{\sqrt{n}}\xi_u + 1} du.$$

The last term in the right hand side of this representation tends uniformly to zero as $n \to \infty$ on any finite time interval. By the Birkhoff-Khinchin ergodic theorem, as $n \to \infty$

$$\frac{1}{n} \int_0^{nt} \xi_u^2 du \to t\mathbf{E}\xi_0^2 \quad \mathbf{P}\text{-a.s.}$$

Hence, by an analogue of Pólya's Theorem (see Liptser and Shiryaev 1986, Problem 5.4.2), as $n \to \infty$ we have **P**-a.s.

$$\sup_{t \leq T} \left| \frac{1}{n} \int_0^{nt} \xi_u^2 du - t\mathbf{E}\xi_0^2 \right| \to 0, \quad T > 0.$$

Finally, by Theorem 4.32 $X^n \overset{\mathcal{L}}{\to} aW$ with $a = \left\{ 2 \int_0^\infty \mathbf{E}(\xi_t \xi_0) dt \right\}^{1/2}$.

Consequently,
$$Y_t^n \overset{\mathcal{L}}{\to} Y,$$
where $Y_t = aW_t - t\mathbf{E}\xi_0^2$.

1.7. Markov Processes. Let $(\Omega, \mathcal{F}, \mathbf{F} = (\mathcal{F}_t)_{t\geq 0}, \mathbf{P})$ be a stochastic basis on which is given a progressively measurable homogeneous Markov process $\zeta = (\zeta_t)_{t\geq 0}$ with state space $(E, \mathcal{E})$. Let $f = f(x)$ be an $\mathcal{E}$-measurable function such that
$$\int_0^t |f(\zeta_s)| ds < \infty \ \mathbf{P}\text{-a.s.}, \ t > 0.$$

The invariance principle in the present instance will be formulated for the sequence $X^n = (X_t^n)_{t \geq 0}$, $n \geq 1$, with

$$X_t^n = \frac{1}{\sqrt{n}} \int_0^{nt} f(\zeta_s) ds. \qquad (4.123)$$

First we give the requisite information from the theory of Markov processes.

We denote by $p = p(t, x, dy)$ the transition function of the Markov process, and by $q = q(dx)$ the initial distribution. Associated with the transition function p is the operator T_t, acting according to the formula

Chapter 4.II. The Invariance Principle and Diffusion Approximation

$$T_t h(x) = \int_E h(y) p(t, x, dy)$$

for any $\mathcal{E}$-measurable function $h = h(y)$ with $\int_{E \times E} |h(y)| p(t, x, dy) q(dx) < \infty$. The family $(T_t)_{t>0}$ is a semigroup with the operation $T_t T_s = T_{t+s}$.

The Markov characterization of the process ζ with respect to the filtration **F** is given in the following manner in terms of the subgroup $(T_t)_{t>0}$: for each $\mathcal{E}$-measurable function $h = h(y)$ with $\int_{E^1 \times E^1} |h(y)| p(t, x, dy) q(dx) < \infty$

$$\mathbf{E}(h(\zeta_t)|\mathcal{F}_s) = T_{t-s} h(\zeta_s) \text{ P-a.s., } t > s. \tag{4.124}$$

A probability measure $r = r(dx)$ on $(E, \mathcal{E})$ is called *invariant* if for any bounded $\mathcal{E}$-measurable function $h(x)$

$$\int_E T_t h(x) r(dx) = \int_E h(x) r(dx), \, t > 0.$$

If $q(dx) = r(dx)$, then the homogeneous Markov process ζ is a process that is stationary in the strict sense.

Let $L_2(E, \mathcal{E}, r)$ be the Hilbert space of functions $h = h(y)$ with norm $\|h\|_2 = \left(\int_E h^2(y) r(dy)\right)^{1/2}$. We denote by B_0 the center of the semigroup $(T_t)_{t>0}$:

$$B_0 = \{h \in L_2(E, \mathcal{E}, r) : \lim_{t \to 0} \|T_t h - h\|_2 = 0\}.$$

By the *infinitesimal* operator of the semigroup $(T_t)_{t>0}$ with respect to $L_2(E, \mathcal{E}, r)$ we mean the operator A with domain of definition:

$$D_A = \{h \in B_0 : \exists g \in B_0, \lim_{t \to 0} \left\|\frac{T_t h - h}{t} - g\right\|_2 = 0\},$$

given by the relation:

$$\lim_{t \to 0} \left\|Ah - \frac{T_t h - h}{t}\right\|_2 = 0.$$

We set $R_A = \{Ah : h \in D_A\}$, that is, R_A is the set of values of the operator A.

For a function $h \in D_A$ and any $t > 0$ we have the Dynkin formula

$$T_t h(x) = h(x) + \int_0^t T_u A h(x) du \text{ r-a.s..} \tag{4.125}$$

1.8. We suppose that

$$q(dx) = r(dx).$$

In this case ζ is a process stationary in the strict sense. Let us also assume that ζ is an ergodic process.

In the proof of the invariance principle we can suppose without loss of generality that ζ is a coordinate process and extend its definition to $t < 0$.

We set
$$\xi_t = f(\zeta_t).$$
Clearly the process $\xi = (\xi_t)_{t \in \mathbb{R}^1}$ is an ergodic process that is stationary in the strict sense. If
$$\mathbf{E}\xi_0^2 < \infty, \quad \mathbf{E}\xi_0 = 0$$
and the process ξ satisfies condition (4.119), then by Theorem 4.32 the sequence X^n, $n \geq 1$, satisfies the invariance principle:
$$X^n \xrightarrow{\mathcal{L}} aW,$$
where W is a Wiener process,
$$a = \left\{ 2 \int_0^\infty \mathbf{E}(f(\zeta_t)f(\zeta_0))dt \right\}^{1/2}.$$

The Markov property of the process ζ allows one to give a sufficient condition for (4.119). Namely, in view of the inequality
$$\|\mathbf{E}(\xi_t|\mathcal{F}_0^\xi)\|_2 \leq \|\mathbf{E}(\xi_t|\mathcal{F}_0)\|_2,$$
which follows from the inclusion $\mathcal{F}_0^\xi \subseteq \mathcal{F}_0$, and (4.124), we have
$$\|\mathbf{E}(\xi_t|\mathcal{F}_0^\xi)\|_2 \leq \left(\int_E (T_t f(x))^2 r(dx) \right)^{1/2} = \|T_t f\|_2.$$
This means that the invariance principle for X^n, $n \geq 1$, holds under the satisfaction of the condition
$$\int_0^\infty \|T_t f\|_2 dt < \infty.$$

1.9. One can obtain deeper results by using the properties of the infinitesimal operator A.

Let $f = f(x)$ be a function belonging to R_A — the set of values of A. Then

1) $$\int_E f^2(x) r(dx) < \infty, \quad \int_E f(x) r(dx) = 0;$$

2) the equation
$$Ag = f$$
has a solution with a function $g = g(x)$ such that $\int_E g^2(x) r(dx) < \infty$
$$\left(\text{when } \int_0^\infty \|T_t f\|_2 dt < \infty, \ g(x) = -\int_0^\infty T_t f(x) dt \right);$$

3) if 0 is a simple eigenvalue of the operator A and $q = r$, then under the condition $\int_E g(x) r(dx) = 0$ the equation $Ag = f$ has a unique solution;

4) if 0 is a simple eigenvalue of the operator A, then the Markov process ζ is ergodic.

Theorem 4.33. *Let $q(dx) = r(dx)$, and 0 a simple eigenvalue of the operator A. Then*
$$X^n \xrightarrow{\mathcal{L}} aW,$$
where W is a Wiener process,
$$a = \left\{ -2 \int_E g(x) f(x) r(dx) \right\}^{1/2},$$
g is a solution of the equation $Ag = f$.

The proof of this theorem rests on the Dynkin formula (4.125) applied to the solution g of the equation $Ag = f$. By virtue of this formula, one can define a square-integrable martingale $M = (M_t)_{t \geq 0}$ with respect to the filtration $\mathbf{F}$ that is a process with increments that are stationary in the strict sense such that
$$M_t = g(\zeta_t) - g(\zeta_0) - \int_0^t f(\zeta_u) du \ \mathbf{P}\text{-a.s.}, \ t > 0$$
with
$$\mathbf{E} M_t^2 = -2t \int_E f(x) g(x) r(dx).$$
In this case
$$X_t^n = \frac{1}{\sqrt{n}} [g(\zeta_{nt}) - g(\zeta_0)] - \frac{1}{\sqrt{n}} M_{nt}$$
and the following facts are used:
$$\sup_{t \leq T} \frac{1}{\sqrt{n}} |g(\zeta_{nt})| \xrightarrow{\mathbf{P}} 0, \ \forall T > 0.$$
$$\frac{1}{\sqrt{n}} M_n. \xrightarrow{\mathcal{L}} aW,$$
where
$$a = \sqrt{\mathbf{E} M_1^2}.$$

1.10. If $q(dx) \neq r(dx)$, then the Markov process ζ is no longer a stationary process in the strict sense. In this case, one can approximate it, in some sense, by a stationary ergodic process with $q(dx) = r(dx)$.

To this end we consider for fixed t, x the signed measure
$$\mu_{t,x}(dy) = p(t, x, dy) - r(dy)$$
and denote by $\text{Var}(\mu_{t,x})$ the total variation of this measure. An essential role is played here by the condition: for each $x \in E$
$$\lim_{t \to \infty} \text{Var}(\mu_{t,x}) = 0. \tag{4.126}$$

If condition (4.126) is satisfied, then 0 is a simple eigenvalue of the operator A and hence for $q = r$, (4.126) ensures the ergodicity of the process ζ.

Theorem 4.34. *Let $q \neq r$, $f \in \mathbb{R}_A$ and suppose that condition (4.126) holds. Then the assertion of Theorem 4.33 remains valid.*

Proof. If $q = r$, then ζ is an ergodic process stationary in the strict sense and the assertion of Theorem 4.33 holds. We define the process $\tilde{X}^n = (\tilde{X}^n_t)_{t \geq 0}$ with

$$\tilde{X}^n_t = \frac{1}{\sqrt{n}} \int_{n^{1/4}}^{n^{1/4}+nt} f(\zeta_s) ds.$$

If $q = r$, then $\tilde{X}^n \overset{\mathcal{L}}{=} X^n$ and hence $\tilde{X}^n \overset{\mathcal{L}}{\to} aW$.

Let $\psi = \psi(X)$, $X \in D$, be a function bounded by a constant K and continuous in the Skorokhod topology. We denote by $\mathbf{E}_q$ the expectation corresponding to the distribution of the process ζ with initial distribution q. We have the estimate

$$|\mathbf{E}_q \psi(\tilde{X}^n) - \mathbf{E}_r \psi(\tilde{X}^n)| \leq K \operatorname{Var}(\mu_{n^{1/4},x});$$

hence by (4.126),

$$\lim_n \mathbf{E}_q \psi(\tilde{X}^n) = \lim_n \mathbf{E}_r \psi(\tilde{X}^n) = \mathbf{E} \psi(aW),$$

where $\mathbf{E}$ is the expectation corresponding to the Wiener measure.

Thus, $\tilde{X}^n \overset{\mathcal{L}}{\to} aW$ for $q \neq r$. We now define the process $\hat{X}^n = (\hat{X}^n_t)_{t \geq 0}$ with

$$\hat{X}^n_t = \tilde{X}^n_{(t-n^{-3/4}) \vee 0}.$$

The proof of the theorem is completed by verifying the relations

$$\hat{X}^n \overset{\mathcal{L}}{\to} aW \text{ and } \sup_{t \leq T} |\hat{X}^n_t - X^n_t| \overset{\mathbf{P}}{\to} 0.$$

$\square$

§2. The Stochastic Averaging Principle in Models without Diffusion

2.1. The Stochastic Averaging Principle of Bogolyubov. Let us consider a differential equation with a small parameter ϵ ($\epsilon > 0$):

$$\dot{x}^\epsilon_t = \epsilon b(x^\epsilon_t, \xi_t), \quad x^\epsilon_0 = x_0, \tag{4.127}$$

where $b = b(x, y)$ is a continuous function in (x, y) that is of linear growth and Lipschitzian in x uniformly with respect to y; $\xi = (\xi_t)_{t \geq 0}$ is a measurable stochastic process with the ergodic property:

$$\frac{1}{t}\int_0^t h(x,\xi_s)ds \xrightarrow{P} \int_{E^1} h(x,y)\rho(dy), \ t\to\infty, \ x\in E^1 \quad (4.128)$$

for any measurable function $h = h(x,y)$ with $\int_{E^1}|h(x,y)|\rho(dy) < \infty$, where $\rho = \rho(dy)$ is some probability measure on E^1.

Since ϵ is a small parameter, x_t^ϵ differs from x_0 by a small amount. It is therefore natural to study the behavior of the process (x_t^ϵ) on time intervals of type $[0, \epsilon^{-1}]$. To this end, let us consider a process with "stretched time" $X^\epsilon = (X_t^\epsilon)_{t\geq 0}$ with

$$X_t^\epsilon = x_{t/\epsilon}^\epsilon.$$

The function $X^\epsilon = (X_t^\epsilon)_{t\geq 0}$ is a solution of the differential equation

$$\dot{X}_t^\epsilon = b(X_t^\epsilon, \xi_{t/\epsilon}), \quad X_0^\epsilon = x_0 \quad (4.129)$$

with a random perturbation in "fast time" $(\xi_{t/\epsilon})$.

The *stochastic averaging principle of Bogolyubov* consists in approximating X^ϵ by a deterministic function $X = (X_t)_{t\geq 0}$ defined by the differential equation

$$\dot{X}_t = \bar{b}(X_t), \ X_0 = x \quad (4.130)$$

with averaged right hand side

$$\bar{b}(x) = \int_{E^1} b(x,y)\rho(dy) \quad (4.131)$$

in the following sense: for any $T > 0$

$$\sup_{t\leq T}|X_t^\epsilon - X_t| \xrightarrow{P} 0, \ \epsilon\to 0. \quad (4.132)$$

The proof of (4.132) follows from the integral inequality for $\Delta_t^\epsilon = |X_t^\epsilon - X_t|$ derived from (4.129) and (4.130): $(t \leq T)$

$$\Delta_t^\epsilon \leq \int_0^t |b(X_s^\epsilon, \xi_{s/\epsilon}) - b(X_s, \xi_{s/\epsilon})|ds$$
$$+ \left|\int_0^t [b(X_s, \xi_{s/\epsilon}) - \bar{b}(X_s)]ds\right|$$
$$\leq L\int_0^t \Delta_s^\epsilon ds + \sup_{t\leq T}\left|\int_0^t [b(X_s, \xi_{s/\epsilon}) - \bar{b}(X_s)]ds\right|,$$

where L is the Lipschitz constant.

Hence, taking into account the Gronwall-Bellman inequality, we obtain

$$\sup_{t\leq T}\Delta_t^\epsilon \leq e^{LT}\sup_{t\leq T}\left|\int_0^t [b(X_s, \xi_{s/\epsilon}) - \bar{b}(X_s)]ds\right|.$$

The required relation (4.132) is derived from this, using a result which is of interest in its own right.

Theorem 4.35. *Let ξ be a stochastic process satisfying condition (4.128). If the function $h = h(x,y)$ in (4.128) satisfies a Lipschitz condition in x uniformly with respect to y, then for any $T > 0$*

$$\sup_{t \leq T}\left|\int_0^t [h(x_s, \xi_{s/\epsilon}) - \bar{h}(x_s)]ds\right| \xrightarrow{\mathbf{P}} 0, \quad \epsilon \to 0, \qquad (4.133)$$

where $\bar{h}(x) = \int_{E^1} h(x,y)\rho(dy)$, $(x_t)_{t \geq 0}$ is a continuous function.

Remark. If the convergence in (4.128) holds **P**-a.s., then the convergence in (4.133) is also **P**-a.s. In particular, if $\xi = (\xi_t)_{t \geq 0}$ is an ergodic process that is stationary in the strict sense, then Theorem 4.35 generalizes the Birkhoff-Khinchin ergodic theorem.

2.2. The Averaging Principle in a Queuing Model.

Suppose that we have a finite source of customers of size n and one serving station. The queue to the serving station Q_t at time t is given by the balancing relation

$$Q_t = Q_0 + A_t - D_t,$$

where Q_0 is the initial queue, $A = (A_t)_{t \geq 0}$, $D = (D_t)_{t \geq 0}$ are counting processes with disjoint times of jumps of arrival and service of the customers. With respect to the filtration generated by $(Q_t)_{t \geq 0}$ the processes A and D have compensators

$$\tilde{A}_t = \int_0^t \lambda(n - Q_s)ds, \quad \tilde{D}_t = \int_0^t I(Q_s > 0)nf\left(\frac{Q_s}{n}\right)ds,$$

where $f = f(x)$ is a non-negative function satisfying a Lipschitz condition, $\lambda > 0$.

The random variable Q_t takes $n+1$ values. Therefore, for large n, the analysis of the given Markov model encounters certain difficulties. In connection with this, let us consider the normalized queue $q^n = (q_t^n)$ with

$$q_t^n = \frac{1}{n}Q_t.$$

We study the question of the limit behavior of q^n as $n \to \infty$.

From the balancing relation for Q_t and the definition of the compensators $\tilde{A}_t$ and $\tilde{D}_t$, we have

$$q_t^n = q_0^n + \int_0^t \lambda(1 - q_s^n)ds - \int_0^t I(q_s^n > 0)f(q_s^n)ds + M_t^n,$$

where $M^n = (M_t^n)_{t \geq 0}$ is a square-integrable martingale with

$$M_t^n + \frac{1}{n}(A_t - \tilde{A}_t) - \frac{1}{n}(D_t - \tilde{D}_t),$$

and quadratic characteristic

$$\langle M^n\rangle_t = \frac{1}{n}\int_0^t [\lambda(1-q_s^n) + I(q_s^n > 0)f(q_s^n)]ds.$$

Since $0 \leq q_t^n \leq 1$, it follows that for any $t > 0$ $\langle M^n\rangle_t \to 0$, $n \to \infty$. Hence $\sup_{s\leq t}|M_s^n| \xrightarrow{P} 0$, $n \to \infty$, $\forall t > 0$. The question of the approximation of the processes q^n, $n \to \infty$ by a deterministic function $q = (q_t)_{t\geq 0}$ is therefore solved if q_t is a strictly positive solution of the differential equation

$$\dot{q}_t = \lambda(1-q_t) - f(q_t), \qquad (4.134)$$

in the sense that under the condition $q_0^n \xrightarrow{P} \mathbf{q}_0$ ($\mathbf{q}_0$ is the initial condition for (4.134))

$$\sup_{t\leq T}|q_t^n - q_t| \xrightarrow{P} 0, \ n \to \infty, \ \forall T > 0.$$

§3. Diffusion Approximation of Semimartingales. The Averaging Principle in Models with Diffusion

3.1. For each $n \geq 1$ let $X_n = (X_i^n)_{i\leq d}$ be a semimartingale with values in $\mathbb{R}^d$ on the stochastic basis $(\Omega^n, \mathcal{F}^n, \mathbf{F}^n, \mathbf{P}^n)$ with triple of predictable characteristics $T^n = (B^n, C^n, \nu^n)$ and truncation function

$$h(x) = xI(|x| \leq 1).$$

From the general theorem on the convergence of a sequence of semimartingales to a semimartingale (Theorem 4.24) we can derive the weak convergence condition

$$X^n \xrightarrow{\mathcal{L}} X, \qquad (4.135)$$

where $X = (X_i)_{i\leq d}$ is the unique weak solution of the Itô stochastic equation

$$X_t = X_0 + \int_0^t b(s,X)ds + \int_0^t c(s,X)dW_s \qquad (4.136)$$

with respect to a Wiener process $W = (W_i)_{i\leq d}$ with independent components, that is, X is a semimartingale with triple of predictable characteristics $T(X) = (B(X), C(X), 0)$:

$$B_t(X) = \int_0^t b(s,X)ds, \ C_t(X) = \int_0^t c(s,X)c^*(s,X)ds,$$

(* denotes the transpose).
 We assume that X is a canonical process

$$D = D[0,\infty)(\mathbb{R}^d), \quad \mathcal{D}_t = \sigma\{X_s, \ s\leq t\},$$

Q is a distribution and X is the unique weak solution of equation (4.136). The elements of the vector $b(t, X)$ and of the matrix $c(t, X)$ are assumed to be $\mathcal{B}(\mathbb{R}_+) \otimes \mathcal{D}_\infty$-measurable and $\mathcal{D}_t$-measurable for each t. Thus X is a semimartingale on the stochastic basis $(D, \mathcal{D}_\infty^Q, \mathbf{D}_+^Q, \mathbf{Q})$, where $\mathcal{D}_\infty^Q$ is the completion of $\mathcal{D}_\infty$ with respect to the measure $\mathbf{Q}$, $\mathbf{D}_\infty^Q = (\mathcal{D}_{t+} \vee \mathcal{N})_{t \geq 0}$, $\mathcal{D}_{t+} = \bigcap_{\epsilon > 0} \mathcal{D}_{t+\epsilon}$, and $\mathcal{N}$ is the collection of sets of $\mathcal{D}_\infty^Q$ of $\mathbf{Q}$-measure zero.

The corresponding result on weak convergence (4.135) is stated as follows.

Theorem 4.36. *Let the following conditions be satisfied.*

1) *If $h(t, X)$ denotes any of the elements of the vector $b(t, X)$ or the matrix $c(t, X)$, then*

$$|h(t, X)| \leq L(1 + \sup_{s < t} |X(s)|), \quad (|X(s)| = \sum_{i \leq d} |X_i(s)|),$$

for each $t \in S$ (S is an everywhere dense subset of $\mathbb{R}_+$ $\int_{\mathbb{R}_+ \setminus S} dt = 0$), $h(t, X)$ is continuous in the Skorokhod metric at the point $X \in C$;

2) $X_0^n \xrightarrow{d} X_0$;

3) *for any $L > 0$, $\epsilon > 0$ and $a \in (0, 1]$*

(A) $$\lim_n \mathbf{P}^n \Big(\int_0^L \int_{|x| > a} d\nu^n > \epsilon \Big) = 0,$$

(sup B) $$\lim_n \mathbf{P}^n \Big(\sup_{t \leq L} \Big| B_t^n - \int_0^t b(s, X^n) ds \Big| > \epsilon \Big) = 0,$$

(sup C) $$\lim_n \mathbf{P}^n \Big(\sup_{t \leq L} \Big| \langle M^{na} \rangle_t - \int_0^t c(s, X^n) c^*(s, X^n) ds \Big| > \epsilon \Big) = 0$$

$$\Big(\langle M^{na} \rangle_t = C_t^n + \int_0^t \int_{|x| \leq a} xx^* d\nu^n$$

$$- \sum_{s \leq t} \int_{|x| \leq a} x\nu^n(\{s\}, dx) \Big(\int_{|x| \leq a} x\nu^n(\{s\}, dx) \Big)^* \Big).$$

Then the weak convergence (4.135) holds, where X is the unique weak solution of equation (4.136).

3.2. We present two examples in which Theorem 4.36 is used.

Example 1. We consider the second approximation in the problem of Bogolyubov's stochastic averaging principle. Let X^ϵ and X be defined by the differential equations (4.129) and (4.130).

We define the process $Y^\epsilon = (Y_t^\epsilon)_{t \geq 0}$ from the relation

$$X_t^\epsilon = X_t + \sqrt{\epsilon} Y_t^\epsilon.$$

If $Y^\epsilon \xrightarrow{\mathcal{L}} Y$, then we say that there exists a second approximation of order $\sqrt{\epsilon}$ in the Bogolyubov stochastic averaging problem.

Chapter 4.II. The Invariance Principle and Diffusion Approximation   231

The existence of the second approximation is established under additional assumptions on the process ξ. Namely, $\xi = (\xi_t)_{t\in E}$ is a stationary ergodic process in the strict sense for which (4.128) is satisfied by the Birkhoff-Khinchin theorem. Furthermore, the following conditions are assumed:

$$\sup \|b(x,\xi_0)\|_p < \infty, \ p < 2,$$
$$\int_0^\infty \operatorname{ess\,sup}\|\mathbf{E}(b(x,\xi_t) - \bar{b}(x)|\mathcal{F}_0^\xi)\|_p dt < \infty, \ p > 2 \qquad (4.137)$$

($\|\alpha\|_p = (\mathbf{E}|\alpha|^p)^{1/p}$, where the sup and the ess sup are taken over the set $\{x : |x| \leq \sup_{t\geq 0} |X_t|\}$, X_t is a solution of the differential equation (4.130).

Theorem 4.37. *Let $\xi = (\xi_t)_{t\in\mathbb{R}}$ be an ergodic process that is stationary in the strict sense, for which conditions (4.137) are satisfied with some $p > 2$, and let $b = b(x,y)$ be a function that is continuous in (x,y) and differentiable in x for each y; the function $b_x = b_x(x,y)$ is uniformly bounded and satisfies a Lipschitz condition in x uniformly with respect to y.*
Then
$$Y^\epsilon \xrightarrow{\mathcal{L}} Y, \ \epsilon \to 0,$$

where $Y = (Y_t)_{t\geq 0}$ is the diffusion process defined by the Itô stochastic equation
$$Y_t = \int_0^t a(X_s)Y_s ds + \int_0^t \sigma(X_s)dW_s$$

with respect to the Wiener process $W = (W_t)_{t\geq 0}$; here
$$a(x) = \mathbf{E}b_x(x,\xi_0),$$
$$\sigma(x) = \left\{2\int_0^\infty [\mathbf{E}(b(x,\xi_t)b(x,\xi_0)) - \bar{b}^2(x)]ds\right\}^{1/2},$$

X_t is a solution of equation (4.130).

The proof of this theorem uses Theorem 4.36 and the following generalization of the invariance principle for stationary processes.

Theorem 4.38. *Suppose that we are given:*
1) a function $b = b(x,y)$ that is measurable with respect to both variables and satisfies a Lipschitz condition in x uniformly with respect to y;
2) a continuous function $Z = (Z_t)_{t\geq 0}$;
3) an ergodic process $\xi = (\xi_t)_{t\in\mathbb{R}}$ that is stationary in the strict sense and satisfies conditions (4.137) with a given function $b = b(x,y)$ for some $p > 2$, in which the sup and ess sup are taken over the set $\{x : |x| \leq \sup_{t\geq 0} |Z_t|\}$.
For $\epsilon > 0$ let
$$Y_t^\epsilon = \frac{1}{\sqrt{\epsilon}} \int_0^t [b(Z_s, \xi_{s/\epsilon}) - \mathbf{E}b(Z_s, \xi_0)]ds.$$

Then as $\epsilon \to 0$

$$Y^\epsilon \stackrel{\mathcal{L}}{\to} Y,$$

where $Y = (Y_t)_{t\geq 0}$ is a stochastic integral with respect to a Wiener process

$$Y_t = \int_0^t \sigma(Z_s)dW_s,$$

$$\sigma(z) = \left\{2\int_0^\infty [\mathbf{E}(b(z,\xi_t)b(z,\xi_0)) - (\mathbf{E}b(z,\xi_0))^2]dt\right\}^{1/2}.$$

Example 2. We consider the second approximation for the queueing model (see Sect. 2). We define the process $X^n = (X^n_t)_{t\geq 0}$ from the relation

$$q^n_t = q_t + \frac{1}{\sqrt{n}}X^n_t.$$

If $X^n \stackrel{\mathcal{L}}{\to} X$, then we say that in the queueing model there is a second approximation of the queue of order $\frac{1}{\sqrt{n}}$.

By Theorem 4.36 under the assumption that the function $f = f(x)$ is continuously differentiable and its derivative satisfies a Lipschitz condition, $X^n_0 \stackrel{d}{\to} X_0$, a second approximation of the queue exists of order $\frac{1}{\sqrt{n}}$. Moreover, X is a solution of the Itô stochastic equation

$$X_t = X_0 - \int_0^t (\lambda + f'(q_s))X_s ds + \int_0^t \{\lambda(1-q_s) + f(q_s)\}^{1/2}dW_s$$

with respect to the Wiener process $W = (W_t)_{t\geq 0}$, q_t is a solution of the differential equation (4.134).

3.3. The Averaging Principle in Models with Diffusion.

We consider the stochastic differential equation with a small parameter $\epsilon > 0$

$$dx^\epsilon_t = \epsilon b(x^\epsilon_t, \xi_t)dt + \sqrt{\epsilon}\sigma(x^\epsilon_t, \xi_t)dW_t, \quad x^\epsilon_0 = x \quad (4.138)$$

with respect to a Wiener process $W = (W_t)_{t\geq 0}$ that is independent of the stochastic process $\xi = (\xi_t)_{t\geq 0}$ satisfying the ergodic property (4.128). The change of variables t/ϵ leads to the Itô stochastic equation for $X^\epsilon_t = x^\epsilon_{t/\epsilon}$

$$dX^\epsilon_t = b(X^\epsilon_t, \xi_{t/\epsilon})dt + \sigma(X^\epsilon_t, \xi_{t/\epsilon})dW^\epsilon_t, \quad X^\epsilon_0 = x_0$$

with respect to the Wiener process $W^\epsilon_t = \sqrt{\epsilon}W_{t/\epsilon}$.

Let us assume that the functions $b = b(x,y)$ and $\sigma = \sigma(x,y)$ are continuous in (x,y) and satisfy Lipschitz and linear growth conditions with respect to x uniformly with respect to y.

Chapter 4.II. The Invariance Principle and Diffusion Approximation   233

We set
$$\bar{b}(x) = \int_{E^1} b(x,y)\rho(dy)$$
$$\bar{\sigma}(x) = \left\{\int_{E^1} \sigma^2(x,y)\rho(dy)\right\}^{1/2} \quad (4.139)$$

By Theorem 4.36 one can establish the averaging principle:
$$X^\epsilon \xrightarrow{\mathcal{L}} X, \ \epsilon \to 0,$$
where $X = (X_t)_{t\geq 0}$ is the diffusion process determined by the Itô stochastic equation with "averaged parameters $\bar{b} = \bar{b}(x)$ and $\bar{\sigma} = \bar{\sigma}(x)$" (see (4.139)):
$$dX_t = \bar{b}(X_t)dt + \bar{\sigma}(X_t)d\bar{W}_t, \ X_0 = x_0 \quad (4.140)$$
with respect to some Wiener process $\bar{W} = (\bar{W}_t)_{t\geq 0}$.

§4. Diffusion Approximation for Systems with Physical White Noise

4.1. In the consideration of physical systems, interest often arises in the study of the properties of the stochastic process defined by the ordinary differential equation
$$\dot{x}_t = a(t, x_t) + b(t, x_t)\dot{W}_t,$$
where $\dot{W}_t$ is physical white noise. In such a situation, it is natural to approximate the process (x_t) by some diffusion Markov process, the appropriateness of which is justified by the fact that the analysis of the properties of a diffusion Markov process, as a rule, is simpler than the study of the properties of (x_t).

In connection with this let us consider the sequence $X^n = (X^n_t)_{t\geq 0}$, $n \geq 1$, of solutions of the differential equations
$$\dot{X}^n = a(t, X^n_t) + b(t, X^n_t)\dot{W}^n_t, \quad X^n_0 \equiv X^0, \quad (4.141)$$
where $\dot{W}^n = (\dot{W}^n_t)_{t\geq 0}$, $n \geq 1$, is a sequence of physical white noises. We set
$$W^n_t = \int_0^t \dot{W}^n_s ds$$
and assume that the sequence of stochastic processes $W^n = (W^n_t)_{t\geq 0}$, $n \geq 1$, converges weakly in the Skorokhod topology to $\sigma W = (\sigma W_t)_{t\geq 0}$, where $W = (W_t)_{t\geq 0}$ is a Wiener process, σ is a constant.

In this case it is natural to understand by the approximation of the processes X^n, $n \geq 1$, weak convergence
$$X^n \xrightarrow{\mathcal{L}} X$$

to a diffusion process X determined by an Itô stochastic equation with respect to the Wiener process W.

There are two approaches for such an approximation.

4.2. The first approach is based on the idea of vibro-correctness due to Krasnosel'skij and Pokrovskij (Krasnosel'skij and Pokrovskij 1983), which is that X^n be a continuous mapping of W^n.

We state the corresponding result.

Theorem 4.39. *Suppose that*

1) $a = a(t,x)$ is a function that is continuous in (t,x) and satisfies Lipschitz and linear growth conditions with respect to x uniformly with respect to t;

2) $b = b(t,x)$ is a function continuous in (t,x), along with its partial derivatives $b_t = b_t(t,x)$ and $b_x = b_x(t,x)$;

3) the functions $b_t(t,x)$, $b_x(t,x)$, $b(t,0)$ are uniformly bounded;

4) $b_x = b_x(t,x)$ satisfies a Lipschitz condition with respect to x uniformly with respect to t.

Then

$$W^n \xrightarrow{\mathcal{L}} \sigma W \Longrightarrow X^n \xrightarrow{\mathcal{L}} X,$$

where $X = (X_t)_{t \geq 0}$ is a solution of the Itô stochastic equation

$$\begin{aligned} X_t = X_0 &+ \int_0^t \left[a(s, X_s) + \frac{\sigma^2}{2} b_x(s, X_s) b(s, X_s) \right] ds \\ &+ \int_0^t \sigma b(s, X_s) dW_s. \end{aligned} \quad (4.142)$$

The proof is based on the vibro-correct transformation:

$$X_t^n = Q(W_t^n, 0, Z_t^n, t),$$
$$\dot{Z}_t^n = a(t, Q(W_t^n, 0, Z_t^n, t)), \quad Z_0^n = X_0,$$

where

$$Q(s, s_0, y_0, t) = y_s$$

and y_s is a solution of the differential equation

$$\frac{dy_s}{ds} = b(t, y_s), \quad y_{s_0} = y_0.$$

This transformation enables us to establish that

$$W^n \xrightarrow{\mathcal{L}} \sigma W \Longrightarrow (X^n, Z^n, W^n) \xrightarrow{\mathcal{L}} (X, Z, \sigma W),$$

where

$$X_t = Q(\sigma W_t, 0, Z_t, t),$$
$$\dot{Z}_t = a(t, Q(\sigma W, 0, Z_t, t)), \quad Z_0 = X_0.$$

Here, the stochastic equation for X is derived from this by the Itô formula.

The merit of the vibro-correctness method is that it does not require any kind of additional assumptions on the structure of the processes $\dot{W}^n$, $n \geq 1$. However, this method, generally speaking, does not work in the vector case without an additional Frobenius condition which severely restricts the structure of the matrix $b(t,x)$ (in the vector case, $b(t,x)$ is a matrix, W_t^n is a vector).

4.3. The second method of approximating $X^n \xrightarrow{\mathcal{L}} X$ works both in the scalar and vector cases. However, it requires additional assumptions regarding $\dot{W}^n$.

Theorem 4.40. *Suppose that*

1) $a = a(t,x)$ *is a jointly continuous function satisfying Lipschitz and linear growth conditions with respect to x uniformly with respect to t;*

2) $b = b(t,x)$ *is a bounded function continuous in (t,x), along with its partial derivatives $b_t = b_t(t,x)$, $b_x = b_x(t,x)$, $b_{xt} = b_{xt}(t,x)$, $b_{xx} = b_{xx}(t,x)$;*

3) $\dot{W}_t^n = \sqrt{n}\xi_{nt}$, *where $\xi = (\xi_t)_{t \in E^1}$ is a stationary ergodic process in the strict sense such that*

$$\mathbf{E}\xi_0^2 < \infty, \quad \mathbf{E}\xi_0 = 0,$$

$$\int_0^\infty \|\mathbf{E}(\xi_t|\mathcal{F}_0^\xi)\|_2 dt < \infty \quad (\mathcal{F}_0^\xi = \sigma\{\xi_s, -\infty < s \leq 0\}).$$

Then

$$X^n \xrightarrow{\mathcal{L}} X,$$

where $X = (X_t)_{t \geq 0}$ is a solution of the stochastic equation (4.142) with respect to some Wiener process $W = (W_t)_{t \geq 0}$ and $\sigma = \left\{ 2 \int_0^\infty \mathbf{E}(\xi_t \xi_0) dt \right\}^{1/2}$.

Proof. First we note that

$$W_t^n = \frac{1}{\sqrt{n}} \int_0^{nt} \xi_s ds.$$

In view of condition 3) of the theorem and Lemma 4.2, the process $W^n = (W_t^n)_{t \geq 0}$ admits the decomposition

$$W_t^n = \frac{1}{\sqrt{n}} M_{nt} + \frac{1}{\sqrt{n}}(V_0 - V_{nt}), \qquad (4.143)$$

where $M = (M_t)_{t \geq 0}$ is a square-integrable martingale that is a process with stationary increments in the strict sense, $V = (V_t)_{t \geq 0}$ is a stationary process in the strict sense such that $(V_t, \xi_t)_{t \geq 0}$ is a stationary process in the strict sense. We set

$$M_t^n = \frac{1}{\sqrt{n}} M_{nt}, \quad V_t^n = \frac{1}{\sqrt{n}} V_{nt}.$$

Hence and from (4.141) and (4.143),

$$X_t^n = X_0 + \int_0^t a(s, X_s^n)ds + \int_0^t b(s, X_s^n)dM_s^n - \int_0^t b(s, X_s^n)dV_s^n,$$

where the integrals with respect to dM^n and dV^n are stochastic integrals with respect to a square-integrable martingale and a semimartingale, respectively. Integration by parts in $\int_0^t b(s, X_s^n)dV_s^n$ leads to the relation

$$\int_0^t b(s, X_s^n)dV_s^n = -\int_0^t b_x(s, X_s^n)b(X_s^n)V_s^n\dot{W}_s^n ds + \alpha_t^n,$$

where all the remaining terms are gathered in α_t^n.

Thus,

$$X_t^n = X_0 + \int_0^t [a(s, X_s^n) + b_x(s, X_s^n)b(s, X_s^n)V_s^n\dot{W}_s^n]ds$$
$$+ \int_0^t b(s, X_s^n)dM_s^n + \alpha_t^n.$$

Since ξ is stationary and ergodic in the strict sense, it follows that $(V_t^n, \dot{W}_t^n)_{t\geq 0}$ is as well. Moreover,

$$\mathbf{E} V_t^n \dot{W}_t^n = \mathbf{E}\Big(\frac{1}{\sqrt{n}}V_{nt}\sqrt{n}\xi_{nt}\Big) = \mathbf{E} V_0 \xi_0.$$

It follows from Lemma 4.2 that $V_0 = \int_0^\infty \mathbf{E}(\xi_u|\mathcal{F}_0^\xi)du$. Consequently, $\mathbf{E} V_0 \xi_0 = \int_0^t \mathbf{E}(\xi_u \xi_0)du = \sigma^2/2$. Further,

$$\int_0^t V_s^n \dot{W}_s^n ds = \int_0^t V_{ns}\xi_{ns}ds = \frac{1}{n}\int_0^{nt} V_s \xi_s ds$$

and hence by the Birkhoff-Khinchin ergodic theorem, as $n \to \infty$ we have **P**-a.s.

$$\int_0^t V_s^n \dot{W}_s^n ds \to t\frac{\sigma^2}{2}.$$

Hence, taking into account the analogue of Pólya's Theorem (see Liptser and Shiryaev 1986, Problem 5.4.2) it is not hard to deduce that for any $T > 0$ as $n \to \infty$ we have **P**-a.s.

$$\sup_{t\leq T}\Big|\int_0^t V_s^n \dot{W}_s^n ds - t\frac{\sigma^2}{2}\Big| \to 0.$$

Next, the relative compactness of the family of distributions X^n, $n \geq 1$, is established, which enables us to verify that

$$\sup_{t\leq T}|\alpha_t^n| \xrightarrow{\mathbf{P}} 0, \quad \forall T > 0.$$

This relation makes it possible to establish weak convergence $\bar{X}^n \xrightarrow{\mathcal{L}} X$ instead of $X^n \xrightarrow{\mathcal{L}} X$, where $\bar{X}_t^n = X_t^n - \alpha_t^n$. The weak convergence

$$\bar{X}^n \xrightarrow{\mathcal{L}} X$$

is established via Theorem 4.36. □

§5. Diffusion Approximation for Semimartingales with Normal Reflection in a Convex Domain

5.1. The Skorokhod Problem. Let $D = D_{[0,\infty)}(E^d)$ be the space of right-continuous vector-valued functions $X = (X_t)_{t \geq 0}$, $X_t = (X_1(t), \ldots, X_d(t))$ with left limits. We set

$$D(O) = \{X \in D : X_t \in \bar{O}, \ t \geq 0\},$$

where $\bar{O}$ is the closure of the convex domain $O \in E^d$ (∂O is the boundary of the domain O).

Definition. By a solution of the Skorokhod normal reflection problem inside a domain O for $X \in D$ with $X_0 \in \bar{O}$ is meant a function $Y = (Y_t)_{t \geq 0} \in D(O)$ such that the function

$$\phi = Y - X$$

possesses the following properties:

1) $\phi = (\phi_t)_{t \geq 0} \in D$, $\phi_0 = 0$,
2) the total variation $\text{Var}(\phi)_t$ of the function ϕ_s, $0 \leq s \leq t$, is finite for each $t > 0$,
3) for any continuous bounded function $f : \bar{O} \to E^d$ with $f(y) = 0$ for $y \in \partial O$

$$\int_0^t (f(Y_s), d\phi_s) = 0, \ t > 0$$

$((\cdot, \cdot)$ is the inner product),
4) for any function $\tilde{Y} = (\tilde{Y}_t)_{t \geq 0} \in D(0)$ the function $U_t = \int_0^t (\tilde{Y}_s - Y_s, d\phi_s)$, $t \geq 0$, is non-decreasing,
5) the function ϕ is adapted with the filtration $\mathbf{D} = (\mathcal{D}_t)_{t \geq 0}$ with $\mathcal{D}_t = \bigcap_{\epsilon > 0} \sigma\{X_s, \ s \leq t + \epsilon\}$.

The function ϕ with the properties indicated above is said to be associated with the function $Y \in D(0)$. This function gives a "reflection" along the normal inside the domain O of the function X (see property 4) for ϕ).

The fact of the existence of a solution of the Skorokhod problem is established under two conditions.

(α) There exist a unit vector e and a constant $c > 0$ such that $(e, n) \geq c$ for any vector $n \in \bigcup_{y \in \partial O} \mathcal{N}_y(O)$, where $\mathcal{N}_y(O)$ is the set of all internal vectors at the point $y \in \partial O$.

(β) There exist constants $\epsilon > 0$ and $\delta > 0$ which define at each point $x \in \partial O$ an open ball $B_\epsilon(x_0)$ with center at x_0 and $|x - x_0| \leq \delta$

$$B_\epsilon(x_0) = \{y \in E^d : |y - x_0| < \epsilon\}$$

such that $B_\epsilon(x_0) \subset O$.

Theorem 4.41. *Let O be a convex domain in E^d.*

1) If condition (α) is satisfied, then for any function $X \in D$ with $X_0 \in \bar{O}$ there exists a unique solution of the Skorokhod normal reflection problem

$$Y = X + \phi,$$

where the function $\phi = \phi(X)$ associated with $Y \in D(O)$ is continuous in the Skorokhod topology at each point $X \in D$ and

$$\operatorname{Var}(\phi(X))_t \leq L \sup_{s \leq t} |X_s|, \quad t > 0, \tag{4.144}$$

where the constant L depends only on the c in condition (α).

2) If condition (β) is satisfied, then assertion 1) remains true with the exception of inequality (4.144).

Remark 1. Every bounded domain O satisfies condition (β).

Remark 2. If X is a continuous function, then $\phi(X)$ is a continuous function.

5.2. Semimartingale with Normal Reflection. Suppose that the convex domain $O \subseteq E^d$ satisfies one of the conditions (α) or (β). Then for each $X \in D = D_{[0,\infty)}(E^d)$ with $X_0 \in \bar{O}$ there is a solution of the normal reflection problem:

$$Y_t = X_t + \phi_t(X). \tag{4.145}$$

We set $\mathcal{D}_t = \sigma\{X_s,\ s \leq t\}$ and define a probability measure $\mathbf{Q}$ on $(\mathbf{D}, \mathcal{D}_\infty)$. We denote by $\mathcal{D}^{\mathbf{Q}}_\infty$ the completion of $\mathcal{D}_\infty$ with respect to $\mathbf{Q}$ and also define the filtration $\mathbf{D}^{\mathbf{Q}}_+ = (\mathcal{D}^{\mathbf{Q}}_{t+})_{t \geq 0}$, where $\mathcal{D}^{\mathbf{Q}}_{t+} = \cap_{\epsilon > 0} \mathcal{D}_{t+\epsilon} \vee \mathcal{N}$, $\mathcal{N}$ is the collection of sets of $\mathcal{D}^{\mathbf{Q}}_\infty$ of zero $\mathbf{Q}$ measure. It follows from the properties of the function $\phi(X) = (\phi_t(X))_{t \geq 0}$ that $\phi(X)$ is a $\mathbf{D}^{\mathbf{Q}}_+$-adapted process and this means that $\phi(X)$ is a semimartingale with respect to $\mathbf{D}^{\mathbf{Q}}_+$.

Let us assume that $X = (X_t)_{t \geq 0}$ is a semimartingale on a stochastic basis $(\mathbf{D}, \mathcal{D}^{\mathbf{Q}}_\infty, \mathbf{D}^{\mathbf{Q}}_+, \mathbf{Q})$. We define the filtration $\mathbf{F}^Y_+ = (\mathcal{F}^Y_{t+})_{t \geq 0}$ with

$$\mathcal{F}^Y_{t+} = \bigcap_{\epsilon > 0} \sigma\{Y_s,\ s \leq t + \epsilon\} \vee \mathcal{N},$$

($\mathcal{N}$ is the collection of sets in $\mathcal{D}^{\mathbf{Q}}_\infty$ of zero $\mathbf{Q}$ measure).

If X is a continuous semimartingale, then in its decomposition

$$X_t = X_0 + A_t + M_t \tag{4.146}$$

the process of locally bounded variation $A = (A_t)_{t \geq 0}$ and the local martingale $M = (M_t)_{t \geq 0}$ have continuous trajectories. Furthermore, in accordance

Chapter 4.II. The Invariance Principle and Diffusion Approximation

with Remark 2 to Theorem 4.41, $\phi(X) = (\phi_t(X))_{t \geq 0}$ also has continuous trajectories. It follows from (4.146) and (4.145) that Y is a continuous process with

$$Y_t = X_0 + A_t + M_t + \phi_t(X).$$

The process Y is a semimartingale with respect to $\mathbf{D}_+^\mathbf{Q}$. Since $\mathbf{F}_+^Y \subset \mathbf{D}_+^\mathbf{Q}$, it follows from Theorem 3.22 that Y is a semimartingale with respect to $\mathbf{F}_+^Y$. The decomposition of Y with respect to $\mathbf{F}_+^Y$ has the form

$$Y_t = X_0 + \bar{A}_t + \bar{M}_t, \tag{4.147}$$

where $\bar{A} = (\bar{A}_t)_{t \geq 0}$ and $\bar{M} = (\bar{M}_t)_{t \geq 0}$ are continuous $\mathbf{F}_+^Y$-adapted processes of locally bounded variation and a local martingale, respectively. Here

$$\langle \bar{M} \rangle = \langle M \rangle. \tag{4.148}$$

Let us assume that A is an $\mathbf{F}_+^Y$-adapted process. We set

$$\bar{\phi} = A - \bar{A}.$$

Then

$$Y_t = X_0 + A_t + \bar{M}_t + \bar{\phi}_t.$$

Here, $\bar{\phi}$ is an $\mathbf{F}_+^Y$-adapted process which is the associated process for Y in the normal reflection problem for the semimartingale $\bar{X} = (\bar{X}_t)_{t \geq 0}$ with $\bar{X}_t = X_0 + A_t + \bar{M}_t$.

We now give the statement of the martingale problem of diffusion type with normal reflection in the domain O.

Let $b = b(t, X)$ and $c = c(t, X)$, $t \in \mathbb{R}_+$, $X \in D$, be a vector and a symmetric positive semidefinite matrix of dimension d and $d \times d$, respectively, with $\mathcal{B}(\mathbb{R}_+) \otimes \mathcal{D}_\infty$-measurable elements that are $\mathcal{D}_t$-measurable for each $t \geq 0$.

Definition. The probability measure $\mathbf{Q}$ solves the martingale problem of diffusion type with drift coefficient $b(t, Y)$ and diffusion coefficient $c(t, Y)$ and normal reflection in the domain O ($Y = X + \phi(X)$) if

$$\int_0^t (\|b(s, X)\| + \|c(s, X)\|) ds < \infty, \quad \mathbf{Q}\text{-a.s.}, \ t > 0$$

and the stochastic process $\bar{M} = (\bar{M}_t)_{t \geq 0}$ with

$$\bar{M}_t = Y_t - X_0 - \int_0^t b(s, Y) ds - \bar{\phi}_t,$$

where $\bar{\phi} = (\bar{\phi}_t)_{t \geq 0}$ is the function associated with $Y = D(O)$, is a continuous local martingale with respect to $(\mathbf{F}_+^Y, \mathbf{Q})$ with quadratic characteristic

$$\langle \bar{M} \rangle_t = \int_0^t c(s, Y) ds.$$

It follows from Doob's theorem that the stochastic process Y under consideration (defined if necessary on some extended probability space) is a weak solution of the Itô stochastic equation

$$Y_t = X_0 + \int_0^t b(s,Y)ds + \int_0^t c^{1/2}(s,Y)dW_s + \bar{\phi}_t, \tag{4.149}$$

with respect to some d-dimensional Wiener process $W = (W_t)_{t \geq 0}$ with independent components.

The given martingale problem has a unique solution if the set of measures solving this problem with the same restrictions on the σ-algebra $\mathcal{D}_0$ consists of one point. The uniqueness of the solution of the martingale problem is equivalent to the weak uniqueness of the solution of equation (4.149).

Suppose that equation (4.149) has a weak solution. We set

$$Z_t = X_0 + \int_0^t b(s,Y)ds + \int_0^t c^{1/2}(s,Y)dW_s.$$

Then

$$Y_t = Z_t + \phi(Z_t),$$

where $\phi(Z)$ is the function associated with Y in the problem of normal reflection in the domain O of the function $Z = (Z_t)_{t \geq 0}$, that is, $\phi_t(Z) = \bar{\phi}_t$. Hence it follows that Z is a weak solution of the equation

$$Z_t = X_0 + \int_0^t b(s, Z + \phi(Z))ds + \int_0^t c^{1/2}(s, Z + \phi(Z))dW_s. \tag{4.150}$$

The converse is also valid, namely, if Z is a weak solution of (4.150), then Y is a weak solution of (4.149), where both equations possess or do not possess weak uniqueness simultaneously.

5.3. Let us now consider the problem of the diffusion approximation of a sequence of semimartingales with normal reflection in a convex domain $O \subseteq E^d$.

For each $n \geq 1$ let $X^n = (X_t^n)_{t \geq 0}$ be a semimartingale defined on a stochastic basis $(\Omega^n, \mathcal{F}^n, \mathbf{F}^n, \mathbf{P}^n)$ with $X_t^n \in E^d$ and triple of predictable characteristics $T^n = (B^n, C^n, \nu^n)$, corresponding to which there is the canonical representation

$$X_t^n = X_0^n + B_t^n + X_t^{nc} + \int_0^t \int_{|x|>1} x d\mu^n + \int_0^t \int_{|x| \leq 1} x d(\mu^n - \nu^n)$$

(μ^n is the jump measure of X^n) and

$$X_0^n \in \bar{O}.$$

For X^n there is a solution of the normal reflection problem:

Chapter 4.II. The Invariance Principle and Diffusion Approximation 241

$$Y_t^n = X_t^n + \phi_t(X^n).$$

We give the corresponding results concerning weak convergence

$$Y^n \xrightarrow{\mathcal{L}} Y, \qquad (4.151)$$

where Y is a process of diffusion type with reflection in the domain determined by equation (4.149).

Theorem 4.42. *Let condition* (α) *hold for the domain* O.

If $g(t, Y)$ denotes any of the elements of the vector $b(t, Y)$ or the matrix $c^{1/2}(t, Y)$, then

$$|g(t, Y)| \leq L(1 + \sup_{s<t} |Y_s|). \qquad (4.152)$$

For each $t \in S$ (S is an everywhere dense subset of $\mathbb{R}_+$, $\int_{\mathbb{R}\setminus S} dt = 0$) the elements of $b(t, Y)$ and $c(t, Y)$ are continuous in the Skorokhod metric at each point $Y \in D(O) \cap C$. Equation (4.149) has a unique weak solution. Furthermore,

$$X_0^n \xrightarrow{d} X_0$$

for any $t > 0$, $\epsilon > 0$ and $a \in (0, 1]$

$$\lim_n \mathbf{P}^n \left(\int_0^t \int_{|x|>a} d\nu^n > \epsilon \right) = 0,$$

$$\lim_n \mathbf{P}^n \left(\sup_{s \leq t} \left| B_s^n - \int_0^s b(u, Y^n) du \right| > \epsilon \right) = 0,$$

$$\lim_n \mathbf{P}^n \left(\sup_{s \leq t} \left| \langle M^{na} \rangle_s - \int_0^s c(u, Y^n) du \right| > \epsilon \right) = 0,$$

where

$$\langle M^{na} \rangle_t = C_t^n + \int_0^t \int_{|x| \leq a} xx^* d\nu^n - \sum_{s \leq t} \int_{|x| \leq a} x\nu^n(\{s\}, dx) \left(\int_{|x| \leq a} x\nu^n(\{s\}, dx) \right)^*,$$

* *denotes the transpose.*

Then (4.151) *holds.*

Theorem 4.43. *The assertion of Theorem 4.42 remains true if instead of condition* (α) *on the domain O condition* (β) *is assumed satisfied, and instead of condition* (4.152) *the condition*

$$|g(t, Y)| \leq L.$$

5.4.

Example. Let $A = (A_t)_{t \geq 0}$, $D = (D_t)_{t \geq 0}$ be point processes, and $Q = (Q_t)_{t \geq 0}$ a solution of the equation

$$Q_t = A_t - \int_0^t I(Q_{s-} > 0) dD_s,$$

where $\int_0^t I(Q_{s-} > 0) dD_s = \sum_{s \le t} I(Q_{s-} > 0) \Delta D_s.$

The processes A and D can be interpreted as processes of arrival and of service, and Q as the queuing process in a queuing model with an autonomous service. In a simple model of such a form, it is assumed that A and D are independent Poisson processes with intensities λ and μ respectively. Let us complicate the model by assuming that the processes A and D possess the property:

$$\sum_{t>0} \Delta A_t \Delta D_t = 0,$$

that is, the jumps of A and D are disjoint. Furthermore, for a fixed value of the queue Q_t at time t, the intensities of the processes A and D depend on Q_t:

$$\lambda = \lambda(\epsilon Q_t), \quad \mu = \mu(\epsilon Q_t),$$

where ϵ is a small parameter, $\lambda = \lambda(x)$, $\mu = \mu(x)$ are non-negative continuously differentiable functions with bounded uniformly continuous derivatives $\lambda'(x)$ and $\mu'(x)$.

If λ and μ are constants and $\lambda = \mu$, then a stationary distribution for Q_t does not exist in such a model. In the general case for

$$\lambda(0) = \mu(0) \tag{4.153}$$

it is also unnatural to assume the existence of a stationary distribution. In connection with this, to study a queue on large time intervals under the assumption (4.153), let us consider the normalized process $Y^\epsilon = (Y_t^\epsilon)_{t \ge 0}$ with

$$Y_t^\epsilon = \sqrt{\epsilon} Q_{t/\epsilon}. \tag{4.154}$$

In this case

$$Y^\epsilon \xrightarrow{\mathcal{L}} Y, \tag{4.155}$$

where $Y = (Y_t)_{t \ge 0}$ is the diffusion process with normal reflection in the domain $O = \{x > 0\}$ determined by the stochastic equation

$$Y_t = \int_0^t [\lambda'(0) - \mu'(0)] Y_s ds + \sqrt{2\lambda(0)} W_t + \bar{\phi}_t, \tag{4.156}$$

($W = (W_t)_{t \ge 0}$ is a Wiener process, $\bar{\phi} = (\bar{\phi}_t)_{t \ge 0}$ is the function associated with Y).

In order to establish this fact, it is sufficient to remark that

$$Q_t = X_t - \inf_{s \le t} X_s, \quad X_t = A_t - D_t$$

Chapter 4.II. The Invariance Principle and Diffusion Approximation   243

and hence Q is a solution of the Skorokhod normal reflection problem in the domain $O = \{x > 0\}$ and $\inf_{s \leq t} X_s = \phi_t(X)$ is the function associated with Q.

Further,

$$Y_t^\epsilon = \sqrt{\epsilon}(X_{t/\epsilon} - \inf_{s \leq t/\epsilon}(X_s))$$
$$= \sqrt{\epsilon}X_{t/\epsilon} + \phi_{t/\epsilon}(\sqrt{\epsilon}X) = \sqrt{\epsilon}X_{t/\epsilon} + \phi_t(\sqrt{\epsilon}X_{\cdot/\epsilon}).$$

An application of Theorem 4.42 with $\epsilon = 1/n$ leads to the weak convergence (4.155) with the diffusion process Y determined by the stochastic equation (4.156) with normal reflection. Moreover,

$$Y \stackrel{\mathcal{L}}{=} |\tilde{Y}|,$$

where $\tilde{Y} = (\tilde{Y}_t)_{t \geq 0}$ is the Gaussian diffusion process that is the solution of the linear Itô equation

$$\tilde{Y}_t = \int_0^t [\lambda'(0) - \mu'(0)]\tilde{Y}_s ds + \sqrt{2\lambda(0)}\tilde{W}_t$$

with respect to the Wiener process $\tilde{W} = (\tilde{W}_t)_{t \geq 0}$.

Commentary to Chapter 4

I. §1. See the book (Shiryaev 1980).

§2. Concerning the different types of convergence see, for example, (Shiryaev 1980). The basic information on the Skorokhod topology is contained in the book (Billingsley 1968); see also (Lindvall 1973) and (Stone 1963).

§3. Theorem 4.1 was established in (Lindeberg 1922). Concerning Theorems 4.2, 4.4, see (Gnedenko and Kolmogorov 1949); (Liptser and Shiryaev 1982); (Jacod and Mémin 1980).

§4. The proof of Theorems 4.5 and 4.7 is in the books (Lipster and Shiryaev 1986); (Jacod and Shiryaev 1987). Theorems 4.8–4.10 are due to Jacod (Jacod 1983); see also (Jacod and Shiryaev 1987).

§5. Regarding Theorem 4.11, see (Jacod and Shiryaev 1987). Theorems 4.12, 4.13 are established in (Liptser and Shiryaev 1986), and Theorems 4.14, 4.15 in (Liptser and Shiryaev 1986) and (Jacod and Shiryaev 1987). Theorem 4.16 is in Chap. 8 of (Liptser and Shiryaev 1986), Theorem 4.18 is in (Jacod and Shiryaev 1987), Theorem 4.19 is in (Liptser and Shiryaev 1986).

§6. Necessary and sufficient conditions for the relative compactness of a family of probability measures (Theorem 4.20) are presented in (Billingsley 1968). Theorem 4.21 is due to Kolmogorov-Chentsov and is also presented in (Billingsley 1968). Theorem 4.22, which is one of the most effective instruments for proving relative compactness, is due to Aldous (Aldous 1978), (Aldous 1981).

§7. Theorems 4.23 and 4.24 can be found in (Liptser and Shiryaev 1986); (Jacod and Shiryaev 1987); see also (Butov 1983); (Gregelionis and Mikulyavichus 1981a),

(Gregelionis and Mikulyavichus 1981b); (Gikhman and Skorokhod 1982). Regarding Theorem 4.25, see (Lebedev 1981) and (Jacod and Shiryaev 1987). Theorems 4.26–4.28 can be found in (Jacod and Shiryaev 1987).

§8. The material of this section is borrowed from the book (Jacod and Shiryaev 1987, Chap. 3.)

II. §1. The results of this section are classical. Different versions of them occur in the papers (Rosenblatt 1956), (Volkonskij and Rozanov 1959), (Volkonskij and Rozanov 1961); (Statulyavichus 1960); (Serfling 1968); (Gordin 1969); (McLeish 1975). Additional bibliographical information is given in the books (Ibragimov and Linnik 1965) and (Hall and Heyde 1980). Theorems 4.31 and 4.32 can be found in the book (Liptser and Shiryaev 1989) and are due to Chikin (Chikin 1984); see also (Dürr and Goldstein 1986). The Markov case was considered in the papers (Bhattacharya 1982) and (Touati 1983); see also (Gordin and Lifshits 1978).

§2. The model of the Bogolyubov stochastic averaging principle was borrowed from (Venttsel' and Freidlin 1979). A similar queueing model in the vector case was studied in the paper (Kogan, Liptser and Smorodinskij 1986).

§3. The examples of this section come from (Liptser and Shiryaev 1989, Chaps. 8, 9); (Kogan, Liptser and Smorodinskij 1986). Concerning averaging in models with diffusion, see (Venttsel' and Freidlin 1979).

§4. The results of this section make use of the works (Krasnosel'skij and Pokrovskij 1983); (Doss 1977); (Matskyavichus 1985); (Gyöngy 1988).

§5. The Skorokhod problem was first formulated for half-spaces by Skorokhod (Skorokhod 1961), (Skorokhod 1963) and solved by Tanaka (Tanaka 1979) and Anulova and Liptser (Anulova and Liptser 1990) for a convex domain. Limit theorems in a domain were studied in (Grigelionis and Mikulyavichus 1983); (Anulova and Liptser 1990). The diffusion approximation with reflection with an application to queueing theory was considered in (Borovkov 1980); (De Zelicourt 1981); (Grigelionis and Mikulevicius 1984).

References[*]

Aldous, D.J. (1978): Stopping times and tightness. Ann. Probab. *6*, No. 2, 335–340. Zbl. 391.60007

Aldous, D.J. (1981): Weak convergence and the general theory of processes. Dept. of Statistics. Univ. California, Berkeley

Anulova, S.V., Liptser, R.Sh. (1990): Diffusion approximation for processes with normal reflection. Teor. Veroyatn. Primen. *35*, No. 2, 417–430. [English transl.: Theory Probab. Appl. *35*, No. 3, 411–423 (1990)] Zbl. 713.60081

Bhattacharya, R.N. (1982): On the functional limit theorem and the law of the iterated logarithm for Markov processes. Z. Wahrscheinlichkeitstheorie Verw. Gebiete *60*, 185–201. Zbl. 468.60034

Billingsley, P. (1968): Convergence of Probability Measures. Wiley, New York. Zbl. 172,212

Borovkov, A.A. (1980): Asymptotic Methods in Queueing Theory. Nauka, Moscow. [English transl.: Wiley, New York, 1984] Zbl. 479.60006

---

[*] For the convenience of the reader, references to reviews in Zentralblatt für Mathematik (Zbl.), compiled using the MATH database, have, as far as possible, been included in this bibliography.

Butov, A.A. (1983): On the question of weak convergence of a sequence of semi-martingales to a process of diffusion type. Usp. Mat. Nauk *38*, No. 5, 181–182. [English transl.: Russ. Math. Surv. *38*, No. 5, 135–136 (1983)] Zbl. 532.60040

Chikin, D.O. (1989): Functional limit theorem for stationary processes: martingale approach. Teor. Veroyatn. Primen. *34*, No. 4, 731–741. [English transl.: Theory Probab. Appl. *34*, No. 4, 668–678 (1989)] Zbl. 693.60024

De Zelicourt, C. (1981): Une méthode de martingales pour la convergence d'une suite de processus de sauts markoviens vers une diffusion associée à une condition frontière. Application aux systèmes de files. Ann. Inst. Henri Poincaré, N. Ser. Sect. B *17*, No. 4, 351–375. Zbl. 476.60078

Donsker, M. (1951): An invariance principle for certain probability limit theorems. Mem. Am. Math. Soc. *6*. Zbl. 42,376

Doss, H. (1977): Liens entre équations différentielles stochastiques et ordinaires. Ann. Inst. Henri Poincaré, N. Ser. Sect. B *13*, No. 2, 99–125. Zbl. 359.60087

Dürr, D., Goldstein, S. (1986): Remarks on the central limit theorem for weakly dependent random variables. Lect. Notes Math. *1158*, pp. 104–118. Zbl. 582.60036

Gikhman, I.I., Skorokhod, A.V. (1982): Stochastic Differential Equations and their Applications. Naukova Dumka, Kiev (Russian). Zbl. 557.60041

Gnedenko, B.V. (1939): On a theory of limit theorems for sums of independent random variables. Izv. Akad. Nauk SSSR, Ser. Mat., 181–232, 643–647. Zbl. 24,263

Gnedenko, B.V., Kolmogorov, A.N. (1949): Limit Distributions for Sums of Independent Random Variables. Gostekhizdat, Moscow-Leningrad. [English transl.: Addison-Wesley, Cambridge, Mass, 1954] Zbl. 56,360

Gordin, M.I. (1969): On the central limit theorem for stationary processes. Dokl. Akad. Nauk SSSR *188*, No. 4, 739–741. [English transl.: Sov. Math. Dokl. *10*, 1174–1176 (1969)] Zbl. 212,500

Gordin, M.I., Lifshits, B.A. (1978): The central limit theorem for stationary Markov processes. Dokl. Akad. Nauk SSSR *239*, No. 4, 766–767. [English transl.: Sov. Math. Dokl. *19*, No. 2, 392–394 (1978)] Zbl. 395.60057

Grigelionis, B.I., Mikulyavichus, R. (1981a): On the weak convergence of semimartingales. Litov. Mat. Sb. *21*, No. 3, 9–24. [English transl.: Lith. Math. J. *21*, No. 3, 213–224 (1982)] Zbl. 498.60005

Grigelionis, B.I., Mikulyavichus, R. (1981b): On the weak convergence of stochastic point processes. Litov. Mat. Sb. *21*, No. 4, 41–55. [English transl.: Lith. Math. J. *21*, No. 4, 297–301 (1982)] Zbl. 491.60033

Grigelionis, B.I., Mikulyavichus, R. (1983): On weak convergence of random processes with boundary conditions. Lect. Notes Math. *972*, pp. 260–275. Zbl. 511.60033

Grigelionis, B.I., Mikulevicius, R. (1984): On the diffusion approximations in queueing theory. Proc. Conf. on Teletraffic, Moscow.

Gyöngy, I. (1988): On the approximation of stochastic differential equations. Stochastics *23*, 331–352. Zbl. 635.60071

Hall, P., Heyde, C.C. (1980): Martingale Limit Theory and its Application. Academic Press, New York. Zbl. 462.60045

Ibragimov, I.A., Linnik, Yu.V. (1965): Independent and Stationary Sequences of Random Variables. Nauka, Moscow. [English transl.: Walters-Noordhoff, Groningen, 1971] Zbl. 154,422

Jacod, J. (1983): Processus à accroissements indépendants: une condition nécessaire et suffisante de convergence en loi. Z. Wahrscheinlichkeitstheorie Verw. Gebiete *63*, 109–136. Zbl. 526.60065

Jacod, J., Mémin, J. (1980): Sur la convergence des semimartingales vers un processus à accroissements indépendants. Lect. Notes Math. *784*, pp. 227–248. Zbl. 433.60034

Jacod, J., Shiryaev, A.N. (1987): Limit Theorems for Stochastic Processes. Springer, Berlin Heidelberg New York. Zbl. 635.60021

Khinchin, A.Ya. (1938): Limit Laws for Sums of Independent Random Variables. Gostekhizdat: Moscow

Kogan, Ya.A., Liptser, R.Sh., Smorodinskij, A.V. (1986): The Gaussian diffusion approximation of closed Markov models for computer communication networks. Probl. Peredachi Inf. 22, No. 1, 49–65. [English transl.: Probl. Inf. Transm. 22, 38–51 (1986)] Zbl. 594.60043

Krasnosel'skij, M.A., Pokrovskij, A.V. (1983): Systems with Hysterisis. Nauka: Moscow. [English transl.: Springer, Berlin Heidelberg New York, 1989] Zbl. 665.47038

Lebedev, V.A. (1981): On relative compactness of families of distributions of semimartingales. Teor. Veroyatn. Primen. 26, No. 1, 143–151. [English transl.: Theory Probab. Appl. 26, No. 1 140–148 (1981)] Zbl. 471.60056

Lévy, P. (1939): Sur une loi de probabilité analogue à celle de Poisson et sur un sousgroupe important du groupe des lois indéfiniment divisibles. Bull. Sci. Math., II. Ser. 63, 247–268. Zbl. 23,57

Lindeberg, J.W. (1922): Eine neue Herleitung des Exponentialgesetzes in der Wahrscheinlichkeitsrechnung. Math. Z. 15, 211–225. Jbuch FdM 48,602

Lindvall, T. (1973): Weak convergence of probability measures and random functions in the function space $D[0, \infty]$. J. Appl. Probab. 10, 109–121. Zbl. 258.60008

Liptser, R.Sh., Shiryaev, A.N. (1982): On weak convergence of semimartingales to stochastically continuous processes with independent and conditionally independent increments. Mat. Sb., Nov. Ser. 116 (158), No. 3 (11), 331–358. [English transl.: Math. USSR Sbornik 44 (1983), No. 3, 299–323 (1983)] Zbl. 484.60024

Matskyavichus, V. (1985): Stability of solutions of symmetric stochastic differential equations. Litov. Mat. Sb. 25, No. 4, 72–84 [English transl.: Lit. Math. J. 25, 343–352 (1985)] Zbl. 588.60050

McLeish, D.L. (1975): Invariance principle for dependent variables. Z. Wahrscheinlichkeitstheorie Verw. Gebiete 32, 165–178. Zbl. 288.60034

Rosenblatt, M.A. (1956): A central limit theorem and a strong mixing condition. Proc. Natl. Acad. Sci. USA 42, No. 1, 43–47. Zbl. 70,138

Rotar', V.I. (1978): On a non-classical estimate of the accuracy of approximation in the central limit theorem. Mat. Zametki 23, No. 1, 143–154. [English transl.: Math. Notes 23, No. 1, 77–83 (1978)] Zbl. 386.60025

Sam Lazaro, J. de, Meyer, P.A. (1975): Questions de théorie des flots. Sémin. Probab. IX. Univ. Strasbourg. Lect. Notes Math. 465, pp. 2–96. Zbl. 311.60019

Serfling, R.J. (1968): Contributions to central limit theory for dependent variables. Ann. Math. Stat. 39, 1158–1175. Zbl. 176,480

Shiryaev, A.N. (1980): Probability. Nauka, Moscow. [English transl.: Springer, Berlin Heidelberg New York, 1984] Zbl. 536.60001

Shiryaev, A.N. (1986): The Theory of Martingales. Nauka, Moscow

Shiryaev, A.N. (1989): Theory of Martingales. Kluwer, Dordrecht

Skorokhod, A.V. (1961): Stochastic equations for diffusion processes in a bounded domain. Teor. Veroyatn. Primen. 6, No. 3, 287–298. [English transl.: Theory Probab. Appl. 6, 264–274 (1961)] Zbl. 215,535

Skorokhod, A.V. (1963): Stochastic equations for diffusion processes in a bounded domain. II. Teor. Veroyatn. Primen. 7, No. 1, 5–25. [English transl.: Theory Probab. Appl. 7, 3–23 (1962)] Zbl. 201,493

Statulyavichus, V.A. (1960): Some new results for sums of weakly dependent random variables. Teor. Veroyatn. Primen. 5, No. 2, 233–234.

Stone, C. (1963): Weak convergence of stochastic processes defined on semi-infinite time intervals. Proc. Am. Math. Soc. 14, 694–696. Zbl. 116,356

Tanaka, H. (1979): Stochastic differential equations with reflecting boundary condition in convex regions. Hiroshima Math. J. 9, No. 1, 163–177. Zbl. 423.60055

Touati, A. (1983): Théorèmes de limite centrale fonctionelle pour les processus de Markov. Ann. Inst. Henri Poincaré, N. Ser. Sect. B *19*, 43–59. Zbl. 511.60029

Venttsel' A.D., Freidlin, M.I. (1979): Random Perturbations of Dynamical Systems. Nauka, Moscow. [English transl.: Springer, Berlin Heidelberg New York, 1984] Zbl. 522.60055

Volkonskij, V.A., Rozanov, Yu.A. (1959): Some limit theorems for random functions. I. Teor. Veroyatn. Primen. *4*, No. 2, 186–207. [English transl.: Theory Probab. Appl. *4*, 178–197 (1960)] Zbl. 92,335

Volkonskij, V.A., Rozanov, Yu.A. (1961): Some limit theorems for random functions. II. Teor. Veroyatn. Primen. *6*, No. 2, 202–214. [English transl.: Theory Probab. Appl. *6*, 186–198 (1961)] Zbl. 108,313

Zolotarev, V.M. (1967): A generalization of the Lindeberg-Feller theorem. Teor. Veroyatn. Primen. *12*, No. 4, 666–667. [English transl.: Theory Probab. Appl. *12*, 608–618 (1967)] Zbl. 234.60031

Author Index

Aldous, D.J.   201, 243, 244
Alyushina, L.A.   76, 91
Anulova, S.V.   35, 244

Bachelier, L.   7
Baklan, V.V.   76, 90
Barlow, M.T.   49, 50, 72
Bass, R.F.   72
Bell, D.R.   91, 101, 107, 108
Bernoulli, J.   166
Bernshtein, S.N.   26, 36, 111, 154
Bhattacharya, R.N.   244
Bichteler, K.   91, 108
Billingsley, P.   52, 72, 243, 244
Birkhoff, G.D.   217
Bismut, J.M.   91, 99, 100, 103, 109
Blagoveshchenskij, Yu.N.   54, 56, 72
Bobrov, A.A.   174
Bogolyubov, N.N.   226
Bohlmann, G.   111, 154
Borovkov, A.A.   244
Brown, R.   7
Butov, A.A.   243, 245

Cartan, H.   47, 72
Chaleyot-Maurel, M.   101
Chebyshev, P.L.   169, 170, 171
Chentsov, N.N.   243
Chikin, D.O.   244
Chitashvili, R.Ya.   42, 48, 72, 74
Conway, E.D.   51, 72
Courrège, P.   153, 154

Daletskij, Yu.L.   76, 90
Dellacherie, C.   125, 152, 153, 154
De Moivre, A.   167, 168
De Zelicourt, C.   244, 245
Doléans-Dade, C.   153, 155
Donsker, M.   13, 185, 245
Doob, J.   152, 153, 155
Doss, H.   47, 72, 244, 245
Dunford, N.   77, 90
Dürr, D.   244, 245

Einstein, A.   7
Elliott, R.J.   153, 155

Fedorenko, I.V.   46, 72
Fisk, D.L.   153, 155
Freidlin, M.I.   36, 54, 56, 72, 244, 247

Gaveau, B.   109
Genis, I.L.   35, 36
Gikhman, I.I.   26, 27, 35, 36, 43, 153, 155, 244, 245
Girsanov, I.V.   30, 31, 36, 141, 154, 155
Gnedenko, B.V.   176, 177, 243, 245
Goldstein, S.   244, 245
Gordin, M.I.   243, 244
Grigelionis, B.I.   153, 154, 155 243, 244, 245
Gyöngy, I.   47, 76, 90, 244, 245

Hall, P.   244, 245
Harrison, M.   65, 66, 72
Haussmann, U.   101, 109
Hellinger, E.   143, 144, 145
Heyde, C.C.   244, 245
Hida, T.   92, 109
Hilbert, D.   111, 155
Hörmander, L.   91, 101, 109

Ibragimov, I.A.   244
Ikeda, N.   27, 35, 36, 41, 48, 49, 51, 59, 61, 64, 66, 70, 72, 99, 105, 107, 109
Itô, K.   19, 27, 32, 36, 72, 92, 109, 129, 153, 155

Jacod, J.   91, 108, 152, 153, 154, 155, 177, 243, 245

Kabanov, Yu.M.   152, 153, 154
Kakutani, S.   154, 156
Kazamaki, N.   36
Khinchin, A.Ya.   174, 175, 214, 215, 216, 246
Kleptsyna, M.L.   44, 45, 46, 48, 72
Kogan, Ya.A.   244, 246
Kolmogorov, A.N.   27, 37, 111, 152, 156, 175, 177, 243
Krasnosel'skij, M.A.   47, 48, 60, 72, 234, 244, 246
Krylov, N.V.   35, 37, 42, 45, 48, 54, 55, 59, 73, 75, 76, 80, 82, 84, 85, 87, 89, 90, 92
Kunita, H.   153, 154, 156
Kusuoka, S.   91, 101

Ladyzhenskaya, O.A.   43, 73
Laplace, P.S.   168, 169
Lebedev, V.A.   51, 73, 246
Lebesgue, H.   140

Author Index

Lévy, P.   179, 181, 246
Liese, F.   156
Lifshits, B.A.   244, 245
Lindeberg, J.W.   173, 174, 243, 246
Lindvall, T.   246
Linnik, Yu.V.   244, 245
Lions, P.L.   35, 37, 69, 73
Liptser, P.Sh.   37, 39, 64, 67, 69, 73, 152, 153, 156, 177, 243, 244, 246
Lomnicki, A.   111, 156
Lyapunov, A.M.   171, 173

Malliavin, P.   91, 92, 109
Malyutov, M.B.   35, 37
Manabe, S.   45, 73
Markov, A.A.   171
Matskyavichyus, V.   47, 73, 244, 246
McKean, H.P.   64, 72, 73
McLeish, D.L.   244
Mel'nikov, A.V.   46, 73
Mémin, J.   153, 154, 177, 245
Métivier, M.   83, 90, 153, 156
Meyer, P.A.   152, 153, 154, 156, 246
Michel, D.   91, 101, 109
Mikulyavichyus, P.   243, 244, 245

Nakao, S.   41, 73
Nikol'skij, S.M.   37, 85, 90
Nisio, M.   43, 73
Norris, N.   91, 101, 107, 109
Novikov, A.A.   30, 37
Nualart, D.   91, 109

Olejnik, O.A.   91, 109
Orey, S.   153, 156

Pardoux, E.   76, 80, 90
Pettis, B.J.   77
Poisson, S.D.   169, 182
Pokrovskij, A.V.   47, 48, 60, 72, 234, 244
Portenko, N.I.   35, 37
Prokhorov, Yu.V.   170

Radkevich, E.A.   91, 109
Rajkov, D.A.   174
Rao, K.M.   153, 156
Reiman, M.   65, 66, 72
Rosenblatt, M.   244, 246
Rotar', V.I.   174, 246
Rozanov, Yu.A.   244, 247
Rozovskij, B.L.   76, 80, 82, 84, 85, 87, 89

Safonov, M.V.   59, 73

Sam Lazaro, J.   246
Schrödinger, E.   91
Schwartz, J.T.   77, 90
Serfling, R.J.   243, 246
Sevast'yanov, B.A.   58, 73
Shiga, T.   45, 73
Shigekawa, I.   91, 96, 109
Shiryaev, A.N.   37, 39, 64, 70, 73, 152, 153, 154, 177, 243, 244, 246
Skorokhod, A.V.   27, 35, 36, 43, 45, 51, 57, 64, 73, 91, 109, 153, 155, 243, 244, 246
Smoluchowski   7
Smorodinskij, A.V.   244
Solonnikov, V.A.   43, 73
Statulyavichus, V.A.   244, 246
Stirling   168
Stone, C.   243, 246
Stricker, C.   154, 156
Stroock, D.W.   43, 53, 61, 62, 67, 70, 74, 91, 101, 107, 109
Sussman, H.J.   47, 74
Sznitman, A.S.   35, 37, 69, 73

Tanaka, H.   35, 37, 48, 69, 74, 244
Tikhonov, A.N.   17
Torondzhadze, T.A.   42, 72, 74
Touati, A.   244, 247
Trauber, P.   109
Tsirel'son, B.S.   40, 49

Ural'tseva, N.N.   43, 73

Van Schuppen, J.H.   154, 156
Varadhan, S.R.S.   43, 63, 67, 71
Venttsel', A.D.   33, 37, 154, 157, 244
Veretennikov, A.Yu.   29, 37, 42, 43, 44, 45, 48, 58, 74, 92, 101, 110
Viot, M.   83, 90
Vishik, M.I.   87, 90
Volkonskij, V.A.   243, 247
von Mises, R.   111, 156
Watanabe, S.   27, 35, 36, 40, 41, 42, 43, 48, 49, 51, 59, 64, 72, 91, 99, 105, 107, 109, 110, 153, 154, 182
Wiener, N.   7, 92, 153
Wong, E.   154, 156

Yamada, T.   40, 41, 43, 45, 74
Yoeurp, Ch.   153, 157

Zakai, M.   91, 92, 93, 94, 95, 96, 97, 98, 100, 103, 109
Zolotarev, V.M.   174, 247
Zvonkin, A.K.   29, 37, 40, 42, 43, 49, 75

Subject Index

absolute continuity
– local   140
– (and singularity) of measures   147
– – in the case of discrete time   148
– – for Markov processes with a countable set of states   151
– – for point processes   149
– – for processes with independent increments   150
– – for semimartingales with Gaussian martingale part   149
– – for semimartingales with local uniqueness condition   152
asymptotic negligibility   173

bilinear form   94
Bogolyubov stochastic averaging principle   226
Brownian bridge   28
Brownian motion   7

canonical representation   125, 180, 190
characteristic function   171
coefficient
– of strong mixing   220
– of uniformly strong mixing   220
comparison theorems   45
compensator   120
– predictable   120
– of a process   120
– of a random measure   121
complete integrability condition   46
completely measurable subset   77
convergence, rate of   58
cumulant   193

decomposition
– Doob-Meyer   119
– Lebesgue   140
– of local martingales   123
differentiation
– in L_p norms   54
– with respect to initial data   54
– with respect to a parameter   54
– pointwise   54
differential, stochastic   22
direct analysis   172
distributions of processes   12
Doléans-Dade equation   130
drift, singular   63

exponential
– generalized   179
– martingale   29
– stochastic   130

filtration   112
functional limit theorem   13

Girsanov theorem   141
graph   113
– of stopping time   113
group of measure-preserving transformations   215

Harnak's inequality   59
heat equation   15
Hellinger integral   143

infinitesimal operator   223
– of a semigroup   223
integration-by-parts formulae   99
invariance principle   213, 215
Itô change of variable formula   22, 23, 24, 129

$\mathcal{L}$-limit   55
$\mathcal{L}$-derivative   55
$\mathcal{L}B$-limit   55
$\mathcal{L}B$-derivative   55
Langevin equation   28
law of iterated logarithm   10
limit process, existence of   26
Lindeberg condition   173
Lyapunov condition   173
local density   140
local time   33

Malliavin derivative   92
– directional   93
mapping
– measurable   77
– strongly measurable   77
– weakly measurable   77
martingale   77, 116, 158, 188
– continuous with values in H 81
– exponential   29
– in Hilbert space   77
– locally integrable representation   137
– – continuous part   124
– – purely discontinuous part   124

- locally square-integrable 123
- orthogonal 123
  problems 50, 135
- square-integrable 117
- uniformly integrable 117
measure
- extended Poisson 122
- invariant 57, 223
- locally absolutely continuous 140
- random 120
-- integer-valued 121
-- integrable 121
-- optional 121

non-regularity of trajectories 10

pathwise-unique solution 39, 48
- of an SDE in a domain 68
Polish space 161
probability space 111
product of characteristic functions 172
projection
- dual predictable 120
- optional 116
- predictable 116
process
- arbitrary 199
- completely measurable 77
- diffusion 27
- **F**-adapted 113
- general Markov 57
- Hellinger, of order α 145
-- zero-order 145
- increasing 119
- indistinguishable stochastic processes 115
- left quasicontinuous 116
- of local density 141
- optional 114, 116
- Ornstein-Uhlenbeck 27
- point 149
- predictable 114, 115, 116, 190
- real simple stochastic 20
- standard 7
- stochastic 113
-- simple 20
- Wiener 7, 20, 123
-- one-dimensional 7
-- recurrent 18
purely discontinuous part of a local martingale 123

quadratic characteristic 123
quadratic covariation 126

- predictable (quadratic characteristic) 123
quadratic variation 125
- theorem 10
quasimartingale 126

random change of time 139
random variable (terminal) 77, 116
reflection principle 13
rules of the Malliavin calculus 96

semimartingale 125, 189
- problem 135
sequence
- announcing, of Markov times 114
- weakly convergent, of measures 13
σ-algebra
- of invariant sets 213
- optional 114
- predictable 114, 190
Skorokhod problem 64
Skorokhod topology 163
Sobolev space 85
Stirling formula 167
stochastic basis 112, 189
stochastic differential equation (SDE) 27, 38
- in a domain 64
- with reflection 65
- with aftereffect 38
stochastic exponentials, method of 194
stochastic integral 127, 131
stochastic intervals 113
strict solution of an SDE 39
strong Markov property 16
strong parabolicity condition 87
strong solution of an SDE 38
- existence theorems 45
submartingale 116
supermartingale 116, 158

time
- Markov 24, 112
-- accessible 115
-- totally inaccessible 115
- stopping 24, 112, 115
trajectory, most probable 62, 63
transition function 222
triple of characteristics 134
- predictable 191
truncation function 176

uniqueness of solution of an SDE

– in a domain   69
– – strong   68
– – weak   71
– of distribution   39
– pathwise   39, 68
– – strong   39, 68

– – weak   39
– of a stationary measure   57

Wald identities   24
weak solution of an SDE   40
Wiener process, existence of   7

Encyclopaedia of Mathematical Sciences
Editor-in-Chief: R.V. Gamkrelidze

Dynamical Systems

Volume 1: D. V. Anosov, V. I. Arnol'd (Eds.)
Dynamical Systems I
Ordinary Differential Equations and Smooth Dynamical Systems
2nd printing 1994. IX, 233 pages. 25 figures.
ISBN 3-540-17000-6

Volume 2: Ya. G. Sinai (Ed.)
Dynamical Systems II
Ergodic Theory with Applications to Dynamical Systems and Statistical Mechanics
1989. IX, 281 pages. 25 figures.
ISBN 3-540-17001-4

Volume 3: V. I. Arnold (Ed.)
Dynamical Systems III
Mathematical Aspects of Classical and Celestial Mechanics
2nd ed. 1993. XIV, 291 pages. 81 figures.
ISBN 3-540-57241-4

Volume 4: V. I. Arnol'd, S.P. Novikov (Eds.)
Dynamical Systems IV
Symplectic Geometry and its Applications
1990. VII, 283 pages. 62 figures.
ISBN 3-540-17003-0

Volume 5: V. I. Arnol'd (Ed.)
Dynamical Systems V
Bifurcation Theory and Catastrophe Theory
1994. IX, 271 pages. 130 figures.
ISBN 3-540-18173-3

Volume 6: V. I. Arnol'd (Ed.)
Dynamical Systems VI
Singularity Theory I
1993. V, 245 pages, 55 figures.
ISBN 3-540-50583-0

Volume 16: V. I. Arnol'd, S. P. Novikov (Eds.)
Dynamical Systems VII
Nonholonomic Dynamical Systems. Integrable Hamiltonian Systems
1994. VII, 341 pages. 9 figures.
ISBN 3-540-18176-8

Volume 39: V. I. Arnol'd (Ed.)
Dynamical Systems VIII
Singularity Theory II Applications
1993. V, 235 pages. 134 figures.
ISBN 3-540-53376-1

Volume 66: D. V. Anosov (Ed.)
Dynamical Systems IX
Dynamical Systems with Hyperbolic Behavior
1995. VII, 235 pages. 39 figures.
ISBN 3-540-57043-8

Partial Differential Equations

Volume 30: Yu. V. Egorov, M. A. Shubin (Eds.)
Partial Differential Equations I
Foundations of the Classical Theory
1991. V, 259 pages. 4 figures.
ISBN 3-540-52002-3

Volume 31: Yu. V. Egorov, M. A. Shubin (Eds.)
Partial Differential Equations II
Elements of the Modern Theory. Equations with Constant Coefficients
1995. VII, 263 pages. 5 figures.
ISBN 3-540-52001-5

Volume 32: Yu. V. Egorov, M. A. Shubin (Eds.)
Partial Differential Equations III
The Cauchy Problem. Qualitative Theory of Partial Differential Equations
1991. VII, 197 pages.
ISBN 3-540-52003-1

Volume 33: Yu. V. Egorov, M. A. Shubin (Eds.)
Partial Differential Equations IV
Microlocal Analysis and Hyperbolic Equations
1993. VII, 241 pages. 6 figures.
ISBN 3-540-53363-X

Volume 34: M. V. Fedoryuk (Ed.)
Partial Differential Equations V
Asymptotic Methods for Partial Differential Equations
Due 1997. Approx. 240 pages. 21 figures.
ISBN 3-540-53371-0

Volume 63: Yu. V. Egorov, M. A. Shubin (Eds.)
Partial Differential Equations VI
Elliptic andParabolic Operators
1994. VII, 325 pages. 5 figures.
ISBN 3-540-54678-2

Volume 64: M.A. Shubin (Ed.)
Partial Differential Equations VII
Spectral Theory of Differential Operators
1994. V, 272 pages.
ISBN 3-540-54677-4

Volume 65: M. A. Shubin (Ed.)
Partial Differential Equations VIII
Overdetermined Systems. Dissipative Singular Schrödinger Operator. Index Theory
1996. VII, 258 pages.
ISBN 3-540-57036-5

> **Please order from**
> Springer-Verlag Berlin
> Fax: + 49 / 30 / 8 27 87- 301
> e-mail: orders@springer.de
> or through your bookseller

■ ■ ■ ■ ■ ■ ■ ■ ■

Springer-Verlag, P. O. Box 31 13 40, D-10643 Berlin, Germany.

Springer for mathematics

Grundlehren der mathematischen Wissenschaften

Volume 293: D. Revuz, M. Yor
Continuous Martingales and Brownian Motion
2nd ed. 1994. XI, 560 pages. 8 figures.
Hardcover DM 168,–
ISBN 3-540-57622-3

Volume 312: K. L. Chung, Z. Zhao
From Brownian Motion to Schrödinger's Equation
1995. XII, 287 pages. 7 figures.
Hardcover DM 148,–
ISBN 3-540-57030-6

Graduate Texts in Mathematics

Volume 157: P. Malliavin
Integration and Probability
1995. XXI, 322 pages.
Hardcover DM 74,–
ISBN 0-387-94409-5

Volume 143: J. L. Doob
Measure Theory
1994. XII, 210 pages.
Hardcover DM 88,–
ISBN 3-540-94055-3

Universitext

B. Oksendal
Stochastic Differential Equations
An Introduction with Applications
4th ed. 1995. XVI, 272 pages.
Softcover DM 48,–

Springer Textbook

Ya. G. Sinai
Probability Theory
An Introductory Course
1992. VIII, 140 pages. 14 figures.
Softcover DM 48,–
ISBN 3-540-53348-6

Applications of Mathematics

Volume 33: P. Embrechts, C. Klüppelberg, T. Mikosch
Modelling Extremal Events
for Insurance and Finance
1997. XV,, 644 pages. 100 figures.
Hardcover DM 118,–
ISBN 3-540-60931-8

Applied Mathematical Sciences

Volume 122: Yu. Gliklikh
Global Analysis in Mathematical Physics
Geometric and Stochastic Models
1997. Approx. 220 pages.
Hardcover DM 88,–
ISBN 0-387-94867-8

Please order from
Springer-Verlag Berlin
Fax: + 49 / 30 / 8 27 87- 301
e-mail: orders@springer.de
or through your bookseller

Prices subject to change without notice.
In EU countries the local VAT is effective.

Springer-Verlag, P. O. Box 31 13 40, D-10643 Berlin, Germany.

Printing: Mercedesdruck, Berlin
Binding: Buchbinderei Lüderitz & Bauer, Berlin